Lecture Notes in Physics

Edited by J. Ehlers, München, K. Hepp, Zürich, and
H. A. Weidenmüller, Heidelberg
Managing Editor: W. Beiglböck, Heidelberg

25

Constructive Quantum Field Theory

The 1973 "Ettore Majorana" International School of Mathematical Physics

Edited by G. Velo and A Wightman

Instituto di Fisica, A. Righi, Bologna/Italy
Princeton University, Princeton, NJ/USA

Springer-Verlag Berlin Heidelberg GmbH 1973

ISBN 978-3-540-06608-8 ISBN 978-3-540-37912-6 (eBook)
DOI 10.1007/978-3-540-37912-6

 Library of Congress Catalog Card Number 73-21055.

Originally published by Springer-Verlag Berlin Heidelberg New York in 1973

Offsetprinting and bookbinding: Julius Beltz, Hemsbach/Bergstr.

TABLE OF CONTENTS

INTRODUCTION

The present volume collects lecture notes from the session of the International School of Mathematical Physics "Ettore Majorana" on Constructive Quantum Field Theory that took place at Erice (Sicily) July 26 to August 5, 1973.

The School was a NATO Advanced Study Institute sponsored by the Italian Ministry of Public Education, the Italian Ministry of Scientific and Technological Research, and the Regional Sicilian Government.

The book contains introductory material on functional analysis and probability theory, as well as detailed discussions of the existing state of knowledge of quantum field theory models.

In the opinion of the Editors, it can serve both as a review for experts of a rapidly developing subject and as an introduction for those with only a basic knowledge of field theory.

Unfortunately, the lecture notes of K. Symanzik were not prepared in time to be published in this volume. We hope they will be published elsewhere. In any case, the reader is referred to his Cargèse Lectures 1973, on related subjects, which are in course of publication.

FUNCTIONAL ANALYSIS AND PROBABILITY THEORY

Michael C. Reed, Princeton University

For many years now the standard tools of functional analysis have found application in quantum mechanics and quantum field theory; indeed the fundamental concepts of functional analysis, Hilbert and Banach spaces, bounded and unbounded operators, are the mathematical objects out of which specific models are constructed. By "standard tools" I mean the spectral theorem, Stone's theorem, and various methods of proving self-adjointness and investigating the properties of specific self-adjoint operators. It is not necessary to lecture on these topics since many of you already know them and in any case they are readily available in functional analysis texts and in several introductory lecture series meant especially for physicists (see for example [8] or [12]). Most of you already well-acquainted with quantum field theory know that many different branches of mathematics have been used in attempting to understand and solve the difficult mathematical problems that are involved, among them group representation, distribution theory, several complex variables and Banach algebras. So perhaps you were not too surprised to learn that yet another branch of analysis, probability theory and stochastic processes is now being applied in quantum field theory. The influx of these methods has come from two sources: first from the observation that certain field theory problems are analogous to problems in statistical mechanics; second, Irving Segal and Edward Nelson have often said that probabilistic methods are not just tools but are generic to the problems themselves, that is, that to some extent the problems of field theory are really problems in probability theory.

My purpose in these lectures then is to present an introduction to the probability theory concepts and methods which will be used by the other lecturers. We will start with the fundamentals, so you should not be angry if I say things that you already know (though you are allowed to be impatient).

§1. FUNDAMENTALS

We begin with the basic definitions. A probability space is a triple $\langle\Omega,\Sigma,\mu\rangle$ where Ω is a set, Σ is a σ-algebra of subsets, and μ is a positive measure of mass one on $\langle\Omega,\Sigma\rangle$. A (real) random variable is a measurable real-valued function on Ω. If x is a random variable on Ω, the distribution of x is the probability measure μ_x on the real line given by:

$$\mu_x\{A\} = \mu\{x^{-1}[A]\} = \mu\{\omega \in \Omega \mid x(\omega) \in A\}$$

where A is a Borel set in $\mathbb{R}$. We define the mean of x, $E(x)$, and the variance of x, $\mathrm{Var}(x)$, by

$$E(x) = \int_{\mathbb{R}} \lambda \, d\mu_x(\lambda)$$

and

$$\mathrm{Var}(x) = \int_{\mathbb{R}} (\lambda - E(x))^2 \, d\mu_x(\lambda)$$

if the integrals exist. If f is a measurable function on $\mathbb{R}$ then $f(x)$ is also a random variable on $\langle\Omega,\Sigma,\mu\rangle$ and

$$E(f(x)) = \int_\Omega f(x(\omega)) \, d\mu(\omega) = \int_{\mathbb{R}} f(\lambda) \, d\mu_x(\lambda) .$$

Notice that if $x \in L^2(\Omega,d\mu)$ then

$$E(x) = (x,\mathbb{1})_{L^2(\Omega,d\mu)} = \int_\Omega x \, d\mu(\omega)$$

and

$$\mathrm{Var}(x) = (x - E(x), x - E(x))_{L^2(\Omega,d\mu)} = \int_\Omega (x - E(x))^2 \, d\mu(\omega) .$$

<u>Example 1a</u>. Consider an experiment which consists of tossing an unbiased coin N times. If we denote the result of a toss by 0 if it is tails and by 1 if it is heads, then the set of all outcomes Ω may be represented by $\Omega = \{\omega = \langle n_i \rangle_{i=1}^{N} \mid n_i = 0 \text{ or } 1\}$, i.e. each point ω represents the possible outcome of a sequence of tosses, and each point in Ω has measure $\frac{1}{2^N}$. Let x_i be the random variable on Ω given by

$$x_i = \begin{cases} 1 & \text{if } n_i = 1 \\ 0 & \text{if } n_i = 0 \end{cases}$$

Then each x_i has the distribution $\mu_{x_i} = \frac{1}{2}\delta(\lambda) + \frac{1}{2}\delta(\lambda - 1)$, mean $\frac{1}{2}$, and variance equal to $\frac{1}{4}$.

We now return to our general definitions.

Let x_i, $i = 1,\dots,k$, be random variables on a probability space $\langle \Omega, \Sigma, \mu \rangle$. Then the vector-valued function $x(\omega) = \langle x_1(\omega),\dots,x_k(\omega)\rangle$ defines a measure μ_x on $\mathbb{R}^k$ by

$$\mu_x\{A\} = \mu\{x^{-1}[A]\} = \mu\{\omega \mid \langle x_1(\omega),\dots,x_k(\omega)\rangle \in A\}$$

for each Borel set A in $\mathbb{R}^k$. The measure μ_x is called the <u>joint distribution</u> of $x_1, x_2, \dots, x_k$.

Usually, independence of random variables is defined in terms of joint distributions, but we will use the following definition in terms of σ-algebras. Let T be an index set. A family of measurable sets $\{A_t\}_{t\in T}$, $A_t \subset \Omega$, is called independent if

$$\mu\{\bigcap_{i=1}^{k} A_{t_i}\} = \prod_{i=1}^{k} \mu\{A_{t_i}\}$$

for any finite subfamily, $t_i \in T$. A family of σ-algebras Σ_t, $t \in T$ with $\Sigma_t \subset \Sigma$ is said to be independent if each family of sets $\{A_t\}$ where $A_t \in \Sigma_t$ is independent. Finally, a family of random variables $\{x_t\}_{t\in T}$ is independent if the family of σ-algebras $\{\Sigma_t\}_{t\in T}$ is independent where Σ_t is the σ-algebra generated by x_t (i.e. the smallest σ-algebra on Ω so that x_t is measurable). It follows from this definition that if the random variables $x_1, x_2, \ldots, x_k$ are independent then their joint distribution μ_x is just the product measure $\mu_x = \bigotimes_{i=1}^{k} \mu_{x_i}$ in $\mathbb{R}^k$. This is the usual definition of independence.

If $A, B \in \Sigma$ and $\mu(B) \neq 0$ we define

$$P(A|B) = \frac{\mu(A \cap B)}{\mu(B)}$$

$P(A|B)$ is called the conditional probability of A given B. If A and B are measurable sets in $\mathbb{R}$ and x, y are random variables with $\mu\{x^{-1}[B]\} \neq 0$ then we set

$$P(y \in A \mid x \in B) \equiv P(y^{-1}(A) | x^{-1}(B))$$

$P(y \in A \mid x \in B)$ is called the conditional probability that y is in A, given that x is in B.

Example 1b. Let us return to example 1 and define $S_i^j = \{\omega \in \Omega | \; n_j = i\}$ for $i = 0,1$, and $j = 1,2,\ldots,n$. Let Σ_j be the σ-algebra generated by S_1^j and S_0^j (i.e. it just consists of the four sets $\emptyset$, Ω, S_1^j and S_0^j). The σ-algebras Σ_j are easily checked to be independent (this of course depends on our choice of measure); thus the random variables x_i are independent. If $c(\omega)$ is the random variable which assigns to each ω the integer corresponding to the toss on which heads first appeared or $N+1$ if heads did not appear on any of the tosses then c has the distribution $\sum_{j=1}^{N} \frac{1}{2^j}\,\delta(\lambda - j) + \frac{1}{2^N}\,\delta(\lambda - (N+1))$ and c is not independent of any of the x_i. You can check yourself in this example that the notions of conditional probability which we have defined correspond to your intuition of conditional probability.

Before leaving this elementary example let us change our point of view slightly and ask in what sense the distributions of the random variables x_i

determine μ. Let $\Sigma^{(j)}$ be the σ-algebra generated by $\Sigma_1, \Sigma_2, \ldots, \Sigma_j$. Then, the distribution of x_1 determines μ on $\Sigma^{(1)}$, the joint distribution of x_1 and x_2 determines μ on $\Sigma^{(2)}$ and so forth. The joint distribution of $x_1, \ldots, x_N$ determines μ completely since $\Sigma^{(N)}$ is just the σ-algebra of all subsets of Ω. (Notice that in this special case, since the x_i are independent, we can calculate any of the joint distributions directly from the individual distributions). So we have an increasing family of σ-algebras $\Sigma^{(1)} \subset \Sigma^{(2)} \subset \ldots \subset \Sigma^{(N)}$ generated by larger and larger families of random variables and the restriction of μ to a particular algebra is determined by the joint distributions of the random variables which generate the algebra. I have reformulated this trivial example to provide some intuition about the more difficult things to come.

Before presenting another example we make one more definition. If x_i, $i = 1, \ldots, N$ are random variables, the matrix

$$\{\Gamma_{ij}\} = E((x_i - Ex_i)(x_j - Ex_j))$$

is called the <u>covariance matrix</u> of the variables x_i. Notice that

$$\sum_{i,j=1}^{N} \Gamma_{ij} \bar{\alpha}_i \alpha_j = E(|\Sigma\, \alpha_i (x_i - Ex_i)|^2)$$

so the covariance matrix is positive definite. Also if x_i and x_j have zero mean then $\Gamma_{ij} = (x_i, x_j)_{L^2(\Omega, d\mu)}$.

<u>Example 2</u>. (Gaussian random variables). A random variable x is said to be a <u>Gaussian random variable</u> if its distribution μ_x is given by

$$d\mu_x(\lambda) = \frac{1}{\sigma\sqrt{2\pi}} e^{-\frac{(\lambda-m)^2}{2\sigma^2}} d\lambda$$

for some $\sigma > 0$ and $m \in \mathbb{R}$. It is easy to check that σ and m are the variance and mean respectively of x. A finite collection $x_1,\ldots,x_N$ of random variables is called Gaussian if there is a symmetric positive definite matrix on $\mathbb{R}^N$, Q, and real numbers $m_1,\ldots,m_N$ so that the joint distribution of $x_1,\ldots,x_N$ is

$$(1.1)\qquad \frac{(\mathrm{Det}(Q))^{\frac{1}{2}}}{(2\pi)^{N/2}}\, e^{-\frac{1}{2}(Q(\vec{\lambda}-\vec{m}),(\vec{\lambda}-\vec{m}))}\, d\vec{\lambda}$$

It follows by integrating out variables that the joint distribution of any subset or linear combinations of the x_i is again Gaussian, in particular the x_i themselves are Gaussian random variables. Further, a little linear algebra and explicit computation shows that $E(x_i) = m_i$ and that the covariance matrix of $x_1,\ldots,x_N$ is just Q^{-1}. Thus, if $x_1,\ldots,x_N$ is a Gaussian family their joint distribution is completely determined by their means and covariance matrix. In particular, if the covariance matrix is diagonal (which is the same thing as saying that $x_1 - m_1, \ldots, x_N - m_N$ are mutually orthogonal in $L^2(\Omega,d\mu)$), the joint distribution is a product of one-dimensional distributions. That is, for a Gaussian family of random variables with means equal zero, $(x_i,x_j)_{L^2(\Omega,d\mu)} = 0$ if and only if x_i and x_j are independent. Thus, since a finite linear combination of the x_i is again Gaussian with mean zero, we can (by Gram-Schmidt orthogonalization) find a set $\xi_1,\ldots,\xi_N$ of independent Gaussian random variables of mean zero and variance one (this just means $\|\xi_i\|^2_{L^2(\Omega,d\mu)} = 1$) so that each x_i is a linear combination of the ξ_i.

Finally if $x_1,\ldots,x_N$ are random variables, the function

$$C(\alpha_1,\ldots,\alpha_N) = \int_\Omega e^{i\,\Sigma\,\alpha_i x_i(\omega)}\, d\mu(\omega)$$

$$= \int_{\mathbb{R}^N} e^{i\,\Sigma\,\alpha_i\lambda_i}\, d\mu_{\vec{x}}(\vec{\lambda})$$

is called the <u>characteristic</u> <u>function</u> of $x_1,\ldots,x_N$. Notice that if $d\mu_{\vec{x}}(\lambda)$ is given by (1.1), (i.e., $x_1,\ldots,x_N$ is a Gaussian system), then

$$(1.2) \qquad C(\alpha_1,\ldots,\alpha_N) = e^{i \sum_i \alpha_i m_i - \sum_{i,j} \Gamma_{ij}\alpha_i\alpha_j}$$

where $\Gamma = Q^{-1}$ is just the covariance matrix. Conversely, if C is given in the form (1.2), then the uniqueness of the Fourier transform shows that the joint distribution of $x_1,\ldots,x_N$ is given by (1.1) where $Q = \Gamma^{-1}$ so $x_1,\ldots,x_N$ form a Gaussian system.

We remark that we have introduced real-valued random variables; we will sometimes use complex-valued random variables in which case the distributions of the random variables are just measures on C. In general, a random variable is a measurable map from $<\Omega, \Sigma>$ to another measurable space $<\Omega_1, \Sigma_1>$.

This concludes the elementary introduction I promised. To see the enormously rich applications of these elementary probability notions you should read the books by William Feller [2].

§2. STOCHASTIC PROCESSES

A _stochastic process_ is a family of random variables $\{x_t\}_{t\in T}$ on a probability space $<\Omega,\Sigma,\mu>$ indexed by a set T. For the time being T will just be $(0,\infty)$ so a stochastic process will just be a map $x_{(\cdot)}(\cdot)$ of $(0,\infty)\times\Omega$ into R so that for each $t\in(0,\infty)$, $x_t(\cdot)$ is a measurable function on Ω. Notice that for each fixed $\omega\in\Omega$, $x_t(\omega)$ is just a real-valued function on R. You should think of $x_t(\omega)$ as the value of some observable quantity at time t; for example the rate of arrival of particles at a counter, or the position of a particle undergoing Brownian motion on the x axis.. That is, each fixed ω corresponds to a possible history of the value of some parameter for $t\in(0,\infty)$. A different $\omega_1\in\Omega$ will correspond to a different history since (in general) the functions $x_t(\omega)$ and $x_t(\omega_1)$ will be different.

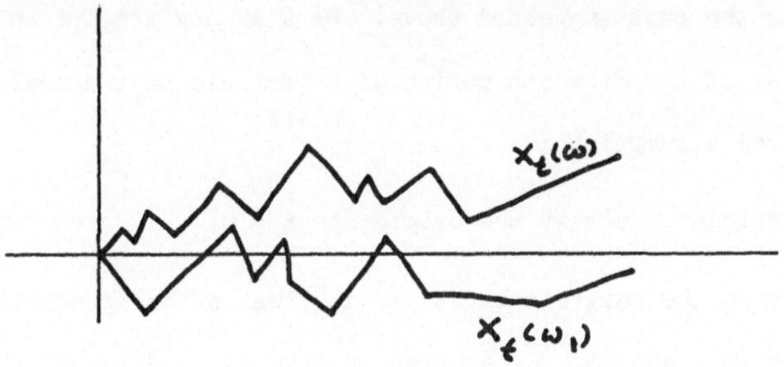

The measure μ on Ω tells us essentially which (collections of) histories are more likely and which are less likely. For example

$$\mu\{\omega| \ \alpha \leq x_{t_0}(\omega) \leq \beta\} = \int_{\alpha}^{\beta} d\mu_{x_{t_0}}$$

is the probability that if we look at time t_0 the value of the parameter will be greater than or equal to α and less than or equal to β. Similarly, the probability that $\alpha_0 \leq x_{t_0} \leq \beta_0$ and $\alpha_1 \leq x_{t_1} \leq \beta_1$ is

$$\mu\{\omega| \ \alpha_0 \leq x_{t_0}(\omega) \leq \beta_0; \alpha_1 \leq x_{t_1}(\omega) \leq \beta_1\} = \int_{\alpha_1}^{\beta_1}\int_{\alpha_0}^{\beta_0} d\mu_{t_0,t_1}$$

where μ_{t_0,t_1} is the joint distribution of $x_{t_0}(\cdot)$ and $x_{t_1}(\cdot)$. As another example, set $M_T(\omega) = \sup_{t\in(0,T)} x_t(\omega)$. Then, assuming that $M_T(\cdot)$ is measurable (see below), $\mu\{\omega| \ M_T(\omega) \leq \beta\}$ gives us the probability that the path or history will not go above β before time T. The point here is that the underlying dynamics of the situation is reflected in the measure μ which allows us to assign probabilities to certain subsets of the set of all paths in which we are especially interested. For any given dynamics, of course, the measure μ is not given a priori but must somehow be constructed out of the physics of the problem being studied. That is, the underlying dynamics should tell us how to construct μ, or better yet how to directly construct a measure on the set of paths for it is this measure which contains the answers to probabilistic questions about the dynamics such as the ones suggested above. We will now attempt such a construction for the case of the Brownian motion of a particle on the real line.

We make three assumptions:

1. The particle starts at the origin at $t = 0$.

2. Given that the particle is at λ at time s the probability that it will be in a set $B \subset \mathbb{R}$ at time $t > s$ is $\frac{1}{\sqrt{2\pi|t-s|}} \int_B e^{-(\xi-\lambda)^2/2|t-s|} d\xi$. For convenience we set

$$P(\lambda,\xi,t) \equiv \frac{1}{\sqrt{2\pi t}} e^{-(\xi-\lambda)^2/2t} .$$

3. (The Markov Condition) Given that the particle is in a set A at time s, the future history of the particle will depend on A but not on the past history before s.

Condition one just says that we are only interested in paths which start at the origin so for our space Ω we take

$$\Omega = \{\text{all real-valued functions } f \text{ on } [0,\infty] \text{ so that } f(0) = 0\} .$$

Our stochastic process will just be evaluation at t, i.e. $x_t(f) = f(t)$ where each f is (a priori) considered as a possible history, $f(t)$ being the position at time t. Since we want $x_t(\cdot)$ to be a measurable function on Ω for each t we want all sets of the form $\{f| \ x_t(f) \in A\}$ to be measurable for each t and each Borel set A in $\mathbb{R}$. Any σ-algebra which contains these sets also contains all sets of the form

$$\{f| \ \langle x_{t_1}(f), x_{t_2}(f), \ldots, x_{t_n}(f)\rangle \in A\}$$

where $t_1 < t_2 < \ldots < t_n$ and $A \subset \mathbb{R}^n$. These sets are called the <u>cylinder sets</u> and we will denote the collection of cylinder sets by Σ_0. Let us see how conditions 2 and 3 determine the measure we want to construct on Σ_0.

If we set $s = 0$ in condition 2 we see that the distribution of x_t is given by

$$d\mu_t = \frac{1}{\sqrt{2\pi t}} e^{-\lambda^2/2t}\, d\lambda = P(0,\lambda,t)\, d\lambda$$

Let $\mu_{t_1t_2}$ denote the joint distribution of x_{t_1} and x_{t_2}, $t_1 < t_2$, and let A and B be Borel sets in $\mathbb{R}$. The probability that x_{t_1} will be in a small interval $\Delta\lambda_1$ is $P(0,\lambda_1,t_1)\Delta\lambda_1$ and given that $x_{t_1} = \lambda_1$, the probability that x_{t_2} will lie in the small interval $\Delta\lambda_2$ is $P(\lambda_1,\lambda_2,|t_2 - t_1|)\Delta\lambda_2$. Thus, the probability that x_{t_1} is in $\Delta\lambda_1$ <u>and</u> x_{t_2} is in $\Delta\lambda_2$ is the product

$$P(0,\lambda_1,t_1)P(\lambda_1,\lambda_2,|t_2 - t_1|)\Delta\lambda_1\Delta\lambda_2 \tag{2.1}$$

Therefore,

$$\mu_{t_1t_2}(A \times B) = \int_{\lambda_1\in A}\int_{\lambda_2\in B} P(0,\lambda_1,t_1)P(\lambda_1,\lambda_2,|t_2 - t_1|)\, d\lambda_1\, d\lambda_2 \tag{2.2}$$

Thus condition 2 also determines the joint distributions of any pairs x_{t_1} and x_{t_2}.

To determine the joint distributions for three or more of the x_t we need condition 3. Let $t_1 < t_2 < t_3$ be given. Then the probability that x_{t_1} has a value in $\Delta\lambda_1$, x_{t_2} in $\Delta\lambda_2$, <u>and</u> x_{t_3} is in $\Delta\lambda_3$ is just (2.1) times the conditional probability that $x_{t_3} \in \Delta\lambda_3$ given that $x_{t_1} \in \Delta\lambda_1$ <u>and</u> $x_{t_2} \in \Delta\lambda_2$. But, by condition 3 this conditional probability is just the conditional probability that $x_{t_3} \in \Delta\lambda_3$ given that $x_{t_2} \in \Delta\lambda_2$. Thus, the probability that $x_{t_1} \in \Delta\lambda_1$, $x_{t_2} \in \Delta\lambda_2$ and $x_{t_3} \in \Delta\lambda_3$ is

$$P(0,\lambda_1,t_1)P(\lambda_1,\lambda_2,|t_2 - t_1|)P(\lambda_2,\lambda_3,|t_3 - t_2|)\Delta\lambda_1\Delta\lambda_2\Delta\lambda_3 \tag{2.3}$$

Using (2.2) we can now easily write down an integral expression analogous to (2.2) for $\mu_{t_1t_2t_3}(A\times B\times C)$, the joint distribution of x_{t_1}, x_{t_2} and x_{t_3}. The other joint distributions work in exactly the same way so we see that conditions

1), 2), and 3) determine a measure μ on Σ_0. Such a measure is called a <u>cylinder measure</u> for obvious reasons.

This brings us to our first real mathematical problem, namely, Σ_0 is an algebra of sets but not a σ-algebra. It is important to extend μ to the smallest σ-algebra Σ containing Σ_0 because, one wants to consider various random variables (say $y(\omega)$) which are limits of combinations of the x_{t_i} and for such functions sets of the form $\{\omega | \alpha \leq y(\omega) \leq \beta\}$ will be in Σ but not in Σ_0 in general. The problem is to extend μ in such a way that the extension is countably additive. There is a general theorem (see Royden [11]) which says that this can be done in a unique way if μ is already countably additive on Σ_0 whenever it can be. That is, if $\{A_n\}$ is a sequence of disjoint sets in Σ_0 <u>and</u> $A = \cup A_n \in \Sigma_0$, then $\mu(A) = \sum_{n=1}^{\infty} \mu(A_n)$. This amounts to the same thing as showing that if $E_1 \supset E_2 \supset \ldots \supset E_n \supset \ldots$ are in Σ_0 and $\cap E_n = \emptyset$, then $\mu(E_n) \longrightarrow 0$. Using a compactness argument this condition may be checked in our case (for the proof see for example [7] or [15]).

<u>Theorem</u>. (Kolmogorov) A finite consistent cylinder measure on $\underset{[0,\infty)}{\times} \mathbb{R}$ can be extended to a unique countably additive measure on the smallest σ-algebra containing the cylinder sets. The word "consistent" just means that cylinder measure has the property that $\mu_{t_1 \ldots t_n}$ arises from $\mu_{s_1 \ldots s_m}$ by integrating out the extra variables whenever $\{t_1, \ldots, t_n\} \subset \{s_1, \ldots, s_m\}$.

The second mathematical problem that now arises is: Where does the measure μ have its support? We have constructed μ so that it is a measure on the set of all real-valued functions on $[0,\infty)$ such that $f(0) = 0$. But surely if our model is a realistic one for Brownian motion the support of the measure must be on a much smaller class of functions, for example on the set of continuous functions since we believe intuitively that Brownian paths are continuous. So one reason to investigate the support is to find the analytical properties of Brownian paths. The second reason is technical but very important. Suppose that we set $M_T(\omega) = \sup_{t \in (0,T)} \{x_t(\omega)\}$. We pointed out before why we might be interested in $M_T(\omega)$ as a random variable. The trouble is that there is no reason for $M_T(\cdot)$ to be measurable since it is the supremum over uncountably many random variable and we are only guaranteed in elementary measure theory that the supremum of countably many measurable functions is measurable. On the other hand, let $\tilde{M}_T(\omega) = \sup_{t_n} \{x_{t_n}(\omega)\}$ where $\{t_n\}$ is a countable dense set in $(0,\infty)$. Then $\tilde{M}_T(\cdot)$ is measurable and $\tilde{M}_T$ and M_T agree on the continuous functions. Therefore if the set $C_0(0,\infty) = \{f \mid f \text{ continuous and } f(0) = 0\}$ has measure one we can take our space of paths to be $C_0(0,\infty)$ in which case the sup function $\tilde{M}_T$ is measurable.

Now, what sort of condition should guarantee that paths are continuous? The condition should say roughly that if $|t - s|$ is small then the distribution of $x_t - x_s$ should be concentrated close to the origin. An example of such a sufficient condition is the requirement that for some $\alpha, \beta > 0$,

$$E(|x_t - x_s|^\beta) \le M|t - s|^\alpha \tag{2.4}$$

For a nice proof see [15]. In our case the joint distribution of x_t and x_s is

$$\frac{1}{\sqrt{2\pi s}\,\sqrt{2\pi|t-s|}}\, e^{-\lambda^2/2s}\, e^{-(\lambda_2-\lambda_1)^2/2|t-s|}\, d\lambda_1\, d\lambda_2$$

Changing to new variables $\xi = \frac{\lambda_1 + \lambda_2}{2}$ and $\eta = \lambda_2 - \lambda_1$ and integrating out ξ we find that the distribution of $x_t - x_s$ is $\frac{1}{\sqrt{2\pi|t-s|}}\, e^{-\eta^2/2|t-s|}\, d\eta$. Thus,

$$\begin{aligned} E(|x_t - x_s|^4) &= \frac{1}{\sqrt{2\pi(t-s)}} \int_{-\infty}^{\infty} \eta^4\, e^{-\eta^2/2|t-s|}\, d\eta \\ &= 3|t - s|^2 \end{aligned}$$

so the condition is satisfied and almost all paths are continuous. In fact, much more is known about Brownian paths. For example, almost all of them are nowhere differentiable. For a proof of this and other facts about the modulus of continuity see [6] or [15].

The stochastic process which we have been using as an example is very special for several reasons. First of all it is a Gaussian process which just means that all of the joint distributions of finitely many $x_{t_1},\ldots,x_{t_m}$ are Gaussian. Secondly, it is a Markov process since condition 3 holds. In §3 we will reformulate condition 3 in more precise analytical terms. Finally, we treated the case of a Brownian particle starting at zero. More generally we could have started with the particle at λ_0 in which case the distribution of x_t would be $P(\lambda_0,\lambda,t)\,d\lambda$ or we might only know that x_0 has some initial distribution $d\mu_0(\lambda)$ in which case the distribution of x_t will be

$$\left(\int_R P(\xi,\lambda,t)\,d\mu_0(\xi)\right)d\lambda$$

The construction of the cylinder measure is similar to what we have done and the extension is again by the Kolmogorov theorem. Of course, the resulting measure μ on the space of paths will be different for different initial distributions.

What we have learned from this simple example is that two non-trivial mathematical problems immediately arise in dealing with stochastic processes, namely the construction of measures on spaces of paths and the support properties of these measures.

§3. CONDITIONAL EXPECTATIONS

The Brownian motion which we constructed in the last section was a very special kind of stochastic process; a Markov process. The Markov character results from the special assumption 3 which we used in calculating explicit expressions for the desired measure on cylinder sets. In order to formulate precisely the notion of Markov process we need to introduce the concept of conditional expectation and to derive some of its properties.

Let $<\Omega,\Sigma,\mu>$ be a probability space and let $\Sigma_0 \subset \Sigma$ be a sub σ-algebra. Let $f \in L^1(\Omega,\Sigma,\mu)$ and define

$$S(B) = \int f(\omega)\chi_B(\omega)\, d\mu(\omega)$$

for each $B \in \Sigma_0$. Then $S(B)$ is a finite signed measure on Σ_0 and $S(B)$ is clearly absolutely continuous with respect to the restriction of μ to Σ_0 (we denote this restriction again by μ). Thus, by the Radon-Nikodyn theorem there is a unique function (up to sets of measure zero) $E(f|\Sigma_0)$ in $L^1(\Omega,\Sigma_0,\mu)$ so that

$$S(B) = \int_B E(f|\Sigma_0)(\omega)\, d\mu(\omega)$$

for all $B \in \Sigma_0$, i.e.

$$\int_B f(\omega)\, d\mu(\omega) = \int_B E(f|\Sigma_0)(\omega)\, d\mu(\omega)$$

The point is that $E(f|\lambda_0)(\omega)$ is measurable with respect to Σ_0 while $f(\omega)$ may not be since Σ_0 is a subalgebra of Σ. $E(f|\Sigma_0)(\omega)$ is called the <u>conditional expectation of</u> f <u>with respect to</u> Σ_0.

<u>Example 1</u>. Let $\Omega = \mathbb{R}^2$, Σ equal the Borel sets in $\mathbb{R}^2$, and suppose (to make things simple) that $d\mu = u(x,y)\, dx\, dy$ where $u(x,y) > 0$ for all $<x,y> \in \mathbb{R}^2$. Let Σ_0 consist of all sets of the form $B \times R$ where B is a Borel set in $\mathbb{R}$. Then the restriction of $u(x,y)\, dx\, dy$ to Σ_0 is just the measure on $\mathbb{R}$ given by

$$\left(\int_R u(x,y)\, dy\right) dx$$

and the conditional expectation of a function $f \in L^1(\mathbb{R}^2,\Sigma,\mu)$ is given by

$$E(f|\Sigma_0)(x) = \frac{\int_{\mathbb{R}} f(x,y)u(x,y)\, dy}{\int_{\mathbb{R}} u(x,y)\, dy}$$

Example 2. Let $\Omega = \mathbb{R}$, let Σ be the Borel sets and suppose that μ is any probability measure on $\mathbb{R}$. Let Σ_t consist of all Borel subsets of the interval $[-t,t]$ plus the complement of $[-t,t]$ and $\mathbb{R}$. The restriction of μ to Σ_t assigns $\mu(B)$ to any Borel subset of $[-t,t]$ and assigns the number $\int_{|x|>t} d\mu(x)$ to the complement of $[-t,t]$. If $f \in L^1(\mathbb{R},\Sigma,d\mu)$ then

$$E(f|\Sigma_t)(x) = \begin{cases} f(x) & |x| \le t \\ \dfrac{\int_{|x|>t} f(x)\, d\mu(x)}{\int_{|x|>t} d\mu(x)} & |x| > t \end{cases}$$

Notice that $E(f|\Sigma_t)(x)$ is constant for $|x| > t$ which it must be if $E(f|\Sigma_t)$ is to be measurable with respect to Σ_t.

Conditional expectations have the following elementary properties.

Proposition. 1. If $f \ge 0$, then $E(f|\Sigma_0) \ge 0$.

2. If Σ_{00} is a sub σ-algebra of Σ_0, then

$$E(E(f|\Sigma_0)|\Sigma_{00}) = E(f|\Sigma_{00}) .$$

3. If $g \in L^\infty(\Omega,\Sigma_0,\mu)$ then

$$E(fg|\Sigma_0) = gE(f|\Sigma_0) .$$

4. If Σ' is a sub σ-algebra of Σ so that Σ_0 and Σ' are independent, then if $f \in L^1(\Omega,\Sigma',\mu)$,

$$E(f|\Sigma_0) = E(f)$$

i.e. $E(f|\Sigma_0)$ is a constant function.

Proof: Property 1 is immediate and property 2 follows from the definition of conditional expectation since

$$\int_B E(f|\Sigma_0)\, d\mu = \int_B f\, d\mu = \int_B E(f|\Sigma_{00})\, d\mu$$

for all $B \in \Sigma_{00}$. Suppose $B \in \Sigma_0$, then by definition

$$\int f\chi_B \, d\mu = \int E(f|\Sigma_0)\chi_B \, d\mu$$

so if $B_i \in \Sigma_0$, $i = 1,\ldots,N$

$$\int f \sum^N c_i\chi_{B_i} \, d\mu = \int E(f|\Sigma_0) \Sigma \, c_i\chi_B \, d\mu \ .$$

By the dominated convergence theorem this implies that

$$\int fg \, d\mu = \int E(f|\Sigma_0)g \, d\mu \tag{3.1}$$

for all $g \in L^\infty(\Omega,\Sigma_0,d\mu)$. In particular, if $B \in \Sigma_0$,

$$\int_B fg \, d\mu = \int_B E(f|\Sigma_0)g \, d\mu$$

so by the uniqueness statement of the Radon-Nikodyn theorem $E(fg|\Sigma_0) = gE(f|\Sigma_0)$. This proves (3). Finally, suppose Σ' and Σ_0 are independent and $A \in \Sigma'$, $B \in \Sigma_0$. Then, $\mu(A \cap B) = \mu(A)\mu(B)$ so

$$\int_B \chi_A \, d\mu = \mu(A \cap B) = \mu(A)\mu(B) \ .$$

Thus, if $A_i \in \Sigma'$, $i = 1,\ldots,N$,

$$\int_B \Sigma \, c_i\chi_{A_i} \, d\mu = (\Sigma \, c_i\mu(A_i))\mu(B) \ .$$

So, using the dominated convergence theorem again we have

$$\int_B f \, d\mu = (\int_\Omega f \, d\mu)\mu(B)$$

for all $f \in L^1(\Omega,\Sigma',d\mu)$ which proves (4).

Given a sub σ-algebra $\Sigma_0 \subset \Sigma$, the conditional expectation is a map

$$f \xrightarrow{E_{\Sigma_0}} E(f|\Sigma_0)$$

of Σ-measurable functions into Σ_0-measurable functions. E_{Σ_0} is clearly linear and has the following important additional properties:

Proposition. 5. E_{Σ_0} is a contraction from $L^p(\Omega,\Sigma,d\mu)$ to $L^p(\Omega,\Sigma_0,d\mu)$ for each $1 \leq p \leq \infty$.

6. E_{Σ_0} is the orthogonal projection of $L^2(\Omega,\Sigma,d\mu)$ onto $L^2(\Omega,\Sigma_0,d\mu)$.

Proof: If $f \in L^1(\Omega,\Sigma,d\mu)$ and $f \geq 0$ then

$$\|E_{\Sigma_0} f\|_1 = \int E_{\Sigma_0} f \, d\mu = \int f \, d\mu = \|f\|_1$$

For a general f we write $f = f_+ - f_-$ where $f_+, f_- \geq 0$, and f_+ and f_- have disjoint support. Then,

$$\|E_{\Sigma_0} f\| \leq \|E_{\Sigma_0} f_+\|_1 + \|E_{\Sigma_0} f_-\|_1$$

$$= \|f_+\|_1 + \|f_-\|_1 = \|f\|_1$$

so E_{Σ_0} is a contraction on L^1. If $f \in L^\infty(\Omega,\Sigma,d\mu)$, then for all $B \in \Sigma_0$,

$$-\|f\|_\infty \mu(B) \leq \int_B f \, d\mu \leq \|f\|_\infty \mu(B)$$

so,

$$-\|f\|_\infty \mu(B) \leq \int_B E_{\Sigma_0} f \, d\mu \leq \|f\|_\infty \mu(B)$$

which implies that $\|E_{\Sigma_0} f\|_\infty \leq \|f\|_\infty$. Thus E_{Σ_0} is also a contraction on L^∞. (5) now follows immediately from the Riesz-Thorin theorem.

To prove (6) notice that $L^2(\Omega,\Sigma_0,d\mu)$ is a closed subspace of $L^2(\Omega,\Sigma,d\mu)$, $E_{\Sigma_0}^2 = E_{\Sigma_0}$, and $\text{Ran}\, E_{\Sigma_0} = L^2(\Omega,\Sigma_0,d\mu)$, so E_{Σ_0} is a projection onto $L^2(\Omega,\Sigma_0,d\mu)$. For $f_1, f_2 \in L^2(\Omega,\Sigma,d\mu)$

$$(E_{\Sigma_0} f_1, f_2) = (E_{\Sigma_0} f_1, E_{\Sigma_0} f_2) = (f_1, E_{\Sigma_0} f_2)$$

so $E_{\Sigma_0}^* = E_{\Sigma_0}$ and E_{Σ_0} is an orthogonal projection.

Typically, conditional expectations arise by taking Σ_0 to be the σ-algebra generated by a collection $\{x_\alpha\}_{\alpha\in I}$ of random variables where I is some index set; in this case $E(x|\Sigma_0)$ is often denoted by $E(x|\{x_\alpha\}_{\alpha\in I})$. Consider the case where $I = \{1,2,\ldots,N\}$; then we are conditioning with respect to an algebra Σ_0 generated by finitely many random variables, $x_1,x_2,\ldots,x_N$. Given a random variable x, $E(x|\{x_i\}_{i=1}^N)$ is then a function on Ω measurable with respect to Σ_0. From this it follows that there is a Borel function φ on $\mathbb{R}^n$ so that $E(x|\{x_i\}_{i=1}^N)(\omega) = \varphi(x_1(\omega),\ldots,x_N(\omega))$. Therefore, if $B \in \mathbb{R}^N$,

$$\int_{x^{-1}(B)} E(x|\{x_i\}_{i=1}^N)(\omega)\, d\mu(\omega) = \int_B \varphi(\lambda_1,\ldots,\lambda_N)\, d\mu_{x_1,\ldots,x_N}(\lambda_1,\ldots,\lambda_N)$$

where $\mu_{x_1,\ldots,x_N}$ is the joint distribution of $x_1,\ldots,x_N$. In the case where $x,x_1,\ldots,x_N$ is a Gaussian family the conditional expectation is especially simple.

Proposition. Let $x,x_1,\ldots,x_N$ be a Gaussian family of random variables with means equal to zero and $x_1,\ldots,x_N$ orthonormal. Then

$$E(x|\{x_i\}_{i=1}^N) = \sum_{i=1}^N (x,x_i)x_i$$

Proof: Since $x,x_1,\ldots,x_N$ are Gaussian they are all in $L^2(\Omega,\Sigma,d\mu)$. Further $x - \sum_{i=1}^N (x,x_i)x_i$ is orthogonal to each x_i so $x - \sum_{i=1}^N (x,x_i)x_i$ is independent of the x_i since the random variables are Gaussian (see §1). Thus, by property 4,

$$\begin{aligned} E(x - \sum_{i=1}^N (x,x_i)x_i|\{x_i\}_{i=1}^N) &= E(x - \sum_{i=1}^N (x,x_i)x_i) \\ &= E(x) - \sum_{i=1}^N (x,x_i)E(x_i) \\ &= 0 \end{aligned}$$

so,

$$E(x\,|\,\{x_i\}_{i=1}^N) = E(\sum_{i=1}^N (x,x_i)x_i\,|\,\{x_i\}_{i=1}^N)$$

$$= \sum_{i=1}^N (x,x_i)x_i \ .$$

We now introduce conditional probabilities as a special case of conditional expectations. Let y be a random variable, A a Borel set in $\mathbb{R}$, and set $x = \chi_{y^{-1}(A)}$, where χ_S always denotes the characteristic function of the set S ($\chi_S(\omega) = 1$ for $\omega \in S$, $\chi_S(\omega) = 0$ if $\omega \notin S$). Then $E(\chi_{y^{-1}(A)}\,|\,\{x_i\}_{i=1}^N)(\omega)$ is denoted by $P(y \in A\,|\,\{x_i\}_{i=1}^N)$ and is called the <u>conditional probability that</u> y <u>is in</u> A <u>given</u> $x_1,\ldots,x_N$. The corresponding $\varphi(\lambda_1,\ldots,\lambda_N)$ (see above) is denoted by $P(y \in A\,|\,x_1 = \lambda_1, \ldots, x_N = \lambda_N)$ and called the <u>conditional probability that</u> $y \in A$ <u>given that</u> $x_1 = \lambda_1, x_2 = \lambda_2, \ldots, x_N = \lambda_N$.

We are now ready to give our precise definition of Markov process. Let $\{x(t)\}$ be a stochastic process indexed by $[0,\infty)$. Let $\Sigma_{t_1,\ldots,t_n}$ denote the σ-algebra generated by $x(t_1),\ldots,x(t_n)$; $\Sigma_{(a,b)}$ denotes the σ-algebra generated by $x(t)$, $a < t < b$; and denote the corresponding conditional expectations by $E_{t_1,\ldots,t_n}$ and $E_{(a,b)}$. $\{x(t)\}_{t\in(0,\infty)}$ is called a Markov process if for each $s > 0$ and random variable $y \in L^1(\Omega,\Sigma_{(s,\infty)},\mu)$,

$$E_{[0,s]}y = E_s y \tag{3.2}$$

i.e., $E_{[0,s]}y$ is already measurable with respect to Σ_s. If s is the present time, $\Sigma_{(s,\infty)}$ corresponds to future events, so the Markov condition says intuitively that the expectation of some event in the future given some information about the present <u>and</u> the past is the same as the expectation just given the present. To illustrate that the fancy definition is just a formal statement of this intuitive idea, let us return briefly to the Brownian motion constructed in §2. Let $r < s < t$ and let A, B, C be Borel sets in $\mathbb{R}$. Then

$$\mathrm{Prob}\{x_t \in A,\ x_s \in B,\ x_r \in C\} = E(\chi_{x_t^{-1}(A)}\,\chi_{x_s^{-1}(B)}\,\chi_{x_s^{-1}(C)})$$

$$= E((E_{r,s}\chi_{x_t^{-1}(A)})\chi_{x_s^{-1}(B)}\,\chi_{x_r^{-1}(C)})$$

$$= \int_{x_r^{-1}(C)} \int_{x_s^{-1}(B)} (E_{r,s}\chi_{x_t^{-1}(A)})(\omega)\, d\mu(\omega)$$

$$= \int_C \int_B \varphi(\xi,\eta)\, d\mu_{r,s}(\xi,\eta)$$

where $\varphi(\xi,\eta)$ is the function expressing $E_{r,s}\chi_{x_t^{-1}(A)}$ in terms of x_r and x_s and $\mu_{r,s}$ is the joint distribution of x_r and x_s. On the other hand we calculated in §2 that

$$\mathrm{Prob}\{x_t \in A,\ x_s \in B,\ x_r \in C\} = \int_C \int_B \left(\int_A e^{-(\eta-\lambda)^2/2(t-s)}\, d\lambda\right) d\mu_{r,s}(\xi,\eta)$$

Thus, $\varphi(\xi,\eta) = \int_A e^{-(\eta-\lambda)^2/2(t-s)}\, d\lambda$ is a function of η only, so $E_{r,s}\chi_{x_t^{-1}(A)}$ is already measurable with respect to Σ_s, i.e.

$$E_{r,s}\chi_{x_t^{-1}(A)} = E_s\chi_{x_t^{-1}(A)}$$

The same argument shows that if $r_1 < r_2 < \dots < r_n < s < t_1 < t_2 < \dots < t_m$, and $A_i \in R$, then

$$E_{r_1,r_2,\dots,r_n,s} \sum_{i=1}^{m} c_i \chi_{x_{t_i}^{-1}(A_i)} = E_s \sum_{i=1}^{m} c_i \chi_{x_{t_i}^{-1}(R)}$$

and now the dominated convergence theorem and a measure theory argument show that in fact (3.2) holds.

We now return briefly to the general case. Given a Markov process $\{x_t\}$ we define the <u>transition probabilities</u> $P(s,t,\lambda,A)$ by

$$P(s,t,\lambda,A) = \mathrm{Prob}(x_t \in A \mid x_s = \lambda)\ .$$

That is,

$$P(s,t,x_t(\omega),A) = E_s \chi_{x_t^{-1}(A)}$$

For each s, t, and λ, $P(s,t,\lambda,\cdot)$ is a probability measure on $\mathbb{R}$ and for each s, t, and A, $P(s,t,\cdot,A)$ is a positive Borel function on $\mathbb{R}$. Just as in the case of Brownian motion we can write down the joint distributions in terms of the transition probabilities and the initial distribution, for example:

$$\mu\{x_t \in A,\ x_s \in B\} = \int_B \int_R P(s,t,\eta,A)P(0,s,\xi,d\eta)\ d\mu_s(\xi)$$

Let $r < s < t$, then

$$\mu\{x_t \in A,\ x_s \in R,\ x_r \in B\} \tag{3.3}$$

$$= \int_B \int_R P(s,t,\eta,A)P(r,s,\xi,d\eta)\ d\mu_r(\xi)$$

But,

$$(3.3) = \mu\{x_t \in A,\ x_r \in B\}$$

$$= \int_B P(r,t,\xi,A)\ d\mu_r(\xi)$$

so by the uniqueness of the conditional expectation we have

$$P(r,t,\xi,A) = \int_R P(s,t,\eta,A)P(r,s,\xi,d\eta) \tag{3.4}$$

almost everywhere w.r.t. μ_n. Equation (3.4) is called the Chapman-Kolmogorov equation and it holds because given a measure μ on $\langle\Omega,\Sigma\rangle$ the family of joint distributions must be consistent. Conversely, given a set of transition probabilities which satisfy (3.4) identically one can use the Kolmogorov construction exactly as we did in the case of Brownian motion to construct a Markov process with the given transition probabilities.

§4. SEMI-GROUPS

In the last section we saw that studying Markov processes can in some sense be reduced to studying transition probabilities which satisfy the Chapman-Kolmogorov equations. Such a system of transition probabilities is called stationary (and the corresponding Markov process is called homogeneous) if $P(s,t,\lambda,A)$ only depends on $|t - s|$. That is, for each t, we have a function $P(t,\lambda,A)$ which is a probability measure in A for λ fixed and a positive Borel function in λ for A fixed, so that the transition probabilities are just given by

$$P(s,t,\lambda,A) = P(t-s,\lambda,A) .$$

In this case the Chapman-Kolmogorov equations take the form:

$$(4.1) \qquad P(t+s,\lambda,A) = \int_{\mathbb{R}} P(t,\eta,A)P(s,\lambda,d\eta)$$

We now define

$$(4.2) \qquad (T_t f)(\lambda) = \int_{\mathbb{R}} P(t,\lambda,d\xi)f(\xi)$$

Then $T_t\colon L^\infty \to L^\infty$, $\|T_t f\|_\infty \leq \|f\|_\infty$, and property (4.1) immediately implies that the semigroup property holds.

$$T_{t+s}f = T_s T_t f = T_t T_s f$$

Thus, in studying Markov processes with stationary transition probabilities we are led naturally to the study of an associated semi-group of contraction operators.

In the lectures by Nelson semi-groups on the space $L^\infty(\Omega,\Sigma_0,\mu)$ will arise directly from the conditional expectations, so it is useful to do the analogous construction here. Each $u \in L^\infty(\Omega,\Sigma_0,\mu)$ can be uniquely written $u(\omega) = f(x_0(\omega))$ where f is a bounded Borel function on $\operatorname{supp}\mu_0 \subset \mathbb{R}$. In fact, the map $u \to f$ is an isometry between $L^\infty(\Omega,\Sigma_0,d\mu)$ and $L^\infty(\operatorname{supp}\mu_0, d\mu_0)$. For each $u \in L^\infty(\Omega,\Sigma_0,d\mu)$ we define $\tilde{T}_t$ as follows

$$u = f(x_0(\omega)) \longrightarrow f(x_t(\omega)) \xrightarrow{E_0} E(f(x_t(\omega))|\Sigma_0)$$

$$\underbrace{\qquad\qquad\qquad\qquad\qquad}_{\tilde{T}_t}$$

$\tilde{T}_t$ is clearly a linear contraction on $L^\infty(\Omega,\Sigma_0,d\mu)$. To verify the semi-group property, let A be a Borel subset of $\operatorname{supp}\mu_0$. Then,

$$\begin{aligned}
\tilde{T}_{t+s}\chi_A(x_0(\omega)) &= E_0\chi_A(x_{t+s}(\omega)) \\
&= E_0\chi_{x_{t+s}^{-1}(A)} \\
&= E_0E_{[0,s]}\chi_{x_{t+s}^{-1}(A)} \\
&= E_0E_s\chi_{x_{t+s}^{-1}(A)} && \text{(the Markov property)} \\
&= E_0P(s,t+s,x_s(\omega),A) \\
&= E_0P(t,x_s(\omega),A) && \text{(stationary trans. prob.)} \\
&= \tilde{T}_sP(t,x_0(\omega),A) \\
&= \tilde{T}_sE_0\chi_{x_t^{-1}(A)} \\
&= \tilde{T}_sE_0\chi_A(x_t(\omega)) \\
&= \tilde{T}_s\tilde{T}_t\chi_A(x_0(\omega))
\end{aligned}$$

Since linear combinations of the functions $\chi_A(x_0(\omega))$ are dense in $L^\infty(\Omega,\Sigma_0,\mu)$ and the $\tilde{T}_t$ are bounded this proves the semi-group property. Of course this semi-group on $L^\infty(\Omega,\Sigma_0,\mu)$ is just the result of lifting the semi-group

$$(T_tf)(\lambda) = \int f(\xi)P(t,\lambda,d\xi)$$

on $L^\infty(\text{supp } \mu_0, d\mu_0)$ to $L^\infty(\Omega,\Sigma_0,\mu)$ using the isomorphism between the two spaces. Our computation shows how the Markov property and the assumption of stationary transition probabilities gives rise directly to a semi-group of contractions on $L^\infty(\Omega,\Sigma_0,\mu)$.

I will now give a very brief sketch of the semi-group theory which you will need to know. A contraction semi-group on a Banach space B is a one-parameter family $\{T_t\}_{t\geq 0}$ of contraction operators so that $T_{t+s} = T_sT_t = T_tT_s$ for all $s,t \geq 0$, and $T_0 = I$. $\{T_t\}$ is said to be strongly continuous if $T_tu - u \longrightarrow 0$ as $t \to 0$ for all $u \in B$. To see that the condition of strong continuity is not trivial notice that in the case of Brownian motion the semi-group given by (4.2) is strongly continuous on the space of bounded continuous functions but is not strongly continuous on the bounded measurable functions. This is because $(T_tf)(\lambda)$ is continuous for all $t > 0$ even if f is just bounded and measurable.

We are interested particularly in the case $B = L^p(X,d\mu)$, an L^p space of real or complex functions, where T_t is assumed to have the additional properties that $(T_tf)(x) \geq 0$ if $f(x) \geq 0$ and $T_t\mathbb{1} = \mathbb{1}$ where $\mathbb{1}$ is the function which is identically one. Any such contraction semi-group on the real valued $L^p(X,d\mu)$ functions can be extended uniquely to a contraction semi-group on the complex-valued $L^p(X,d\mu)$ functions; so from now on we will always deal with Banach spaces over the complex numbers.

Let T_t be a strongly continuous semi-group and set $A_t = t^{-1}(I - T_t)$. We now set $D(A) \equiv \{u \in B \mid \lim_{t\to 0} A_tu \text{ exists}\}$ and define $Au = \lim_{t\to 0} A_tu$ for $u \in D(A)$. If we set $u_s = \int_0^s T_tu\, dt$, then

$$T_ru_s = \int_0^s T_{t+r}u\, dt$$

and

$$A_r u_s = \frac{1}{r} \int_0^s (T_t u - T_{t+r} u)\, dt$$

$$= \frac{1}{r} \int_0^r T_t u\, dt - \frac{1}{r} \int_s^{s+r} T_t u\, dt \xrightarrow[r \to 0]{} u - T_s u .$$

Thus, for all $u \in B$, $u_s \in D(A)$. Since $u_s \to u$ as $s \to 0$, we see that $D(A)$ is in fact dense in B. Furthermore, if $u \in D(A)$, then $A_t T_t u = T_t A_t u$ so $T_t: D(A) \to D(A)$ and

$$\frac{d}{dt} T_t u = -AT_t u = -T_t Au .$$

By similar techniques one can show that A is closed. A is called the <u>infinitesimal</u> <u>generator</u> of T_t and we will write $T_t \equiv e^{-tA}$. The generators of contraction semi-groups on Banach spaces are characterized by the Hille-Yosida theorem (see [5]) but we will only need a simple special case.

<u>Proposition</u>. A closed operator A on a Hilbert space $\mathcal{H}$ is the generator of a strongly continuous semi-group of self-adjoint contractions if and only if A is self-adjoint and $A \geq 0$.

<u>Proof</u>: The if part follows immediately from the spectral theorem. We just define

$$e^{-tA} u = \int_0^\infty e^{-\lambda t}\, dE_\lambda u$$

and use the functional calculus and the dominated convergence theorem to prove the semi-group property and strong continuity. e^{-tA} is self-adjoint because $e^{-\lambda t}$ is real valued.

Conversely suppose T_t is a strongly continuous, self-adjoint, contraction semi-group on $\mathcal{H}$ and let A be the generator of T_t. If $u,v \in D(A)$, then

$$(Au,v) = \lim_{t\to 0} (\tfrac{1}{t}(I - T_t)u,v)$$
$$= \lim_{t\to 0} (u,\tfrac{1}{t}(I - T_t)v)$$
$$= (u,Av)$$

so A is symmetric. Further since T_t is a self-adjoint contraction $(T_t u,u)$ is real and

$$(T_t u,u) \leq \|T_t u\|\|u\| \leq (u,u)$$

Thus $\frac{1}{t}((I - T_t)u,u) \geq 0$ for all t, so $(Au,u) = \lim_{t\to 0} \frac{1}{t}((I - T_t)u,u) \geq 0$. Now, let $\lambda > 0$; to show that A is self-adjoint we need only prove that $\mathrm{Ran}\,(A + \lambda) = \mathcal{H}$ or $\mathrm{Ker}\,(A^* + \lambda) = \{0\}$. Suppose $(A^* + \lambda)v = 0$ and let $u \in D(A)$. Then

$$\frac{d}{dt}(T_t u,v) = -(AT_t u,v)$$
$$= -(T_t u,A^*v)$$
$$= \lambda(T_t u,v)$$

so $(T_t u,v) = (u,v)e^{\lambda t}$. Since $\lambda > 0$ and $\{T_t\}$ are contractions, this is a contradiction unless $(u,v) = 0$. But $(u,v) = 0$ for all $u \in D(A)$ means $v = 0$ since $D(A)$ is dense. Thus $\mathrm{Ker}\,(A^* + \lambda) = \{0\}$ and A is self-adjoint.

Now, let e^{-tA} be the strongly continuous self-adjoint semi-group generated by A as in the first part of the proof and set $w(t) = T_t u - e^{-tA}u$. Since $w(0) = 0$, $\|w(t)\|^2 \geq 0$, and

$$\frac{d}{dt}(w(t),w(t)) = -2(Aw(t),w(t)) \leq 0$$

we see that $w(t) = 0$ for all t, i.e. $T_t = e^{-tA}$.

The problem which we want to investigate is this. Suppose $<\Omega,\mu>$ is a probability space and $A \geq 0$ is self-adjoint on $L^2(\Omega,d\mu)$. Under what circumstances is e^{-tA} a contraction semi-group on $L^p(\Omega,d\mu)$ for $p \neq 2$. Before stating and (partially) proving a theorem about this we need some definitions.

A strongly continuous, bounded semi-group T_t on a Banach space $\mathcal{B}$ is called a <u>bounded holomorphic semi-group of angle</u> θ, $0 < \theta < \frac{\pi}{2}$ if

(1) T_t is the restriction to the positive real axis of a family of operators T_z, $z \in S_\theta = \{z \mid |\arg z| < \theta\}$, so that $T_z u$ is a holomorphic vector-valued function for all $u \in \mathcal{B}$, $z \in S_\theta$ and $T(z + z') = T(z)T(z')$ for $z, z' \in S_\theta$.

(2) For each $\theta_1 < \theta$, T_z is uniformly bounded in the sector S_{θ_1} and $T(z)u \to u$ as $z \to 0$ in S_{θ_1}.

Notice that on $L^2(\Omega, d\mu)$ we may define

$$e^{-zA}u = \int_0^\infty e^{-z\lambda}\, dE_\lambda u \quad , \qquad \operatorname{re} z > 0 .$$

Using properties of the functional calculus and the dominated convergence theorem one can easily show that (1) and (2) hold so e^{-tA} is a bounded holomorphic semi-group of angle $\frac{\pi}{2}$ on $L^2(\Omega, d\mu)$.

We will call e^{-tA} an <u>L^p-contractive semi-group</u> if $\|e^{-tA}u\|_p \leq \|u\|_p$ for all $u \in L^2 \cap L^p$ and <u>all</u> $p \in [1, \infty]$. If the map $t \to e^{-tA}$ is strongly continuous for all $p < \infty$, we will call e^{-tA} a <u>continuous</u> L^p-contractive semi-group.

<u>Theorem</u>. (Stein) Let $<\Omega, \mu>$ be a finite measure space $(\mu(M) = 1)$ and A a positive self-adjoint operator on $L^2(M, d\mu)$. Then,

(a) If e^{-tA} is positivity preserving (i.e., $(e^{-tA}f)(x) \geq 0$ if $f(x) \geq 0$) and $e^{-tA}\mathbb{1} = \mathbb{1}$ (where $\mathbb{1}$ is the function which is identically one) then e^{-tA} is an L^p-contractive semi-group.

(b) Every L^p-contractive semi-group is automatically continuous. Moreover, $\operatorname{Ker}(e^{-tA} \restriction L^p) = \{0\}$ for all $p > 1$ and $\operatorname{Ran}(e^{-tA} \restriction L^q)$ is dense in L^q for all $q < \infty$.

(c) For $1 < p < \infty$, e^{-tA} is a bounded holomorphic semi-group in the sector

$$S^{(p)} = \{z \mid |\arg z| \le \tfrac{\pi}{2}(1 - |\tfrac{2}{p} - 1|)\}$$

<u>Proof</u>: We begin by showing that e^{-tA} is a contraction on all the L^p spaces. First, suppose $f \in L^2$ and $f \ge 0$. Then,

$$\|e^{-tA}f\|_1 = (1\!\!1, e^{-tA}f) = (e^{-tA}1\!\!1, f) = (1\!\!1, f) = \|f\|_1$$

If $f \in L^2$ is real-valued, then we write $f = f_+ - f_-$ where $f_+, f_- \ge 0$ and $(f_+, f_-) = 0$. Then

$$\|e^{-tA}f\|_1 \le \|e^{-tA}f_+\|_1 + \|e^{-tA}f_-\|_1$$

$$= \|f_+\|_1 + \|f_-\|_1 = \|f\|$$

Finally, suppose $f(x) \in L^2$ is complex-valued. Then

$$|(e^{-tA}f)(x)| = \sup_{\eta \text{ rational}} \{\mathrm{Re}\,[e^{-i\eta}(e^{-tA}f(x)]\}$$

$$= \sup_{\eta \text{ rational}} \{\mathrm{Re}\,[(e^{-tA}(e^{-i\eta}f))(x)]\}$$

$$= \sup_{\eta \text{ rational}} \{(e^{-tA}(\mathrm{Re}\; e^{-i\eta}f))(x)\}$$

for almost all x, where we have used the fact that e^{-tA} takes real functions into real functions since e^{-tA} is positivity preserving. Also, for each real $g \in L^2$,

$$(e^{-tA}g)(x) = e^{-tA}g_+ - e^{-tA}(g_-)$$

$$\le e^{-tA}g_+ + e^{-tA}g_-$$

$$= e^{-tA}|g(x)|$$

almost everywhere. Therefore,

$$|(e^{-tA}f)(x)| \le e^{-tA}|f(x)| \quad \text{a.e.}$$

which implies that $\|e^{-tA}f\|_1 \le \|e^{-tA}|f(x)|\|_1 = \|f\|_1$. Thus, for all $f \in L^2$, $\|e^{-tA}f\|_1 \le \|f\|_1$.

If $f \in L^\infty \subset L^2$, then

$$\|e^{-tA}f\|_\infty = \sup_{\substack{\|g\|_1=1 \\ g \in L^2}} (e^{-tA}f,g)$$

$$= \sup_{\substack{\|g\|_1=1 \\ g \in L^2}} (f,e^{-tA}g)$$

$$\le \|f\|_\infty$$

so e^{-tA} is a contraction on L^∞ also. Thus by the Riesz-Thorin Theorem, e^{-tA} is a contraction on all the L^p spaces.

To prove that e^{-zA} on $L^p(\Omega,d\mu)$, $1 < p < \infty$, is holomorphic in the sectors $S^{(p)}$ and bounded in the sub-sectors one needs Stein's generalization of the Riesz-Thorin Theorem. The reader can find the details in [13]. We will conclude by proving the strong continuity. If $1 \le p \le 2$ then

$$\|e^{-tA}f - f\|_p \le \|e^{-tA}f - f\|_2$$

for $f \in L^2$. Since L^2 is dense in L^p strong continuity follows since $\{e^{-tA}\}$ are uniformly bounded on L^p and e^{-tA} is strongly continuous on L^2. Now, suppose $2 < p < \infty$ and let q satisfy $\frac{1}{p} + \frac{1}{q} = 1$. Suppose $e^{-t_0A}\psi = 0$ for some $\psi \in L^q$. Then $e^{-sA}\psi = 0$ for all $s \ge t_0$ and by analyticity $e^{-tA}\psi = 0$ for all $t > 0$. Since e^{-tA} is strongly continuous on L^q, $\psi = 0$. Thus $\mathrm{Ker}\,(e^{-tA}) = \{0\}$ on L^q for each $t > 0$. The reader can easily check that the adjoint of e^{-tA} on L^q is e^{-tA} on L^p. So we conclude that $\mathrm{Ran}\,(e^{-tA})$ is dense in L^p. Let $\psi = e^{-t_0A}\varphi$, $\varphi \in L^p$. Then

$$\|e^{-tA}\psi - \psi\|_p \le \|e^{-(t+t_0)}\varphi - e^{-t_0A}\varphi\|_p \longrightarrow 0$$

as $t \to 0$ by the holomorphicity in the interior of $S^{(p)}$ quoted above. Since $\mathrm{Ran}\,(e^{-t_0 A})$ is dense and the $\{e^{-tA}\}$ are uniformly bounded, we conclude that e^{-tA} is strongly continuous on L^p.

Warning. The semi-group $\tilde{T}_t$ which we constructed earlier (4.2) on $L^\infty(\Omega,\Sigma_0,\mu_0)$ is not L^p-contractive. To see this one need only calculate that

$$\|\chi_A(x_0(\omega))\|_{L^1(\Omega,\Sigma_0,\mu_0)} = \mu_0(A)$$

but

$$\|\tilde{T}_t\chi_A(x_0(\omega))\|_{L^1(\Omega,\Sigma_0,\mu_0)} = \|P(t,x_0(\omega),A)\|_{L^1}$$
$$= \mu_t(A)$$

and for appropriate A, $\mu_t(A) > \mu_0(A)$. If the Markov process which we constructed was stationary (i.e. $\mu_t(A) = \mu_0(A)$ for all t and A) rather than just having stationary transition probabilities, then we would have L^p-contractivity. This is why the processes which Ed Nelson will construct will be generalizations of the so called Ornstein-Uhlenbeck process which is stationary.

§5. GENERALIZED STOCHASTIC PROCESSES

Suppose that x_t, $t > 0$ is a stochastic process of the type we have previously discussed, i.e. for each t, x_t is a random variable on some probability space. Suppose that we write

(5.1)
$$\psi(f) = \int_0^\infty f(t)x_t\,dt.$$

where f is in some suitably nice class of functions E. Then $f \to \psi(f)$ is a linear map from E to the random variables. This suggests the following:

<u>Definition</u>. Let E be a locally convex topological space and $<\Omega,\Sigma,\mu>$ a probability space. A linear map ψ from E to the random variables on $<\Omega,\Sigma,\mu>$ is called a <u>generalized stochastic process over</u> E <u>on</u> $<X,\Sigma,\mu>$.

The space E will usually be some space of smooth functions like $S(0,\infty)$ or $S(\mathbb{R}^n)$. Given a stochastic process one can construct a generalized stochastic process by (5.1) but the converse is not necessarily true. Generalized stochastic processes are "random variable-valued distributions" while stochastic processes are "random variable-valued functions." It is this more general notion of stochastic process which arises in quantum field theory.

In §2 we showed how a stochastic process can be "realized" on the set of all functions on $[0,\infty]$. What this meant was that we could construct a probability measure on the set of all functions so that the stochastic process $\tilde{x}_t$ given by evaluation at t has the same finite dimensional distributions as x_t. We begin this section by making an analogous construction for $\psi(f)$. We will denote by E^* the dual space of E and by E^*_{alg} the algebraic dual space, i.e. the set of all everywhere defined linear functionals. We first introduce a σ-algebra on E^*_{alg}. Let F be a finite dimensional subspace of E and define

$$F_a = \{T \in E^*_{alg} \mid T(f) = 0, \text{ for all } f \in F\}$$

Let ρ_F be the natural projection of E^*_{alg} onto E^*_{alg}/F_a. Since E^*_{alg}/F_a is isomorphic to F^* and F^* is finite dimensional there is a natural notion (from $\mathbb{R}^n$) of Borel set in E^*_{alg}/F_a. A set $A \subset E^*_{alg}$ is called a <u>cylinder set based on</u> F if A is of the form $A = \rho_F^{-1}(B)$ where B is a Borel set in E^*_{alg}/F_a. The collection of cylinder sets based on F will be denoted by Σ_F. The smallest σ-algebra containing all the Σ_F for all finite dimensional subspaces F in E will be denoted by Σ_E. Now for $f \in E$, $T \in E^*_{alg}$, define

$$\varphi(f)(T) = <f,T> \equiv T(f)$$

For each f, $\varphi(f)$ is a function on E^*_{alg} measurable with respect to Σ_E. Thus if μ^* is a probability measure on $\langle E^*_{alg}, \Sigma_E \rangle$, then $f \longrightarrow \varphi(f)$ is a generalized stochastic process on $\langle E^*_{alg}, \Sigma_E, \mu^* \rangle$ over E.

Now, let ψ be a generalized stochastic process over E on $\langle \Omega, \Sigma, \mu \rangle$ and let $\{f_i\}_{i=1}^n$ be a basis for a finite dimensional subspace F of E. Let μ_F be the joint distribution of $\{\psi(f_i)\}_{i=1}^n$, i.e.

$$\mu_F(B) = \mu\{\omega \in \Omega \mid \langle \psi(f_1)(\omega), \ldots, \psi(f_n)(\omega) \rangle \in B\}$$

Then μ_F induces a measure $\hat{\mu}_F$ on E^*_{alg}/F_a by $\hat{\mu}_F(\hat{B}) = \mu_F(B)$ where

$$\hat{B} = \{\sum_{i=1}^n c_i \hat{f}_i \mid \langle c_1, \ldots, c_n \rangle \in B\}$$

and $\{\hat{f}_i\}_{i=1}^n$ is the dual basis to $\{f_i\}_{i=1}^n$. We now define

$$\mu^*_F(A) = \hat{\mu}_F(\rho_F(A))$$

for a cylinder set A in E^*_{alg} based on F. It can be checked that μ^* is independent of the choice of basis $\{f_i\}_{i=1}^n$ and the μ^*_F are consistent. That is, if $F_1 \subset F_2$ then $\mu^*_{F_2} \restriction \Sigma_{F_1} = \mu_{F_1}$. Therefore, setting

$$\mu^*(A) = \mu^*_F(A) \quad , \qquad A \text{ cylinder set based on } F$$

we obtain a measure on the family of cylinder sets in E^*_{alg}. The joint distribution of $\{\varphi(f_i)\}_{i=1}^n$ is

$$\begin{aligned}\nu(B) &= \mu^*_F\{T \mid \langle \varphi(f_1)(T), \ldots, \varphi(f_n)(T) \rangle \in B\} \\ &= \mu^*_F\{T \mid \langle\langle f_1, T\rangle, \langle f_2, T\rangle, \ldots, \langle f_n, T\rangle\rangle \in B\} \\ &= \hat{\mu}_F\{\Sigma\, c_i \hat{f}_i \mid \langle c_1, \ldots, c_n \rangle \in B\} \\ &= \mu_F(B)\end{aligned}$$

so $\{\varphi(f_i)\}_{i=1}^n$ and $\{\psi(f_i)\}_{i=1}^n$ have the same joint distributions. The only step left to show that $\psi(\cdot)$ can be realized as $\varphi(\cdot)$ on $<E^*_{alg},\Sigma_E>$ is to show that the measure μ^* which is presently only defined on the cylinder sets can be extended to a probability measure on Σ_E. This situation is analogous to the extension problem which we faced in §2 and which was solved by the Kolmogorov theorem.

Theorem. (Lenard) Every cylinder measure on E^*_{alg} can be extended to a countably additive measure on $<E^*_{alg},\Sigma_E>$.

Proof: Let $\mu^* = \{\mu^*_F\}$ be the cylinder measure on E^*. By a standard measure theory lemma [11], μ^* can be uniquely extended if and only if for each decreasing sequence, $\ldots A_n \supset A_{n+1} \ldots$, of cylinder sets, $\mu(A_n) > \delta$ for all n implies that $\bigcap_n A_n \neq \emptyset$. Each A_n is of the form

$$A_n = \{T \in E^*_{alg} \mid <T(f_1^{(n)}),\ldots,T(f_{k(n)}^{(n)})> \in B_n \subset \mathbb{R}^{k(n)}\}$$

for some $f_i^{(n)} \in E$, $i = 1,\ldots,k(n)$. We may assume that all the $f_i^{(n)}$ are selected from a linearly independent set $\{g_i\}_{i=1}^{\infty}$, because we can always change B_n by expressing a non-independent f as a finite linear combination of previous ones. Now, consider the map:

$$\theta:\ E^*_{alg} \longrightarrow \prod_{i=1}^{\infty} \mathbb{R}$$

given by

$$\theta(T) = <T(g_1),T(g_2),\ldots>$$

θ is onto since given $<w_i>$ we set $T(g_i) = w_i$, extend by linearity to finite linear combinations of the g_i, and define T to be zero elsewhere. (Note that in general T is not 1-1 and is not onto when restricted to E^*). If S is a cylinder set in $\prod_{i=1}^{\infty} \mathbb{R}$, we define $\tilde{\mu}(S) = \mu^*(\theta^{-1}(S))$. Thus $\tilde{\mu}$ is a consistent cylinder measure on $\prod_{i=1}^{\infty} \mathbb{R}$ and so by the Kolmogorov theorem has a countably additive extension to Σ, the smallest σ-algebra containing such S. Therefore, since $\tilde{\mu}(\theta(A_n)) > \delta$ for all n and $\theta(A_n) \supset \theta(A_{n+1})$ there is a point $<w_i> \in \bigcap_{n=1}^{\infty} \theta(A_n)$. Since θ is onto, there is a $T \in \bigcap_{n=1}^{\infty} A_n$. Thus, μ^* has an extension.

Corollary 1. Every generalized stochastic process ψ over E can be realized as the canonical process φ on $<E^*_{alg}, \Sigma_E, \mu^*>$ where μ^* is the probability measure constructed above.

From now on we will denote measures on $<E^*_{alg}, \Sigma_E>$ by μ instead of μ^*.

Corollary 2. Let $C(\cdot)$ be a complex-valued function on E of positive type such that $C(tf)$ is a continuous function of t for each $f \in E$ and $C(0) = 1$. Then there is a probability measure μ on $<E^*_{alg}, \Sigma_E>$ so that

$$C(f) = \int_{E^*_{alg}} e^{i<f,T>} \, d\mu(T) \tag{5.2}$$

C is called the characteristic function of μ.

Proof: To say that C is of positive type means that for any finite collection $\{f_i\}_{i=1}^n$ of vectors in E, $\{C(f_i - f_j)\}$ is a positive matrix on $\mathbb{C}^n$. Let F be a finite-dimensional subspace in E. Then the restriction of C to F is a continuous positive definite function with $C(0) = 1$, so by the classical theorem of Bochner, there is a unique probability measure $\hat{\mu}_F$ on F^* so that

$$C(f) = \int_{F^*} e^{i<f,T>} \, d\hat{\mu}_F(T)$$

for all $f \in F$. We now lift this measure exactly as we did before to a measure μ_F on Σ_F. $\{\mu_F\}$ is a consistent (by the uniqueness in Bochner's theorem) family and so defines a cylinder measure μ on E^*_{alg}. By the theorem, μ extends to a probability measure on $<E^*_{alg}, \Sigma_E>$.

Since E^*_{alg} is difficult to handle analytically it is important to know when measures on $<E^*_{alg}, \Sigma_E>$ have their support in smaller spaces, in particular on E^*. A sufficient condition is provided by the following theorem of

Minlos ([3], [4], [9]).

Theorem. Let μ be a probability measure on $<E^*_{alg}, \Sigma_E>$. Then if E is a nuclear space and if the characteristic function of μ is continuous on E, μ is supported by E^*.

Corollary. Let ψ be a generalized stochastic process on $<\Omega,\Sigma,\mu>$ over a nuclear space E. Suppose ψ is continuous in the sense that $f_\gamma \xrightarrow{E} f$ implies that $\psi(f_\gamma) \longrightarrow \psi(f)$ in measure on $<\Omega,\Sigma,\mu>$. Then ψ can be realized as the canonical process φ on $<E^*,\Sigma_E,\mu^*>$ where μ^* is a probability measure on $<E^*,\Sigma_E>$.

Proof: We already know that ψ can be realized on $<E^*_{alg},\Sigma_E,\mu^*>$. But the continuity condition implies that the characteristic function of μ^* is continuous on E so the support of μ^* is on E^*.

We remark that if E is metrizable (like $S(\mathbb{R}^n)$) the continuity condition can be replaced by the requirement that $f_n \xrightarrow{E} f$ implies that $\varphi(f_n) \longrightarrow \varphi(f)$ pointwise almost everywhere.

Example (the CCR). Let E be a real nuclear space with a continuous scalar product $(\cdot,\cdot)$. A representation of the Weyl relations over E is a pair of maps $f \longrightarrow U(f)$, $f \longrightarrow V(g)$ from E to the unitary operators on a separable Hilbert space $\mathcal{H}$ so that

(1) $U(f + g) = U(f)U(g)$, $V(f + g) = V(f)V(g)$

(5.3) (2) $V(g)U(f) = e^{i(f,g)}U(f)V(g)$

(3) $f_\gamma \xrightarrow{E} f$ implies that $U(f_\gamma) \longrightarrow U(f)$ and $V(f_\gamma) \longrightarrow V(f)$ strongly.

Gelfand and Vilenkin show in [3] how any representation which has a cyclic vector Ω_0 for $\{U(f)\}_{f\in E}$ can be realized on $L^2(E^*,\Sigma_E,d\mu)$ for an appropriate μ. We will sketch the result to show how the Minlos theorem is used; for details see

[3]. Define

$$L(f) = (U(f)\Omega_0,\Omega_0)_H$$

Then L is a continuous complex-valued function of positive type on E satisfying $L(0) = 1$, so by the Minlos theorem

$$L(f) = \int_{E^*} e^{i\langle f,T\rangle}\, d\mu(T)$$

for some probability measure μ on $\langle E^*,\Sigma_E\rangle$. Define

$$S\colon \sum_{i=1}^{n} \alpha_i U(f_i)\Omega_0 \longrightarrow \sum_{i=1}^{n} \alpha_i e^{i\langle f_i,\cdot\rangle}$$

Then it is easily checked that S extends to a unitary map of H onto $L^2(E^*,\Sigma_E,d\mu)$. If we set

$$\tilde{U}(f) = SU(f)S^{-1}$$

$$\tilde{V}(f) = SV(f)S^{-1}$$

Then $\tilde{U}(f)$ and $\tilde{V}(f)$ act on $h \in L^2(E^*,\Sigma_E,d\mu)$ by

$$(5.4)\qquad \begin{aligned} (\tilde{U}(f)h)(T) &= e^{i\langle f,T\rangle}h(T) \\ (\tilde{V}(f)h)(T) &= (\tilde{V}(f)\mathbb{1})(T)h(T+f) \end{aligned}$$

Conversely, given any probability measure μ on $\langle E^*,\Sigma_E\rangle$ the formulas (5.4) define a representation of the Weyl relations over E on $L^2(E^*,\Sigma_E,d\mu)$. It turns out that two such representations are equivalent if and only if the corresponding measures $d\mu_1$ and $d\mu_2$ are equivalent (i.e., absolutely continuous with respect to each other).

Remark. It is reasonable to ask why (5.3) is the appropriate generalization of the finite dimensional case of the Weyl relations. Instead we could say that a representation of the Weyl relations is a pair of families $\{U_k(t)\}$, $\{V_\ell(t)\}$ of strongly continuous unitary groups on a Hilbert space H so that:

(5.5)

$$(1)\quad [U_k(t),U_\ell(s)] = 0 = [V_k(t),V_\ell(S)]$$

$$(2)\quad V_k(t)U_\ell(s) = e^{ist\delta_{k\ell}}U_\ell(s)V_k(t)$$

Of course, given a representation of the form (5.3) we can always get a representation of the form (5.5) by choosing an orthonormal basis $\{f_k\}$ in the inner product $(\cdot,\cdot)$ on E and defining $U_k(t) = U(tf_k)$, $V_k(t) = V(tf_k)$. However, to go from (5.5) to (5.3) one must show that a certain family of infinite products $\prod_k U_k(t_k)$ and $\prod_k V_k(s_k)$ make sense. In fact, this can be done. Every representation of the form (5.5) can be continuously extended to one of the form (5.3); see [10].

Generalized stochastic processes often arise in the following way. Let $(\cdot,\cdot)$ be a continuous inner product on a nuclear space E. Then $e^{-\frac{1}{2}(f,f)}$ is a continuous characteristic function and thus by the Minlos theorem there is a probability measure μ on $\langle E^*,\Sigma_E\rangle$ so that

$$e^{-\frac{1}{2}(f,f)} = \int_{E^*} e^{i\langle f,T\rangle}\, d\mu(T)$$

Now, let $f_1,\dots,f_N$ be independent vectors in E. Then

$$e^{-\frac{1}{2}\sum\limits_{i,j}\alpha_i\alpha_j(f_i,f_j)} = e^{-\frac{1}{2}(\Sigma\,\alpha_i f_i,\ \Sigma\,\alpha_i f_i)}$$

$$= \int_{E^*} e^{i\sum\limits_{i=1}^{N}\alpha_i\varphi(f_i)(T)}\, d\mu(T)$$

$$= \int_{\mathbb{R}^N} e^{i\,\Sigma\,\alpha_i\lambda_i}\, d\mu_{\varphi(f_1),\dots,\varphi(f_n)}(\vec{\lambda})$$

Therefore, by our remarks in §1, $\varphi(f_1),\dots,\varphi(f_N)$ are a Gaussian system (i.e. $d\mu_{\varphi(f_1),\dots,\varphi(f_n)}$ is a multi-variate Gaussian) with means equal to zero and covariance $\Gamma_{ij} = (f_i, f_j)$. For this reason μ is called a Gaussian measure on E^* and $\varphi(\cdot)$ is called the Gaussian process with mean zero and covariance $= (f,g)$.

The following theorem is essentially a corollary of the proof of the Minlos Theorem.

Theorem. Let E be a nuclear space, μ a measure on E^*, C its characteristic function. Let $(\cdot,\cdot)_0$ be a continuous inner product on E and H_0 the completion of E in $(\cdot,\cdot)_0$. Suppose that C is continuous on H_0. Let T be a Hilbert-Schmidt operator on H_0 satisfying:

(a) T is one to one

(b) $E \subset \operatorname{Ran} T$ and $T^{-1}(E)$ is dense in H_0.

(c) The map $E \xrightarrow{T^{-1}} H_0$ is continuous.

Then the support of μ is on $(T^{-1})^* H_0^* \subset E^*$.

In the statement $(T^{-1})^*$ and H_0^* mean "adjoint" and "dual space" in the pairing between E and E^*, not the inner product on H_0. For a proof of the theorem see [4], [14]. To illustrate the use of this theorem we conclude with an example.

Example. Let μ be the measure on $S'(\mathbb{R}^n)$ corresponding to the Gaussian process with means zero and covariance $(f,(-\Delta+1)^{-1}g)$. Let $P = -\Delta + 1$ and let H_{-1} denote the completion of $S(\mathbb{R}^n)$ in the norm $\|P^{-\frac{1}{2}}f\|_2$. H_{-1} will play the role of H_0 in the theorem. Let $H = L^2(\mathbb{R}^n)$. Since $P^{\frac{1}{2}}$ is a unitary map of H into H_{-1}, $T\colon H_0 \to H_0$ is Hilbert-Schmidt if and only if $T_0 = P^{-\frac{1}{2}}TP^{\frac{1}{2}}$ is Hilbert-Schmidt on H. Since $T = P^{\frac{1}{2}}T_0P^{-\frac{1}{2}}$, we have $T^{-1} = P^{\frac{1}{2}}T_0^{-1}P^{-\frac{1}{2}}$ and $(T^{-1})^* = P^{-\frac{1}{2}}(T_0^{-1})^*P^{\frac{1}{2}}$. Thus if T_0 is Hilbert-Schmidt on H and T satisfies conditions (a), (b), and (c), the support of μ will be on

$$(T^{-1})^* H^*_{-1} = P^{-\frac{1}{2}}(T_0^{-1})^* P^{\frac{1}{2}}(P^{-\frac{1}{2}}H)$$

$$= P^{-\frac{1}{2}}(T_0^{-1})^* H$$

Let $T_0 = P^{-\frac{n}{4}} P_1^{-\alpha} Q^{-\beta}$ where $P_1 = (-\dfrac{\partial^2}{\partial x_1{}^2} + 1)$ and $Q = (x^2 + 1)$. Suppose that

(5.4) $$\beta > \frac{n}{4} \quad , \quad \alpha > 0$$

Then T_0 will be Hilbert-Schmidt since it is an integral operator with kernel

$$[\widehat{((p^2 + 1)^{-\frac{n}{4}}(p_1{}^2 + 1)^{-\alpha})}(x - y)](y^2 + 1)^{-\beta}$$

and the kernel is in $L^2(R^{2n})$ if (5.6) is satisfied. This depends on the fact that

$$\int \frac{d^n p}{(p^2 + 1)^{n/2}(p_1{}^2 + 1)^{\alpha}} < \infty$$

for all $\alpha > 0$. Furthermore, it is easy to check that $T = P^{-\frac{1}{2}} T_0 P^{\frac{1}{2}}$ satisfies the conditions (a), (b), and (c) of the theorem. Thus, μ has support on

$$\{P^{\frac{n}{4} - \frac{1}{2}} P_1{}^{\alpha} Q^{\beta} f \mid f \in L^2(\mathbb{R}^n)\}$$

for all α and β satisfying (5.4). In the case $n = 2$, μ has support on the set

$$\{P_1{}^{\alpha} Q^{\beta} f \mid f \in L^2(\mathbb{R}^2)\}$$

for any $\alpha > 0$. In particular, for almost all (with respect to μ) $g \in S'(\mathbb{R}^n)$, $(-\dfrac{d^2}{dx_1{}^2} + 1)^{-\alpha} g$ is locally in L^2.

The local result, i.e. that the support of μ is on paths which are locally in $P^{\frac{n}{4}-\frac{1}{2}} P_1^{\alpha} L^2$, is contained in the work of J. Cannon [16]. For further results on the L^2 support properties of μ see M. Reed and L. Rosen [17]. The following appendix by P. Colella and O. Lanford contains results on the support of μ in terms of the lim sup properties of the sample paths.

BIBLIOGRAPHY

[1] Breiman, L., Probability, Addison-Wesley, 1968.

[2] Feller, W. F., An Introduction to Probability Theory and Applications, Vol. I ( ), Vol. II (1971), John Wiley & Sons.

[3] Guelfand, I. M. and N. Y. Vilenkin, Les Distributions, T. IV, Dunod, Paris, 1967.

[4] Hida, T., Stationary Stochastic Processes, Mathematics lecture notes, Princeton Univ. Press, 1970.

[5] Hille, E. and R. S. Phillips, Functional Analysis and Semi-groups, Amer. Math. Soc. colloq. pub. 31, 1957.

[6] Ito, K. and H. P. McKean Jr., Diffusion Processes and Their Sample Paths, Academic Press, N. Y., 1965.

[7] Loève, M., Probability Theory, Van Nostrand, 1960.

[8] Mathematics of Contemporary Physics, R. F. Streater, ed., Academic Press, 1972.

[9] Minlos, R. A., "Generalized random processes and their extension to a measure," Trudy. Moscow Mat. Obšč. 8 (1959), 497-518.

[10] Reed, M., "A Garding domain for quantum fields," Comm. Math. Phys. 14 (1969), 336-346.

[11] Royden, H., Real Analysis, Macmillan, 1963.

[12] Statistical Mechanics and Quantum Field Theory, ed. By C. de Witt and R. Stora, Gordon and Breach (1971).

[13] Stein, E., Topics in Harmonic Analysis, Annals of Mathematical Studies 63, Princeton, 1970.

[14] Umemura, Y., "Measures on infinite dimensional vector spaces," and "Carriers of continuous measures in Hilbertian norm," Pub. of Research institute of Kyoto Univ. A.1 (1965), pp. 1-54.

[15] Varadhan, S. R. S., Stochastic Processes, lecture notes from the Courant Institute, N. Y., 1968.

[16] Cannon, J., "Continuous sample paths in quantum field theory," to appear in Comm. Math. Phys.

[17] Reed, M. and L. Rosen, "Global properties of the free Markov measure," to appear.

APPENDIX: SAMPLE FIELD BEHAVIOR FOR THE FREE MARKOV RANDOM FIELD

Phillip Colella and Oscar E. Lanford III*
Department of Mathematics, University of California
Berkeley, California 94720

I. INTRODUCTION

This appendix is concerned with the following question: If μ_0 denotes the Gaussian probability measure on $S'(\mathbb{R}^2)$ with mean zero and covariance $((-\Delta+\mathbb{1})^{-1}f,g)$, what are the properties of "typical" distributions with respect to μ_0? A first result in this direction is given in the final paragraphs of Professor Reed's lectures; he shows that, if $\alpha > 0$, then for almost all $T \in S'(\mathbb{R}^2)$, $\left(\frac{-d^2}{dx_1^2} + 1\right)^{-\alpha} T$ is a locally square-integrable function.

For ease of reference, we will summarize our results here in something less than their full generality:

Theorem 1.1. (a) *The set of distributions* T *having the property that there exists a non-empty open set* U_T *on which* T *is equal to a signed measure is a set of* μ_0*-measure zero.*

*Alfred P. Sloan Foundation Fellow. Also supported in part by NSF Grant GP-15735.

b. *Let* $0 < \alpha \leq \frac{1}{2}$. *The set of distributions* T *such that* $\left(\frac{-d^2}{dx_1^2} + 1\!\!1\right)^{-\alpha/2} T$ *is a locally Hölder continuous function with all exponents* $\alpha' < \alpha$* *is a set of* μ_0*-measure one.*

c. *Again let* $0 < \alpha \leq \frac{1}{2}$. *The set of distributions* T *such that*

$$\limsup_{x\to\infty} \frac{\left(\frac{-d^2}{dx_1^2} + 1\!\!1\right)^{-\alpha/2} T(x)}{\sqrt{\log|x|}} = \frac{1}{\pi}\sqrt{\int_{\mathbb{R}^2} (k^2+1)^{-1}(k_1^2+1)^{-\alpha}\, dk}$$

is a set of μ_0*-measure one.*

Result a) is a negative one; it says that a typical distribution is nowhere sufficiently regular to be a locally integrable function or even a signed measure. Result b) on the other hand says that typical distributions fail to be functions only very slightly. That is, if we regularize so as to add an arbitrarily small fraction of a derivative, typical distributions become Hölder continuous functions. Moreover, it is only necessary to add this fraction of a derivative in one direction, so a typical distribution may be regarded as a continuous function of x_2 with values in the space of distributions in x_1. Thus, although a typical distribution is not a function, it still makes sense to restrict it to a line x_2 = const.

*We say that a function T(x) is locally Holder continuous with exponent α' if, for each bounded open set $\Lambda \subset \mathbb{R}^2$, there is a constant $Q(\Lambda)$ such that

$$|T(x) - T(y)| \leq Q(\Lambda)|x-y|^{\alpha'}$$

for all $x,y \in \Lambda$.

Result c) gives rather detailed information about the behavior of a typical distribution at infinity. It says in particular that, for almost every T,

$$\left(\frac{-d^2}{dx_1^2} + \mathbb{1}\right)^{-\alpha/2} T(x)$$

can be majorized by const. × $\sqrt{\log |x|}$ for large x but not by any less rapidly growing function of x. Moreover, we will prove that, if ρ is any non-zero element of $S(\mathbb{R}^2)$, then, for almost every T, $T*\rho(x)$ can be majorized by const × $\sqrt{\log |x|}$ but not by any less rapidly growing function. Thus, we may say that a typical T grows like $\sqrt{\log |x|}$ at infinity. It should be understood, however, that a typical T does not "behave like $\sqrt{\log |x|}$" for large x; instead, it is usually much smaller than $\sqrt{\log |x|}$ and only occasionally takes values which are this large. Indeed, a typical T behaves in some respects as if it were bounded; for example, if f(x) is any square-integrable function, the set of distributions T such that

$$\left(\frac{-d^2}{dx_1^2} + \mathbb{1}\right)^{-\alpha/2} T(x)\ f(x)$$

is square-integrable is a set of μ_0-measure one.

Since $S'(\mathbb{R}^n)$ is not a locally compact space, measure theory on it is not quite standard. We will not give a systematic investigation of the subject, but there are a few simple remarks which should make it seem less strange. To begin with, $S'(\mathbb{R}^n)$ comes equipped with three σ-algebras which might in principle be distinct:

Σ_0: the σ-algebra generated by the continuous linear functionals

Σ_w: the σ-algebra generated by the weakly closed sets

Σ_s: the σ-algebra generated by the closed sets.

Evidently, $\Sigma_0 \subset \Sigma_w \subset \Sigma_s$. Gaussian measures are defined on Σ_0, but it is frequently easier to verify that sets we are interested in belong to Σ_s. Fortunately, we have in fact $\Sigma_0 = \Sigma_w = \Sigma_s$. There are general theorems which imply this equality, but in the case at hand it is easy to give a direct elementary proof: By using Hermite expansions, we can identify $S(\mathbb{R}^n)$ with the space s of rapidly decreasing sequences and $S'(\mathbb{R}^n)$ with the space s' of polynomially bounded sequences. Let

$$K_n = \{(a_j) \in s' : |a_j| \leqslant n(1+j^n) \quad \text{for all} \quad j\} .$$

It is easy to check that K_n is compact and metrizable in the strong topology on s', that $K_n \in \Sigma_0$, and that $s' = \bigcup_{n=1}^{\infty} K_n$. A subset B of s' belongs to $\Sigma_0(\Sigma_s)$ if and only if $B \cap K_n$ belongs to $\Sigma_0(\Sigma_s)$ for all n. Since each K_n is compact, the strong and weak topologies agree on K_n, and since K_n is metrizable, Σ_0 and Σ_w agree on K_n. Hence, Σ_s and Σ_0 agree on K_n for all n so $\Sigma_0 = \Sigma_w = \Sigma_s$. We will from now on denote this σ-algebra simply by Σ, and when we say that a subset of $S'(\mathbb{R}^n)$ is measurable, we mean that it belongs to Σ. We will usually leave the problem of verifying the measurability of the sets we consider to the reader. The fact that $S'(\mathbb{R}^n)$ is a countable union of compact metrizable subsets is frequently useful; to a large extent, it makes possible the reduction of measure theory on $S'(\mathbb{R}^n)$ to measure theory on compact metric spaces.

We will make one small change in the notation used in the Reed lectures. If μ is a Gaussian measure of mean zero on $S'(\mathbb{R}^n)$, the <u>covariance</u> of μ was defined to be the positive semi-definite bilinear form $(\cdot,\cdot)_\mu$ on $S(\mathbb{R}^n)$ given by

$$(f,g)_\mu = \int \mu(dT) \langle f,T\rangle \langle g,T\rangle .$$

By the nuclear theorem, we can write

$$(f,g)_\mu = \int dx\, dy\, f(x)g(y)\, X_\mu(x,y),$$

where $X_\mu(x,y) = X_\mu(y,x)$ is a tempered distribution on $\mathbb{R}^n \times \mathbb{R}^n$. We will also refer to the distribution $X_\mu(x,y)$ as the covariance of μ. Furthermore, if μ is translation-invariant,* we can write

$$X_\mu(x,y) = \hat{X}_\mu(x-y),$$

where $\hat{X}_\mu$ is a distribution of positive type on $\mathbb{R}^n$. We will drop the ^ and write, for example,

$$X_\mu(x,y) = X_\mu(x-y),$$

again referring to $X_\mu(\xi)$ as the covariance of μ. In the reverse direction, if we start with a distribution X of positive type, there is a uniquely determined translation-invariant Gaussian measure with mean zero and covariance $X(x-y)$.

We make very limited claims to novelty for the results presented here. Although we have not been able to find these results in the literature, the ideas involved in proving them are quite standard and, indeed, old-fashioned. The general theory of Gaussian stochastic processes has undergone vigorous development in recent years--for a good list of references, see the reviewer's remark in MR 42#4994

*We are using "translation invariant" as a synonym for the probabilists' term "stationary", i.e., a measure μ on $S'(\mathbb{R}^n)$ is said to be translation invariant if it is invariant under the natural action on $S'(\mathbb{R}^n)$ of the group of translations in $\mathbb{R}^n$.

(1973)--but recent general results do not seem to be directly applicable to the questions investigated here. In any case, our principal objective in writing this appendix was to provide a reasonably self-contained discussion of the answers to these questions in a language accessible to mathematical physicists.

II. REGULARITY

Let $\alpha > 0$. The mapping

$$T \mapsto \left(\frac{-d^2}{dx_1^2} + \mathbb{1}\right)^{-\alpha/2} T$$

is a continuous linear mapping of $S'(\mathbb{R}^2)$ onto itself. We temporarily denote this mapping by Φ. Schematically, what we want to prove is that

$$\mu_0(\Phi^{-1}\mathcal{X}) = 1$$

where $\mathcal{X}$ is some appropriate space of locally Hölder continuous functions, regarded as a subspace of $S'(\mathbb{R}^2)$. This suggests that we investigate the "support properties" of the measure.

$$\mu = \mu_0 \circ \Phi^{-1} .$$

It is easy to see that μ is again Gaussian with mean zero but has covariance

$$\chi_\mu(x-y) = \frac{1}{(2\pi)^2} \int dp\, e^{ip\cdot x}(p^2+1)^{-1} (p_1^2+1)^{-\alpha} . \qquad (2.1)$$

The Fourier transformation can be understood literally, rather than in the sense of distributions, since $(p^2+1)^{-1}(p_1^2+1)^{-\alpha}$ is integrable. This also implies that X_μ is a continuous function. In fact:

Proposition 2.1 (a). *For* $0 < \alpha < \frac{1}{2}$,

$$|X_\mu(0) - X_\mu(x)| \leq \text{const} \times |x|^{2\alpha}$$

(b) *For* $\alpha = \frac{1}{2}$

$$|X_\mu(0) - X_\mu(x)| \leq \text{const} \times |x| \log\left(\frac{1}{|x|}\right) \quad \textit{for small} \quad |x| .$$

We will return later to the proof of this proposition; assume it for the moment. We are now in the following situation: We have a Gaussian measure μ on $S'(\mathbb{R}^2)$ with mean zero and with a covariance which is Hölder continuous at zero. We want to conclude that μ assigns measure one to the set of locally Hölder continuous functions. We can in fact prove quite a general and precise theorem in this direction; we do not need to assume that the measure is translation invariant, nor are we restricted to a two-dimensional index set. To formulate our result efficiently, we need some notation. Let α,β be real numbers, with $\alpha > 0$. For $0 < \varepsilon < 1$, define

$$\delta_{\alpha,\beta}(\varepsilon) = \varepsilon^\alpha \left[\log\left(\frac{1}{\varepsilon}\right)\right]^{-\beta+\frac{1}{2}} . \tag{2.2}$$

(Note the $\frac{1}{2}$ in the exponent of the logarithm; it is put there for convenience later on.) For any real-valued function T defined on $\mathbb{R}^n$, define

$$\|T\|_{\alpha,\beta} = \sup_{0<|x-y|\leq\frac{1}{2}} \left\{ \frac{|T(x) - T(y)|}{\delta_{\alpha,\beta}(|x-y|)[\log(|x|+2)]^{\frac{1}{2}}} \right\}, \tag{2.3}$$

and let

$$\mathcal{H}_{\alpha,\beta} = \{T: \|T\|_{\alpha,\beta} < \infty\}.$$

Then $\mathcal{H}_{\alpha,\beta}$ may be identified with a subspace of $S'(\mathbb{R}^n)$ by

$$\langle f,T\rangle = \int f(x)T(x)dx, \qquad f \in S(\mathbb{R}^n).$$

Our principal regularity result is the following:

Theorem 2.2. *Let* μ *be a Gaussian measure on* $S'(\mathbb{R}^n)$ *with mean zero and covariance* $X(x,y)$. *Assume* $X(x,y)$ *is continuous and that for some* $\alpha > 0$, β, *and* c

$$|X(x,x)+X(y,y)-2X(x,y)|^{\frac{1}{2}} \leq c|x-y|^{\alpha}\left[\log\left(\frac{1}{|x-y|}\right)\right]^{-\beta}$$

for all x,y *with* $|x-y| \leq \frac{1}{2}$. *Then*

$$\mu(\mathcal{H}_{\alpha,\beta}) = 1.$$

Note that, by combining Theorem 2.2 with Proposition 2.1 we obtain a sharper version of statement b) of Theorem 1.1, i.e., we show that, for $0 < \alpha < \frac{1}{2}$, μ_0-almost every T has the property that

$$\left(\frac{-d^2}{dx_1^2} + 1\right)^{-\alpha/2} T \quad \text{belongs to} \quad \mathcal{H}_{\alpha,0} \ (\mathcal{H}_{\frac{1}{2},-\frac{1}{2}} \quad \text{for} \quad \alpha = \tfrac{1}{2}).$$

Belonging to $\mathcal{H}_{\alpha,0}$ is a slightly weaker local property than Hölder continuity with exponent α but is stronger than Hölder continuity with all exponents $\alpha' < \alpha$.

In outline, the proof of the theorem goes as follows. Let D^n denote the set of points in $\mathbb{R}^n$ with dyadic rational co-ordinates, and let X denote $\prod_{x \in D^n} \mathbb{R}$, the space of all real-valued functions on D^n. We will again denote elements of X by T, and we will define $\|T\|_{\alpha,\beta}$, for $T \in X$, by the right-hand side of (2.3) with x,y restricted to belong to D^n. Let $\hat{\mu}$ be the Gaussian measure on X with mean zero and covariance $X(x,y)$. By this we mean that $\hat{\mu}$ is the uniquely determined probability measure on X such that, for any $x_1,\ldots,x_m \in D^n$, the random variables $T \mapsto T(x_1),\ldots,T(x_m)$ are jointly Gaussian, and such that

$$\int T(x)\hat{\mu}(dT) = 0; \quad \int T(x)T(y)\hat{\mu}(dT) = X(x,y) \quad \text{for all} \quad x,y \in D^n .$$

<u>Lemma 2.3.</u> *Let* $\tilde{X} = \{T \in X: \|T\|_{\alpha,\beta} < \infty\}$. *Then*

$$\hat{\mu}(\tilde{X}) = 1.$$

Assuming the lemma, we can easily prove the theorem:

Since $\hat{\mu}(\tilde{X}) = 1$, we can restrict the random variables $T(x)$ to $\tilde{X}$ without changing their Gaussian character or their covariance. Also, there is a natural norm-preserving bijection i of $\mathcal{H}_{\alpha,\beta}$ onto $\tilde{X}$ given by

$$i(T) = T|_{D^n} .$$

Let $\hat{\hat{\mu}}$ be the probability measure on $\mathcal{H}_{\alpha,\beta}$ which is the image of $\hat{\mu}$ under i^{-1}. The random variables on $\left(\mathcal{H}_{\alpha,\beta-\frac{1}{2}}, \hat{\hat{\mu}}\right)$ given by

$$T \mapsto T(x) \qquad (x \in D^n)$$

are, by the definition of $\hat{\hat{\mu}}$, jointly Gaussian with mean zero and covariance $X(x,y)$. If $x \in \mathbb{R}^n \setminus D^n$, and if we take a sequence (x_j) in D^n converging to x, we have

$$T(x) = \lim_{j\to\infty} T(x_j)$$

for all $T \in \mathcal{H}_{\alpha,\beta}$. This, together with the continuity of $X(x,y)$, implies that the mappings $T \mapsto T(x)$ are $\hat{\hat{\mu}}$-measurable for all x, and are jointly Gaussian with mean zero and covariance $X(x,y)$. For $f \in S(R^n)$, the mapping

$$T \to \langle f,T\rangle = \int f(x)T(x)dx$$

is again $\hat{\hat{\mu}}$-measurable, Gaussian, and of mean zero, and

$$\int \hat{\hat{\mu}}(dT) \ \langle f_1,T\rangle \langle f_2,T\rangle = \int dx\, dy\, f_1(x)\, f_2(y) X(x,y)$$

(approximate the integral defining $\langle f,T\rangle$ by Riemann sums). Hence, if we regard $\hat{\hat{\mu}}$ as a measure on $S'(\mathbb{R}^n)$ by putting

$$\hat{\hat{\mu}}(S'(\mathbb{R}^n) \setminus \mathcal{H}_{\alpha,\beta}) = 0,$$

$\hat{\hat{\mu}}$ is a Gaussian measure with mean zero and covariance $X(x,y)$. Since a Gaussian measure is uniquely determined by its mean and covariance,

$$\hat{\hat{\mu}} = \mu,$$

and the theorem is proved, once we have proved Lemma 2.3.

Now let $\Delta = [0,1]^n \cap D^n$, and define $\|T\|_{\Delta,\alpha,\beta}$ on X by the right-hand side of (2.3) with x,y restricted to lie in Δ. We next reduce Lemma 2.3 to the following local regularity statement:

Lemma 2.4. *There exist strictly positive constants* c_1, c_2 *depending only on* n, c, α, β *such that, for all positive* λ

$$\hat{\mu}\{T: \|T\|_{\Delta,\alpha,\beta} \geqslant \lambda\} \leqslant c_1 e^{-c_2\lambda^2} .$$

To prove Lemma 2.3, it suffices to show that

$$\lim_{\lambda\to\infty} \hat{\mu}\{T: \sup_{a\in\mathbb{Z}^n}\left\{\|T_{a/2}\|_{\Delta,\alpha,\beta} [\log(|a|+2)]^{-\frac{1}{2}}\right\} \geqslant \lambda\} = 0, \tag{2.4}$$

where

$$T_{a/2}(x) = T(x-a/2).$$

By Lemma 2.4 and the translation invariance of the hypotheses of Theorem 2.2

$$\hat{\mu}\{T: \|T_{a/2}\|_{\Delta,\alpha,\beta} \geqslant \lambda [\log(|a|+2)]^{\frac{1}{2}}\} \leqslant c_1(|a|+2)^{-c_2\lambda^2}$$

for any $a \in \mathbb{Z}^n$. Hence, the left-hand side of (2.4) is no larger than

$$c_1 \lim_{\lambda\to\infty} \sum_{a\in\mathbb{Z}^n} (|a|+2)^{-c_2\lambda^2} = 0.$$

We now come to the essential and most difficult step--the proof of Lemma 2.4. The argument we will give is a straightforward modification of the argument used by Ito and McKean to construct Wiener measure. (See K. Ito and H.P. McKean, Jr., Diffusion Processes and Their Sample Paths, Springer-Verlag (1965)pp. 12-15.)

For fixed $x,y \in \Delta$, $T(x) - T(y)$ is a Gaussian random variable on $(X,\hat{\mu})$ with mean zero and variance

$$X(x,x) + X(y,y) - 2X(x,y).$$

By the hypotheses of Theorem 2.2, this variance is no larger than $c^2\delta^2_{\alpha,\beta}(|x-y|)\left[\log\left(\frac{1}{|x-y|}\right)\right]^{-1}$ provided that $|x-y| \leq \frac{1}{2}$. Hence, again for $0 < |x-y| \leq \frac{1}{2}$

$$\hat{\mu}\{T: |T(x)-T(y)| \geq \gamma\, \delta_{\alpha,\beta}(|x-y|)\} \leq c_3 \cdot |x-y|^{c_4\gamma^2} \qquad (2.5)$$

for all positive γ. Here, c_3, c_4 are strictly positive constants depending only on c. This inequality is the only property of $\hat{\mu}$ we will need to complete the proof.

We now proceed by constructing, for each positive γ, a set $X(\gamma) \subset X$ and proving

a) $\hat{\mu}(X(\gamma)) \leq c_1 \exp[-c_5\gamma^2]$ for all positive γ

b) There exists a constant c_6 such that for all γ, all $T \notin X(\gamma)$, and all x,y with $|x-y| \leq \frac{1}{2}$,

$$|T(x)-T(y)| \leq c_6\, \gamma\, \delta_{\alpha,\beta}(|x-y|)\ .$$

The lemma then follows immediately, with $c_2 = c_5 c_6^{-2}$.

For notational simplicity, we consider only $n = 2$ for the remainder of the argument; the extension to arbitrary n is immediate. We will say that a pair x,y of elements of Δ is an <u>elementary pair (of order</u> j) if

1) the components of x are integral multiples of 2^{-j}
2) each component of y differs from the corresponding component of x either by zero or by $\pm 2^{-j-1}$.

We now define $X(\gamma)$ to be the set of $T \in X$ such that, for some $j = 1,2,\dots$ and some elementary pair (x,y) of order j, $|T(x)-T(y)| > 2\gamma\delta_{\alpha,\beta}(2^{-j-1})$. In other words:

$$\text{If } T \notin X(\gamma), \text{ then } |T(x)-T(y)| \leq 2\gamma\delta_{\alpha,\beta}(2^{-j-1}) \text{ for all elementary pairs } x,y \text{ of order } j = 1,2,\dots . \tag{2.6}$$

To prove a), we pick an elementary pair x,y (order j) and define $z = (x_1,y_2)$. Then $|x-z|$ is 0 or 2^{-j-1}, and similarly for $|y-z|$. Thus

$$\begin{aligned}\hat{\mu}\{|T(x)-T(y)| > 2\gamma\delta_{\alpha,\beta}(2^{-j-1})\} &\leq \hat{\mu}\{|T(x)-T(z)| > \gamma\delta_{\alpha,\beta}(2^{-j-1})\} \\ &\quad + \hat{\mu}\{|T(z)-T(y)| > \gamma\delta_{\alpha,\beta}(2^{-j-1})\} \\ &\leq 2\, c_3\, 2^{-c_4\gamma^2(j+1)}\end{aligned} \tag{2.7}$$

by (2.5). To estimate $\hat{\mu}(X(\gamma))$, we sum the right-hand side of (2.7) over all elementary pairs. For a given j, the number of elementary pairs of order j is smaller than $8 \times (2^j+1)^2$, so

$$\hat{\mu}(X(\gamma)) \leq \sum_{j=1}^{\infty} 8 \times (2^j+1)^2 \cdot 2 \cdot c_3 \cdot 2^{-c_4\gamma^2(j+1)} = O(4^{-c_4\gamma^2}) \quad \text{as} \quad \gamma \to \infty .$$

Since $\hat{\mu}(X(\gamma)) \leq 1$ for all γ, this proves a).

Turning now to the proof of b), we let x, y be two distinct points of Δ with $|x-y| \leq \frac{1}{2}$. Let

$$j_0 = \min \{j: \max_{i=1,2} |x_i - y_i| \geq 2^{-j}\} .$$

There exist $x^{(0)}$, $y^{(0)}$, whose components are integral multiples of 2^{-j_0}, such that

$$|x_i^{(0)} - x_i| < 2^{-j_0}, \ |y_i^{(0)} - y_i| < 2^{-j_0}, \ |x_i^{(0)} - y_i^{(0)}| \leq 2^{-j_0} \ (i=1,2).$$

What we want to show is that, for $T \notin X(\gamma)$,

$$|T(x) - T(y)| \leq c_6 \gamma \, \delta_{\alpha,\beta}(|x-y|) .$$

Since $|x-y| \geq 2^{-j_0}$, and since $\delta_{\alpha\,\beta}(\varepsilon)$ is increasing in ε for small ε, it suffices to show

$$|T(x) - T(y)| \leq c_7 \, \gamma \delta_{\alpha,\beta}(2^{-j_0}). \tag{2.8}$$

By the triangle inequality

$$|T(x)-T(y)| \leq |T(x)-T(x^{(0)})| + |T(x^{(0)})-T(y^{(0)})| + |T(y^{(0)})-T(y)|.$$

It follows readily from (2.6) that, if $T \notin X(\gamma)$,

$$|T(x^{(0)}) - T(y^{(0)})| \leq 4 \, \gamma \delta_{\alpha,\beta}(2^{-j_0-1}).$$

(Put $z^{(0)} = \frac{1}{2}x^{(0)} + \frac{1}{2}y^{(0)}$; then $(x^{(0)}, z^{(0)})$ and $(y^{(0)}, z^{(0)})$ are elementary pairs of order j_0.) The estimates of $|T(x)-T(x^{(0)})|$ and $|T(y)-T(y^{(0)})|$ are identical; we will give only the first of them. Recall that $|x_i - x_i^{(0)}| < 2^{-j_0}$. It is easy to see that we can construct inductively a sequence $x^{(1)}, x^{(2)}, \ldots$ such that

i) $x^{(k)}, x^{(k+1)}$ is an elementary pair of order j_0+k $(k=0,1,2,\ldots)$

ii) $|x_i^{(k)} - x_i| < 2^{-j_0-k}$ $(i=1,2)$.

It follows from ii) (and the fact that each x_i is a dyadic rational) that $x^{(k)} = x$ for sufficiently large k and from i) that, if $T \notin X(\gamma)$

$$|T(x^{(k+1)}) - T(x^{(k)})| \leqslant 2\gamma\delta_{\alpha,\beta}(2^{-j_0-k-1}) \quad \text{for all } k,$$

so

$$|T(x^{(0)}) - T(x)| \leqslant 2\gamma \cdot \sum_{k=0}^{\infty} \delta_{\alpha,\beta}(2^{-j_0-k-1}).$$

Hence, for $T \notin X(\gamma)$,

$$|T(x) - T(y)| \leqslant 4\gamma \left[\delta_{\alpha,\beta}(2^{-j_0-1}) + \sum_{k=0}^{\infty} \delta_{\alpha,\beta}(2^{-j_0-k-1})\right].$$

Since $\lim_{j\to\infty} \delta(2^{-j-1})/\delta(2^{-j}) = 2^{-\alpha} < 1$,

the estimate (2.8) follows, completing the proof of Lemma 2.4 and hence of Theorem 2.2.

We note in the following proposition some subsidiary results which follow from the above considerations.

<u>Proposition 2.5 a)</u>. *Let* Λ *be a bounded set in* $\mathbb{R}^n$. *There exist constants* c_8, c_9 *(depending on* Λ*) such that*

$$\mu\{T: \sup_{\substack{x,y\in\Lambda \\ |x-y|\leq\frac{1}{2}}} \left[\frac{|T(x)-T(y)|}{\delta_{\alpha,\beta}(|x-y|)}\right] \geq \lambda\} \leq c_8 e^{-c_9\lambda^2}$$

b) *In addition to the hypotheses of Theorem 2.2, assume* $X(x,x)$ *is bounded. Then for* μ*-almost every* T

$$\sup_{x\in\mathbb{R}^n} \left\{\frac{|T(x)|}{\sqrt{\log(|x|+2)}}\right\} < \infty \quad .$$

To prove a), we combine Lemma 2.4 with the argument deriving Theorem 2.2 from Lemma 2.3. To prove b), we note that a) and the fact that $X(x,x)$ is bounded imply the existence of c_{10}, c_{11} such that

$$\mu\left\{T: \sup_{x\in[0,1]^n} \{|T_a(x)|\} \geq \lambda\right\} \leq c_{10} e^{-c_{11}\lambda^2}$$

for all $a \in \mathbb{Z}^n$; we then argue as in the proof of Lemma 2.3 from Lemma 2.4.

It remains to prove Proposition 2.1. We will consider only $\alpha < \frac{1}{2}$; the proof for $\alpha = \frac{1}{2}$ is similar but slightly messier. We want to estimate:

$$X(0)-X(x) = \frac{1}{(2\pi)^2} \int dp_1 dp_2 \, (p_1^2+p_2^2+1)^{-1}(p_1^2+1)^{-\alpha}\{1-\exp[i(p_1x_1+p_2x_2)]\} \; .$$

Using the fact that the remainder of the integrand is even in p_1 and p_2 separately, we can replace the term in braces by

$$\{1 - \cos(p_1x_1)\cos(p_2x_2)\} \leq \{1 - \cos(p_1x_1)\} + \{1 - \cos(p_2x_2)\}.$$

We give the argument for estimating the contribution from the second of the terms on the right; the first is easier. For $\omega > 0$ define

$$\Phi(\omega) = \omega^{1+2\alpha}\int_{-\infty}^{\infty} dp_1(p_1^2+\omega^2)^{-1}(p_1^2+1)^{-\alpha} = \int_{-\infty}^{\infty} d\sigma(\sigma^2+1)^{-1}(\sigma^2+1/\omega^2)^{-\alpha} ;$$

$\Phi(\omega)$ is continuous for $\omega > 0$ and approaches a finite limit as ω approaches ∞; hence, it is bounded on $[1,\infty)$. Now:

$$\int dp_1 dp_2 (p_1^2+p_2^2+1)^{-1}(p_1^2+1)^{-\alpha}\{1-\cos(p_2x_2)\} =$$

$$= 2\int dp_2(p_2^2+1)^{-\frac{1}{2}-\alpha}\Phi(\sqrt{p_2^2+1})\sin^2(p_2x_2/2)$$

$$\leqslant 2\cdot\sup_{\omega\geqslant 1}\Phi(\omega)\cdot x_2^{2\alpha}\cdot\int_{-\infty}^{\infty} d\tau(\tau^2+x_2^2)^{-\frac{1}{2}-\alpha}\sin^2(\tau/2) .$$

Since

$$\int_{-\infty}^{\infty} d\tau(\tau^2+x_2^2)^{-\frac{1}{2}-\alpha}\sin^2(\tau/2)$$

approaches a finite limit as $|x_2|$ approaches zero, we have the desired estimate.

III. NON-REGULARITY

We will prove in this section an abstract version of statement a) of Theorem 1.1.

Proposition 3.1. *Let* X *be a distribution of positive type on* $\mathbb{R}^n$. *Assume that the Fourier transform of* X *(which must be a positive measure) has infinite total mass; equivalently, assume* X *is not a continuous function. Let* μ *denote the Gaussian measure on* $S'(\mathbb{R}^n)$ *with mean zero and covariance* X. *Then* μ*-almost every distribution*

T has the property that there exists no non-empty open set $U \subset \mathbb{R}^n$ *such that* $T|U$ *is a signed measure.*

Proof. We will need:

Lemma 3.2. *Let* U_1, U_2 *be bounded open sets in* $\mathbb{R}^n$ *with* $\overline{U}_1 \subset U_2$. *Let* ν *be a signed measure of finite total variation on* U_2 *, and let* (φ_n) *be a sequence of continuous functions on* $\mathbb{R}^n$ *with support contained in the open ball of radius* $d(U_1, \mathbb{R}^n \setminus U_2)$ *about* 0, *and with*

$$\sum_m \int |\varphi_m(x)|\,dx < \infty. \tag{3.1}$$

Then

$$\lim_{m\to\infty} \int \nu(dy)\varphi_m(x-y) = 0 \quad \text{for almost all} \quad x \in U_1, \tag{3.2}$$

where "almost all" is to be understood in the sense of Lebesgue measure.

Proof:

$$\int_{U_1} dx \left|\int \nu(dy)\varphi_m(x-y)\right| \leq |\nu|(U_2)\int |\varphi_m(x)|\cdot dx$$

where $|\nu|$ denotes the total variation of ν. By (3.1) and the Monotone Convergence Theorem,

$$\sum_m \left|\int \nu(dy)\ \varphi_m(x-y)\right| < \infty \quad \text{a.e. on} \quad U_1,$$

which implies (3.2).

Now let $f(x)$ be a C^∞ function of compact support on $\mathbb{R}^n$ with $\int f(x)dx \neq 0$ and, for each positive λ, define

$$f_\lambda(x) = \lambda^n f(\lambda x) .$$

For each fixed x, the mapping

$$T \mapsto (T * f_\lambda)(x)$$

is a Gaussian random variable on $S'(\mathbb{R}^n)$, with mean zero and variance

$$\int \tilde{X}(dp)|\tilde{f}(p/\lambda)|^2$$

(where $\tilde{X}$ is the Fourier transform of X and $\tilde{f}$ the Fourier transform of f). Since $\tilde{X}$ is a positive measure of infinite total mass, this variance goes to infinity with λ. Choose a sequence of positive numbers a_m with

$$\sum_m a_m < \infty; \tag{3.3}$$

then a sequence λ_m going to infinity so rapidly that

$$a_m^2 \cdot \int \tilde{X}(dp)| \tilde{f}(p/\lambda_m)|^2 \to \infty ; \tag{3.4}$$

and put

$$\varphi_m(x) = a_m f_{\lambda_m}(x).$$

It follows from (3.3) and Lemma 3.2 that, if the restriction of T to some non-empty open set is a signed measure, there is a set of positive Lebesgue measure in $\mathbb{R}^n$ on which

$$\lim_{m\to\infty} T*\varphi_m(x) = 0.$$

We will complete the proof of the proposition by showing

$$\text{For almost all } T \in S'(\mathbb{R}^n),\ \limsup_{m\to\infty} |T*\varphi_m(x)| = \infty \text{ a.e.} \qquad (3.5)$$

To prove this, we note first that the mapping $(T,x) \mapsto T*\varphi_n(x)$ is continuous and hence Borel from $S'(\mathbb{R}^n) \times \mathbb{R}^n$ to $\mathbb{R}$. Hence,

$$Y \equiv \{(T,x) \in S'(\mathbb{R}^n) \times \mathbb{R}^n : \limsup_{m\to\infty} |T*\varphi_m(x)| < \infty\}$$

is a Borel subset of $S'(\mathbb{R}^n) \times \mathbb{R}^n$. Now for any fixed x, the random variables $T \to T*\varphi_m(x)$ $(m=1,2,3,\ldots)$ are Gaussian with variance going to infinity (by (3.4)), so

$$\limsup_{m\to\infty} |T*\varphi_m(x)| = \infty \quad \text{for almost all } T.$$

In other words, for each x,

$$\{T: (T,x) \in Y\}$$

is a set of μ-measure zero. By Fubini's Theorem* , Y is a set of $\mu \otimes dx$ measure zero. Applying Fubini's Theorem again, we conclude that, for almost all $T \in S'(\mathbb{R}^n)$, $\{x: (T,x) \in Y\}$ is a set of Lebesgue measure zero. This is exactly the desired statement (3.5).

*There is a slight subtlety at this point. To apply Fubini's Theorem, we need to know that Y belongs to the product σ-algebra, which may in principle by strictly smaller than the σ-algebra of Borel sets in the product space. It is, however, not hard to show, using the fact that $S'(\mathbb{R}^n)$ is a countable union of compact metrizable spaces, that in this case the product σ-algebra and the Borel σ-algebra on the product space actually coincide.

IV. BEHAVIOR AT INFINITY

Proposition 2.5b implies that the set of distributions which (in some sense) grow no more rapidly than $\sqrt{\log |x|}$ at infinity is a set of μ_0-measure one. In this section, we prove that μ_0-almost every T does in fact grow as fast as $\sqrt{\log |x|}$. It is certainly reasonable to say that a function T(x) grows as fast as $\sqrt{\log |x|}$ at infinity if

$$\limsup_{|x|\to\infty} T(x)[\log(|x|]^{-\frac{1}{2}} > 0.$$

The meaning of the corresponding statement for a distribution is a little more problematical*, but the following two propositions would seem to cover any reasonable interpretation.

Proposition 4.1. *Let $\alpha > 0$. Then μ_0-almost every T satisfies.*

$$\limsup_{|x|\to\infty}\left[\left(\frac{-d^2}{dx_1^2} + \mathbb{1}\right)^{-\alpha/2} T\right](x)[\log(|x|)]^{-\frac{1}{2}} = \frac{1}{\pi}\left[\int dp(p^2+1)^{-1}(p_1^2+1)^{-\alpha}\right]^{\frac{1}{2}}$$

Proposition 4.2. *Let $\rho \in S(\mathbb{R}^2)$. Then μ_0-almost every T satisfies*

$$\limsup_{|x|\to\infty} [T*\rho(x) [\log(|x|)]^{-\frac{1}{2}} = 2[((-\Delta+\mathbb{1})^{-1}\rho,\rho)]^{\frac{1}{2}}.$$

We will prove these propositions by proving the following more general result.

Theorem 4.3. *Let μ be a translation-invariant Gaussian measure of mean zero on $S'(\mathbb{R}^n)$ with covariance $X(x-y)$ which is Hölder continuous at zero and satisfies*

*Note that $2x\cos(x^2)$ is the derivative of the bounded function $\sin(x^2)$. Is the distribution $2x\cos(x^2)$ bounded?

a. $X(0) = 1$

b. $X(x-y)\,|x-y|^n$ is bounded as $|x-y| \to \infty$.

Then

$$\mu\{T: \limsup_{|x|\to\infty} T(x)\,[\log(|x|)]^{-\frac{1}{2}} = \sqrt{2n}\} = 1.$$

To derive the propositions, we apply the theorem to measures of the form

$$\mu = \mu_0 \circ \Phi^{-1}$$

where for Proposition 4.1 we take

$$\Phi(T) = \text{const} \times \left(\frac{-d^2}{dx_1^2} + \mathbb{1}\right)^{-\alpha/2} T$$

and for Proposition 4.2 we take

$$\Phi(T) = \text{const} \times T*\rho \ ;$$

the constants are chosen to satisfy the normalization condition a. in Theorem 4.3.

Lemma 4.4. *Let* Λ *be any bounded set in* $\mathbb{R}^n$, *and let* $K_2 < \frac{1}{2}$. *There exists a constant* K_1 *such that*

$$\mu\{T: \sup_{x\in\Lambda} |T(x)| \geq \lambda\} \leq K_1 \exp[-K_2\lambda^2] \quad \textit{for all positive } \lambda .$$

Lemma 4.5. *For any* $\delta > 0$,

$$\mu\{T: \limsup_{x\to\infty} |T(x)|[\log(|x|)]^{-\frac{1}{2}} > \sqrt{2n(1+\delta)}\} = 0.$$

Lemma 4.6. *For any* $\delta > 0$, *there exists a sequence* x_i *going to* ∞ *in* $\mathbb{R}^n$ *such that*

$$\mu\{T: \limsup_{i} T(x_i)[\log(|x_i|)]^{-\frac{1}{2}} < \sqrt{2n(1-\delta)}\} = 0 \qquad (4.1)$$

The theorem follows at once by applying Lemmas 4.5 and 4.6 to a sequence of positive δ's converging to zero.

Proof of Lemma 4.4. Let $x \in \mathbb{R}^n$. The normalization condition $X(0) = 1$ implies

$$\mu\{T: |T(x)| \geqslant \lambda\} \leqslant K_3 \exp[-\lambda^2/2].$$

Hence, for any $x_1,\dots,x_k$,

$$\mu\{T: \sup_{1\leqslant i\leqslant k} |T(x_i)| \geqslant \lambda\} \leqslant K_3\cdot k \exp[-\lambda^2/2].$$

By Proposition 2.5a, there exist $\alpha > 0$ and c_8, c_9 such that

$$\mu\{T: |T(x)-T(y)| \leqslant \lambda|x-y|^\alpha \text{ for all } x,y \in \Lambda\} \geqslant 1-c_8 \exp[-c_9\lambda^2]. (4.2)$$

Define ε by

$$(1+\varepsilon)^2 = (2K_2)^{-1}$$

and put $\eta = (2c_9\varepsilon^2)^{\frac{1}{2\alpha}}$. Then it follows from (4.2) that

$$\mu\{T: |T(x)-T(y)| \leqslant \varepsilon\lambda \text{ for all } x,y \text{ with } |x-y| \leqslant \eta\} \geqslant 1-c_8\exp[-\lambda^2/2].$$

If we now choose $x_1,\dots,x_k$ such that each point of Λ is within a distance η of some x_i, we get

$$\mu\{T: \sup_{x\in\Lambda} |T(x)| \geqslant (1+\varepsilon)\lambda\} \leqslant (K_3\cdot k+c_8)\exp[-\lambda^2/2],$$

or (replacing $(1+\varepsilon)\lambda$ by λ and recalling the definition of ε)

$$\mu\{T: \sup_{x\in\Lambda}|T(x)| \geqslant \lambda\} \leqslant (K_3\cdot k+c_8)\exp[-K_2\lambda^2].$$

Proof of Lemma 4.5. Now let $\Lambda = [0,1]^n$. Then

$$\mu\{T: \limsup_{|x|\to\infty} \left(|T(x)|[\log(|x|)]^{-\frac{1}{2}}\right) > \sqrt{2n(1+\delta)} =$$

$$\mu\{T: \limsup_{\substack{|a|\to\infty \\ a\in\mathbb{Z}^n}} \left(\sup_{x\in\Lambda}\{|T(x-a)|\}[\log(|a|)]^{-\frac{1}{2}}\right) > \sqrt{2n\ (1+\delta)}\}$$

$$\leqslant \lim_{A\to\infty} \sum_{\substack{|a|\geqslant A \\ a\in\mathbb{Z}^n}} \mu\{T: \sup_{x\in\Lambda}\{|T(x-a)|\} \geqslant \sqrt{2n(1+\delta)\log(|a|)}\}.$$

Now choose $K_2 < \frac{1}{2}$ so that $2K_2(1+\delta) > 1$, and apply Lemma 4.4, (and translation invariance) to get

$$\mu\{T: \sup_{x\in\Lambda}\{|T(x-a)|\} \geqslant \sqrt{2n(1+\delta)\log(|a|)}\} \leqslant K_1|a|^{-n'},$$

where $n' = 2K_2(1+\delta)n > n$. Hence

$$\lim_{A\to\infty} \sum_{|a|\geqslant A}^{\cdot} \mu\{T: \sup_{x\in\Lambda}\{|T(x-a)|\} \geqslant \sqrt{2n(1+\delta)\log(|a|)}\} = 0$$

proving the lemma.

<u>Proof of Lemma 4.6</u>. Let R be a large number (to be chosen later) and let (x_i) be a enumeration of the vectors

$$|a|^{\delta/2}\cdot a, \quad a \in \mathbb{Z}^n, \quad |a| \geq R.$$

We choose the enumeration in such a way that $|x_i|$ is increasing in i, and we write ℓ_i for $\log(|x_i|)$ and T_i for the random variable $T \mapsto T(x_i)$. By Lemma 4.5

$$\limsup_{i\to\infty} |T_i|\cdot\ell_i^{-\frac{1}{2}} < \sqrt{3n}$$

for almost all T, so to prove (4.1) it suffices to prove

$$0 = \mu\{T: -\sqrt{3n\ell_i} \leq T_i \leq \sqrt{2n(1-\delta)\ell_i} \text{ for all but finitely many } i\},$$

i.e., for each I

$$0 = \mu\{T: -\sqrt{3n\ell_i} \leq T_i \leq \sqrt{2n(1-\delta)\ell_i} \quad \text{for all } i \geq I\}.$$

We will in fact only prove this equation for $I = 1$ and obtain the result for general I be observing that our argument works for all sufficiently large values of R.

For each $k = 1,2,3,\ldots$ define

$$\pi_k = \mu\{T: -\sqrt{3n\ell_i} \leq T_i \leq \sqrt{2n(1-\delta)\ell_i} \quad \text{for } 1 \leq i \leq k\}.$$

What we want to show is that

$$\lim_{k\to\infty} \pi_k = 0.$$

We estimate the π_k's recursively, using the inequality

$$\pi_k \leq \pi_{k-1} \cdot \text{ess sup} \left[\mu\{T_k \leq \sqrt{2n(1-\delta)\ell_k} \mid T_1 = t_1, \ldots, T_{k-1} = t_{k-1}\} \right]. \quad (4.3)$$

Here, $\mu\{.|..\}$ denotes conditional probability and the essential supremum is taken over all $t_1, \ldots, t_{k-1}$ such that $|t_i| \leq \sqrt{3n\ell_k}$ for $1 \leq i \leq k-1$. (We could, of course, take the essential supremum over a smaller set of t's.) Since the measure μ is Gaussian, it is easy to check that the conditional distribution of T_k given $T_1 = t_1, \ldots, T_{k-1} = t_{k-1}$ may be chosen to be Gaussian with mean

$- \sum_{i=1}^{k-1} \frac{a_{ki}}{a_{kk}} t_i$ and variance a_{kk}^{-1},

where

$$A = (a_{ij})$$

is the $k \times k$ matrix with inverse

$$A^{-1} = B = (b_{ij}) = (X(x_i - x_j)).$$

By condition a) in Theorem 4.3,

$$b_{ii} = 1 \quad \text{for all} \quad i;$$

also, by condition b) and the fact that the x_i are sparsely distributed in $\mathbb{R}^n$, we can make

$$\sum_{\substack{j=1 \\ j \neq i}}^{k} |b_{ij}|$$

small, uniformly in i,k, by making R large.

Now if $C = (c_{ij})$ is any $k \times k$ matrix, the operator norm of

C corresponding to the ℓ^∞ vector norm on $\mathbb{R}^k$ is given by

$$\|C\|_\infty = \sup_{1 \leqslant i \leqslant k} \sum_{j=1}^{k} |c_{ij}|$$

Thus, if R is large, $\|B-\mathbb{1}\|_\infty$ is small, so (since $A = B^{-1}$), $\|A-\mathbb{1}\|_\infty$ is also small, so $|a_{kk}-1|$ and $\sum_{i=1}^{k} \left|\frac{a_{ki}}{a_{kk}}\right|$ are small. We can therefore choose R large enough so that, for all k,

$$\mu\{T_k \leqslant \sqrt{2n(1-\delta)\ell_k} \;|T_1 = t_1,\ldots,T_{k-1} = t_{k-1}\} \leqslant 1-\exp[-\ell_k n(1-\delta/2)]$$

when $|t_i| \leqslant \sqrt{3n\ell_k}$ for $1 \leqslant i \leqslant k-1$. If we do this, and if we use the fact that the ℓ_k's are an enumeration of the numbers $(1+\delta/2)\log(|a|)$, with $a \in \mathbf{Z}^n$ and $|a| \geqslant R$, we obtain from (4.3)

$$\lim_{k\to\infty} \pi_k \leqslant \prod_{k=1}^{\infty} \{1-\exp[-\ell_k n(1-\delta/2)]\} = \prod_{\substack{|a|\geqslant R \\ a\in\mathbf{Z}^n}} \{1-|a|^{(1-\delta^2/4)n}\} = 0,$$

completing the proof of the lemma and hence of the theorem.

EUCLIDEAN GREEN'S FUNCTIONS AND WIGHTMAN DISTRIBUTIONS

Konrad Osterwalder*)
Harvard University
Cambridge, Massachusetts

I. INTRODUCTION

The use of Euclidean methods in relativistic quantum field theory has a long history. The idea to pass to imaginary times or energies in order to replace the indefinite Minkowski metric by the Euclidean metric appeared first in Dyson's work on renormalization [Dy 1]. Fradkin [Fr 1], Nakano[Na 1], Wick [Wi 1] and most forcefully Schwinger [Sc 1, 2] proposed to consider the Green's functions of a field theory, continued to imaginary times, and to study their properties as solutions of certain differential equations or as the n-point functions of Euclidean field operators. It was Symanzik [Sy 1, 2] who first advocated a purely Euclidean approach to quantum field theory. He realized that given a formal Lagrangian density the construction of the Green's functions at imaginary times - called Euclidean Green's functions or Schwinger functions - might be simpler than the direct construction of Wightman distributions. It is only in the Euclidean context that Feynman's famous history integral [Fe 1, 2] can be given a mathematically rigorous meaning, see e.g. ref. [Ne 4] for the case of boson theories and [OS 1, 2], [Oz 1] for theories involving bosons and fermions. In his work, Symanzik developed many of the concepts and ideas which these days belong to the basic instruments of the Euclidean approach to field theory. More recently Nelson's mathematically rigorous formulation of Euclidean bose field theories in terms of Markov fields [Ne 1-4], his reconstruction theorem and Guerra's first application of Nelson's scheme [Gu 1] paved the way for all the various recent applications of Euclidean methods in constructive field theory.

Both in Nelson's and in Symanzik's work and also in most other work on Euclidean field theory, Euclidean field operators play an essential role. But the existence of such Euclidean field operators is not automatically guaranteed by what one usually knows or assumes about the relativistic theories. It only follows from the so-called Nelson-Symanzik positivity condition on the Schwinger functions, which is expected to hold in theories describing scalar bosons only, but

*) Supported in part by National Science Foundation Grant GP40354X.

not necessarily in cases where fermions are involved.

It is therefore natural to try to work with the Schwinger functions only and to study their relations to the relativistic theory. Since the Schwinger functions are not necessarily the vacuum expectations of a (Euclidean) field, one does not have a "Euclidean field theory" anymore, but only a "Euclidean formulation of (relativistic!) field theory".

It is the purpose of these lectures to define and to study the Schwinger functions within the framework of the Wightman axioms. The main task will be to determine under which conditions Schwinger functions are indeed the Euclidean Green's functions of a well defined Wightman theory.

The plan of these lectures is as follows: First we shall briefly review some of the main definitions and results of axiomatic quantum field theory and give a precise definition of the Schwinger functions. Second we will inquire which properties of the Schwinger functions can be derived as consequences of the Wightman axioms. Finally we will show how to reconstruct the Wightman distributions from Schwinger functions $\mathfrak{S}_n$ satisfying the following conditions

(E0) A distribution property

(E1) Euclidean covariance

(E2) Positivity

(E3) Symmetry

(E4) Cluster property.

These conditions (E0) - (E4) are <u>necessary and sufficient</u> for the existence of Wightman distributions satisfying all the Wightman axioms. In detail the connection is as follows.

<u>Euclidean</u>		<u>Relativistic</u>
[Distribution Covariance Positivity]	⟷	[Distribution Covariance Positivity Spectrum]
[ ] + Symmetry	⟷	[ ] + Locality
[ ] + Cluster	⟷	[ ] + Cluster

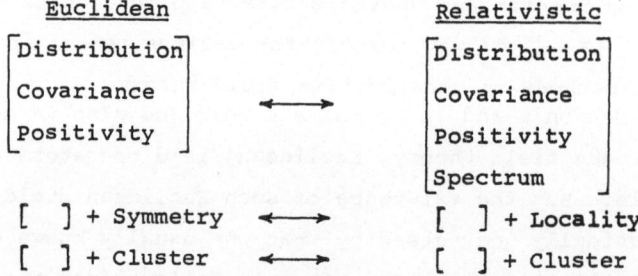

For applications in constructive field theory the distribution property (E0) does not seem to be appropriate. We therefore introduce a second distribution property (E0') and prove that (E0'), (E1) - (E4) are still sufficient for the reconstruction of the Wightman theory. The above chart holds again with all the arrows pointing to the right only. (E0') is a simple temperedness condition with a restriction on the order of the distributions $\mathfrak{S}_n$ (the Euclidean Green's functions), for large n. It is this second result which I think will turn out to be useful in constructive field theory; the equivalence result, though more general, seems to be of purely esthetical interest.

Remark: As discovered first by Schrader and Simon, there is a mistake in the proof of a technical lemma in the original paper on the axioms for Euclidean Green's functions (ref. [OS 3], lemma 8.8). As a consequence, the temperedness condition of ref. [OS 3] seems to be too weak to allow for the reconstruction of the Wightman theory.

In these lectures we restrict our attention to theories involving one neutral scalar field only; the generalization to finitely or countably many fields of general transformation properties requires only minor and mostly obvious modifications, see Osterwalder-Schrader [OS 3, 4]. Most of the material covered in these lectures is taken from references [OS 3, 4].

It is a pleasure to thank Dr. J. Fröhlich for a careful reading of this manuscript and for many critical remarks. I would also like to thank Prof. B. Simon for sending me prior to publication a copy of chapter II of his Zürich lecture notes [Si 1]. It was very useful during the preparation of these notes.

II. WIGHTMAN DISTRIBUTIONS, WIGHTMAN FUNCTIONS, EUCLIDEAN GREEN'S FUNCTIONS

In this section we state the Wightman axioms in terms of the Wightman distributions and summarize some of the classical results of axiomatic field theory. For details see the monographs of Jost [Jo 1] and of Streater and Wightman [SW 1], and references given there.

Let $\varphi(x)$ be a neutral scalar field obeying all the Wightman axioms. Then its vacuum expectation values or Wightman distributions

$$\mathcal{W}_n(\underline{x}) = \mathcal{W}_n(x_1, \dots x_n) = (\Omega, \varphi(x_1) \dots \varphi(x_n)\Omega)$$

have the following properties:

Distribution property: For each n, $\mathcal{W}_n$ is a tempered distribution:

(W0) $\mathcal{W}_n(\underline{x}) \in \mathcal{S}'(\mathbb{R}^{4n})$; $\mathcal{W}_0 = 1$

Notation: $\underline{x} = (x_1, \dots x_n) \in \mathbb{R}^{4n}$, $x_i = (x_i^0, \vec{x}_i) \in \mathbb{R}^4$

Relativistic covariance: For each n, $\mathcal{W}_n$ is Poincaré invariant:

(W1) $\mathcal{W}_n(\underline{x}) = \mathcal{W}_n(\Lambda \underline{x} + a)$

for all $(a, \Lambda) \in \mathcal{P}_+^\uparrow$,

where $\Lambda \underline{x} + a = (\Lambda x_1 + a, \dots \Lambda x_n + a)$.

Positivity: For all finite sequences $f_0, f_1, \dots f_N$ of test functions, $f_0 \in \mathbb{C}$, $f_n \in \mathcal{S}(\mathbb{R}^{4n})$, $n = 1, \dots N$,

(W2) $\sum_{n,m} \mathcal{W}_{n+m}(f_n^* \times f_m) \geq 0$,

where $f_n^* \times f_m$ is defined by $(f_n^* \times f_m)(\underline{x}, \underline{y}) = f_n^*(\underline{x}) f_m(\underline{y})$

and $f_n^*(\underline{x}) = f_n^*(x_1, \dots x_n) = \bar{f}_n(x_n, \dots x_1) \equiv \bar{f}_n(\overleftarrow{\underline{x}})$.

Locality: For any n and k = 1, ... n - 1

(W3) $\mathcal{W}_n(x_1, \dots x_k, x_{k+1}, \dots x_n) = \mathcal{W}_n(x_1, \dots x_{k+1}, x_k, \dots x_n)$

if $(x_k - x_{k+1})^2 < 0$.

Cluster Property: For any spacelike a and k = 1, ... n - 1,

$\underline{x} = x_1, \dots x_k$, $\underline{y} = y_1, \dots y_{n-k}$,

(W4) $\lim_{\lambda \to \infty} \mathcal{W}_n(\underline{x}, \underline{y} + \lambda a) = \mathcal{W}_k(\underline{x}) \mathcal{W}_{n-k}(\underline{y})$.

Spectral condition: Using the translation invariance of the $\mathcal{W}_n$'s we conclude that there exist distributions $W_{n-1} \in \mathcal{S}'(\mathbb{R}^{4(n-1)})$, such that $\mathcal{W}_n(\underline{x}) = W_{n-1}(\underline{\xi})$ where $\underline{\xi} = (\xi_1, \dots \xi_{n-1})$ and $\xi_k = x_{k+1} - x_k$. Then

(W5) $\operatorname{supp} \widetilde{W}_{n-1} \subset \bar{V}_+^{n-1} \equiv \{ \underline{q} \mid q_i \in \bar{V}_+ \ , i = 1, \dots n-1 \}$,

where $\widetilde{W}_{n-1}(\underline{q}) = (2\pi)^{-4(n-1)} \int \exp[-i \sum q_k \xi_k] \, W_{n-1}(\underline{\xi}) \, d^{4(n-1)}\xi$

is the Fourier transform of W_{n-1} , $\bar{V}_+$ is the closed forward light cone, and $q_k \xi_k = q_k^0 \xi_k^0 - \vec{q}_k \vec{\xi}_k$

Remark: Equations (W1) - (W5) have to be interpreted in distributional sense.

From a given set of Wightman distributions satisfying (W0) - (W5) we can reconstruct the physical Hilbert space $\mathcal{H}$, the vacuum vector $\Omega \in \mathcal{H}$, the field operators $\varphi(x)$ and a unitary representation $\mathcal{U}(a, \Lambda)$ of $\mathcal{P}_+^\uparrow$ in $\mathcal{H}$. This is the Wightman reconstruction theorem; see Wightman [Wg 1].

Let us now introduce vector valued distributions formally defined by

$$\Psi_n(x_1, \underline{\xi}) = \varphi(x_1) \dots \varphi(x_n) \Omega \ ,$$

where $\underline{\xi} = (\xi_1, \dots \xi_{n-1})$ and $\xi_k = x_{k+1} - x_k$ for k =1 , ...n-1. Ψ_n is a tempered distribution with values in $\mathcal{H}$. The scalar product of two vectors $\Psi_n(f)$ and $\Psi_m(g)$ for $f \in \mathcal{S}(\mathbb{R}^{4n})$, $g \in \mathcal{S}(\mathbb{R}^{4m})$ is

$$(\Psi_n(f), \Psi_m(g)) = \int \bar{f}(x,\underline{\xi})\, g(x',\underline{\xi}')\, W_{n+m-1}(-\overleftarrow{\underline{\xi}}, -x+x', \underline{\xi}')\, dx\,dx'\,d\underline{\xi}\,d\underline{\xi}'.$$

The Fourier transform of Ψ_n is defined as usual by $\tilde{\Psi}_n(\tilde{f}) = \Psi_n(f)$, where $\tilde{f}$ is the Fourier transform of $f \in \mathcal{S}(\mathbb{R}^{4n})$.

<u>Theorem 1:</u> For n = 1,2,.., the support of $\tilde{\Psi}_n(\underline{q})$ is contained in $\bar{V}_+^n$.

For a proof we just take a test function $f \in \mathcal{S}(\mathbb{R}^{4n})$ with $\tilde{f}(\underline{q}) = 0$ for $\underline{q} \in \bar{V}_+^n$, write down the norm $\|\Psi_n(f)\|$ in terms of the distribution $\tilde{W}_{2n-1}$ and show that it is zero, using the spectrum condition.

We can now study the Laplace transform $\Psi_n(z,\underline{\zeta})$ of $\Psi_n(\underline{q})$. The Laplace transforms of tempered distributions have been extensively studied, e.g. in references [Sz 1, Li 1, Vl 1, SW 1]. The particular theorem we quote below is due to Bros, Epstein and Glaser [BEG].

<u>Theorem 2</u>: Let $\tilde{F}(\underline{q})$ be a distribution in $\mathcal{S}'(\mathbb{R}^{4n})$ with support in $\bar{V}_+^n$. Then there exists an analytic function $G(\underline{z})$, regular in the forward tube $\mathcal{T}_+^n = \{\underline{z} \mid \underline{z} = \underline{x} + i\underline{y} \in \mathbb{C}^{4n},\ \underline{y} \in V_+^n\}$, so that

a) the Fourier transform F of $\tilde{F}$ is the boundary value of G :

$$\lim_{\lambda \downarrow 0} \int G(\underline{x} + i\lambda\underline{y})\, f(\underline{y})\, d^{4n}x = F(f)$$

for all $f \in \mathcal{S}(\mathbb{R}^{4n})$, $\underline{y} \in V_+^n$,

b) For some positive constants c, α, β and for all $\underline{z} \in \mathcal{T}_+^n$,

(1) $$|G(\underline{z})| \leq c\,(1+|\underline{z}|)^{\alpha}\left[1 + \left(\min_{1\leq j\leq n}(y_j^0 - |\vec{y}_j|)\right)^{-\beta}\right]$$

The function $G(\underline{z})$ is called the Laplace transform of $\tilde{F}$ and can be written (heuristically) as

$$G(\underline{z}) = \int \exp(i\textstyle\sum z_i q_i)\, \tilde{F}(\underline{q})\, d^{4n}q .$$

Theorem 2 can be immediately generalized to hold for distributions $\tilde{F}$ with values in some Banach space. Applied to our vector valued distributions $\tilde{\Psi}_n$ it shows that the Laplace transform

$$\Psi_n(z,\underline{\zeta}) = \int \exp[i z q_0 + i \textstyle\sum \zeta_i q_i]\, \tilde{\Psi}_n(q_0, q_1, \ldots q_{n-1})\, d^{4n}q$$

is an analytic vector valued function, regular in $\mathcal{T}_+^n$, with boundary value $\Psi_n(x,\underline{\xi})$.

Applying Theorem 2 to the distributions $\tilde{W}_n(\underline{q})$ we obtain the Wightman functions $W_n(\underline{\zeta})$, which are analytic functions for $\underline{\zeta} \in \mathcal{T}_+^n$. Obviously we have

(2) $$W_{n-1}(\underline{\zeta}) = (\Omega, \Psi_n(z,\underline{\zeta})) ;$$

translation invariance of Ω implies that the right hand side of (2) does not depend on z. The following lemma is also simple to prove.

Lemma 3 (see [Jo 1], p. 74): Let $(z,\underline{\zeta}) \in \mathcal{T}_+^n$, $(z',\underline{\zeta}') \in \mathcal{T}_+^m$. Then

$$(\Psi_n(z,\underline{\zeta}),\ \Psi_m(z',\underline{\zeta}')) = W_{n+m-1}(-\overleftarrow{\underline{\bar{\zeta}}}, -\bar{z}+z', \underline{\zeta}') \ .$$

Using relativistic invariance (W1) and the Bargmann-Hall-Wightman theorem, we obtain a single valued analytic extension of the Wightman functions $W_n(\underline{\zeta})$ - but not of the vectors $\Psi_n(\underline{\zeta})$ - into the extended tube $\mathcal{T}^n_{+,ext} = \{\underline{\zeta} \mid \Lambda\underline{\zeta} \in \mathcal{T}^n_+$ for some $\Lambda \in L_+(\mathbb{C})\}$, where $L_+(\mathbb{C})$ denotes the set of complex Lorentz transformations with determinant 1. The functions $\mathcal{W}_n(\underline{z})$, defined by $\mathcal{W}_n(\underline{z}) = W_{n-1}(\underline{\zeta})$, where $\zeta_k = z_{k+1} - z_k$, are analytic in $\mathfrak{S}^n_{ext} = \{\underline{z} \mid \underline{\zeta} \in \mathcal{T}^n_{+,ext}\}$ and have as boundary values the distributions $\mathcal{W}_n(\underline{x})$. Finally using locality (W3), we obtain a single valued analytic extension of $\mathcal{W}_n(\underline{z})$ into the set $\mathfrak{S}^n_{ext,perm} \equiv \{\underline{z} \mid (z_{\pi(1)},\dots z_{\pi(n)}) \in \mathfrak{S}^n_{ext}$ for some permutation $\pi\}$. We denote this extension again by $\mathcal{W}_n(\underline{z})$. It is invariant under the inhomogeneous complex Lorentz group $iL_+(\mathbb{C})$ and under permutations of the arguments $z_1, \dots. z_n$.

The set $\mathfrak{S}^n_{ext,perm}$ contains $\mathcal{E}^n = \{\underline{z} \mid \underline{z} \in \mathbb{C}^n, \operatorname{Re} z_k^0 = 0, \operatorname{Im}\vec{z}_k = 0$, $z_k \neq z_\ell$ for all $1 \leq k < \ell \leq n\}$. Points in $\mathcal{E}^n$ are called Euclidean points (of non-coinciding arguments).

Definition: The restriction of the Wightman function $\mathcal{W}_n(\underline{z})$ to $\mathcal{E}^{(n)}$ is called the n-point Euclidean Green's function or Schwinger function.

We set $\mathfrak{S}_0 = \mathcal{W}_0 = 1$ and

$$\mathfrak{S}_n(\underline{x}) = \mathfrak{S}_n(x_1,\dots x_n) = \mathcal{W}_n((ix_1^0,\vec{x}_1),\dots (ix_n^0,\vec{x}_n)) ,$$

for $\underline{x} \in \Omega^n = \{\underline{x} \mid x_i \neq x_j$ for all $1 \leq i < j \leq n\}$.

From the invariance properties of the Wightman functions we immediately derive the following lemma.

Lemma 4: The Schwinger functions $\mathfrak{S}_n(\underline{x})$ are invariant under the inhomogeneous proper Euclidean group iSO_4 and under permutations of the arguments $x_1, \dots x_n$.

Let us now introduce the difference variable Schwinger functions $S_{n-1}(\underline{\xi})$, defined by

$$(3) \qquad S_{n-1}(\underline{\xi}) = W_{n-1}((i\xi_1^0,\vec{\xi}_1), \dots (i\xi_{n-1}^0,\vec{\xi}_{n-1})) = \mathfrak{S}_n(\underline{x})$$

where $\xi_k = x_{k+1} - x_k$ and $\underline{x} \in \Omega^n$. We also define the real analytic vector valued functions $\Psi_n^E(\underline{\xi})$ by

$$\Psi_n^E(x,\underline{\xi}) = \Psi_n((ix^0,\vec{x}), (i\xi_1^0,\vec{\xi}_1), \dots (i\xi_{n-1}^0,\vec{\xi}_{n-1})) \qquad ,$$

for $x^0>0$, $\xi_k^0>0$, $1 \le k \le n-1$. Define $\vartheta\underline{\xi}$ by $(\vartheta\underline{\xi})_k = (-\xi_k^0, \vec{\xi}_k)$.
Lemma 3 and the above definitions lead immediately to

Lemma 5. Let all of x^0, ξ_k^0, x'^0, $\xi_k'^0$ be positive. Then

$$(4) \qquad (\Psi_n^E(x,\underline{\xi}), \Psi_m^E(x',\underline{\xi}')) = S_{n+m-1}(-\vartheta\overleftarrow{\underline{\xi}}, -\vartheta x + x', \underline{\xi}') .$$

Lemma 5 of course yields a positivity property for the Schwinger functions, see (E2) below.

The cluster property (W4) implies that for any two vectors Φ and Φ' in $\mathcal{H}$, and spacelike a,

$$(5) \quad \lim_{\lambda\to\infty} (\Phi, U(\lambda a, \mathbb{1})\Phi') = (\Phi, \Omega)(\Omega, \Phi') .$$

Lemma 6. For $0 < x_1^0 < \ldots < x_n^0$, $0 < y_1^0 < \ldots < y_m^0$, and $a = (0, \vec{a})$

$$\lim_{\lambda\to\infty} \mathfrak{S}_{n+m}(\vartheta x_n, \ldots \vartheta x_1, y_1+\lambda a, \ldots y_m+\lambda a) = \mathfrak{S}_n(\vartheta x_n, \ldots \vartheta x_1)\, \mathfrak{S}_m(y_1, \ldots y_m)$$

The lemma follows essentially from substituting $\Phi = \Psi_n^E(\underline{\xi})$, $\Phi' = \Psi_m^E(\underline{\xi}')$, $\xi_k = x_{k+1} - x_k$, $\xi_k' = y_{k+1} - y_k$ in equation (5).

The Schwinger functions are real analytic functions, but they also define distributions. Before we investigate that aspect we have to stop for a mathematical digression. We shall collect some simple lemmas on distributions and Laplace transforms without giving proofs. They are either contained in refs. [SW 1, Vl 1, OS 3] or they are simple consequences of things established there.

Distributions and Laplace Transforms

By $\mathbb{R}_\pm$ we denote the open half intervals $(0, \pm\infty)$, by $\overline{\mathbb{R}}_\pm$ their closure. Let $\mathcal{S}(\mathbb{R}_\pm)$ denote the space of functions $f \in \mathcal{S}(\mathbb{R})$ with supp$f$ in $\mathbb{R}_\pm$, given the induced topology. $\mathcal{S}(\overline{\mathbb{R}}_\pm)$ is the set of all functions, defined on $\overline{\mathbb{R}}_+$, C^∞ on $\mathbb{R}_+$, whose derivatives all have a continuous extension to $\overline{\mathbb{R}}_+$ and are of fast decrease at infinity. The topology on $\mathcal{S}(\overline{\mathbb{R}}_+)$ is defined by the seminorms

$$|g|_{m,+} = \sup_{\substack{x>0\\ \alpha\le m}} (1+x)^m |g^{(\alpha)}(x)| .$$

Lemma 7. The space $\mathcal{S}(\overline{\mathbb{R}}_+)$ is isomorphic to the topological quotient space $\mathcal{S}(\mathbb{R})/\mathcal{S}(\mathbb{R}_-)$.

The main point of this lemma is that any element in $\mathcal{S}(\overline{\mathbb{R}}_+)$ is the restriction to $\overline{\mathbb{R}}_+$ of some element in $\mathcal{S}(\mathbb{R})$.

The dual space of $\mathcal{S}(\mathbb{R})/\mathcal{S}(\mathbb{R}_-)$ is the polar of $\mathcal{S}(\mathbb{R}_-)$, which is the set of distributions $T \in \mathcal{S}'(\mathbb{R})$ with supp$T \in \overline{\mathbb{R}}_+$. Hence by Lemma 7 a distribution in $\mathcal{S}'(\overline{\mathbb{R}}_+)$ can be identified with a distribution in $\mathcal{S}'(\mathbb{R})$

with support in $\overline{\mathbb{R}}_+$. For $f \in \mathcal{S}(\mathbb{R}_+)$ we define $\check{f}$ by

$$\check{f}(p) = \int e^{-px} f(x)\, dx \restriction \overline{\mathbb{R}}_+ .$$

<u>Lemma 8</u>. $f \to \check{f}$ is a continuous map of $\mathcal{S}(\mathbb{R}_+)$ into $\mathcal{S}(\overline{\mathbb{R}}_+)$ whose range is dense and whose kernel is zero.

Let $T \in \mathcal{S}'(\mathbb{R})$ with $\operatorname{supp} T \subset \overline{\mathbb{R}}_+$. Then by Lemma 7, T also defines a distribution in $\mathcal{S}'(\overline{\mathbb{R}}_+)$, again denoted by T, and we conclude that there are contants c and m, such that for all $f \in \mathcal{S}(\mathbb{R}_+)$

(6) $$|T(\check{f})| \leq c\,|\check{f}|_{m,+} \equiv c\,|f|'_m .$$

On the other hand we can use the Laplace transform S of T to define a distribution in $\mathcal{S}'(\mathbb{R}_+)$. For x > 0 we define

$$S(x) = \int e^{-xp}\, T(p)\, dp$$

and for $f \in \mathcal{S}(\mathbb{R}_+)$ we set

(7) $$S(f) = \int f(x)\, S(x)\, dx .$$

(More precisely we first define $S(f)$ for elements in $\mathcal{S}(\mathbb{R}_+)$ which have compact support, then prove continuity of S using a bound on $S(x)$ similar to (1), and finally extend S to all of $\mathcal{S}(\mathbb{R}_+)$.)

The next theorem contains the main result of this mathematical digression.

<u>Theorem 9</u>: Let T be a distribution in $\mathcal{S}'(\mathbb{R})$ with $\operatorname{supp} T \subset \overline{\mathbb{R}}_+$ and define S as in (7). Then for all $f \in \mathcal{S}(\mathbb{R}_+)$

(8) $$S(f) = T(\check{f})$$

(9) $$|S(f)| \leq c\,|f|'_m$$

for some constants c, m depending on T only. Conversely if S is a distribution in $\mathcal{S}'(\mathbb{R}_+)$, satisfying (9) for some c, m, then there exists a unique distribution $T \in \mathcal{S}'(\mathbb{R})$ with support in $\overline{\mathbb{R}}_+$, such that (8) holds.

In our applications we will use a multivariable version of Theorem 9, which due to the nuclear theorem is easy to prove. Let us just introduce the necessary notations and definitions.

By $\mathbb{R}^{4n}_+$ we denote the set $\{\underline{\xi} \mid \xi^0_k > 0 , k = 1, \dots n\}$, by $\overline{\mathbb{R}}^{4n}_+$ its closure. For $f \in \mathcal{S}(\mathbb{R}^{4n}_+)$ we define $\check{f}$ by

$$\check{f}(q) = \int \exp[-\sum_{k=1}^{n} (q_k^0 \xi_k^0 + i\vec{q}_k \vec{\xi}_k)]\, f(\underline{\xi})\, d^{4n}\xi \restriction \overline{\mathbb{R}}_+^{4n} ,$$

and we introduce a set of norms on $\mathcal{S}(\mathbb{R}_+^{4n})$ by

(10) $$|f|_m' = |\check{f}|_{m,+} = \sup_{\substack{\underline{q} \in \mathbb{R}_+^{4n} \\ |\alpha| \le m}} (1+|\underline{q}|)^m \, |\check{f}^{(\alpha)}(\underline{q})| .$$

For $T \in \mathcal{S}'(\mathbb{R}^{4n})$ with $\operatorname{supp} T \subset \overline{\mathbb{R}}_+^{4n}$ we can again define S (formally) by

$$S(\underline{\xi}) = \int \exp[-\sum_{k=1}^{n} (\xi_k^0 q_k^0 + i\vec{\xi}_k \vec{q}_k)]\; T(\underline{q})\, d^{4n}q \restriction \mathbb{R}_+^{4n} ,$$

(Laplace transform with respect to the q_k^0 variables, Fourier transform in the distributional sense for the $\vec{q}_k$ variables). For $f \in \mathcal{S}(\mathbb{R}_+^{4n})$ $S(f) = \int f(\underline{\xi})\, S(\underline{\xi})\, d^{4n}\xi$ again defines a distribution in $\mathcal{S}'(\mathbb{R}_+^{4n})$. We leave it as an exercise to formulate the multivariable version of Theorem 9 for S and T defined as above.

For later use we introduce two more sub-spaces of $\mathcal{S}(\mathbb{R}^{4n})$.

$$\mathcal{S}_0(\mathbb{R}^{4n}) = \{ f \mid f \in \mathcal{S}(\mathbb{R}^{4n}),\ f^{(\alpha)}(\underline{x}) = 0 \text{ if } x_i = x_k \text{ for some } i \neq k, \text{ all } \alpha \}$$

$$\mathcal{S}(\mathbb{R}_<^{4n}) = \{ f \mid f \in \mathcal{S}(\mathbb{R}^{4n}),\ \operatorname{supp} f \subset \mathbb{R}_<^{4n} \},$$

where $\mathbb{R}_<^{4n} = \{ \underline{x} \mid 0 < x_1^0 < \ldots < x_n^0 \}$.

***

Now we are prepared to study the distributional aspects of the Schwinger functions. For $\xi_k^0 > 0$, $S_n(\underline{\xi})$ is just the Fourier-Laplace transform of $\widetilde{W}_n(\underline{q})$, see eq. (3), and the multivariable version of Theorem 9 applies. We obtain

<u>Lemma 10</u>. a) The difference variable Schwinger functions $S_n(\underline{\xi})$ define distributions in $\mathcal{S}'(\mathbb{R}_+^{4n})$ through

$$S_n(f) = \int f(\underline{\xi})\, S_n(\underline{\xi})\, d^{4n}\xi$$

for $f \in \mathcal{S}(\mathbb{R}_+^{4n})$. Furthermore $S_n(f) = \widetilde{W}_n(\check{f})$, and for some c and m

$$|S_n(f)| \le c\, |f|_m' .$$

b) The Schwinger functions $\mathfrak{S}_n(\underline{x})$ define distributions in $\mathcal{S}_0'(\mathbb{R}^{4n})$.

We remark that a) implies that $\mathfrak{S}_n(\underline{x})$ defines a distribution in

$\mathcal{S}'(\mathbb{R}^{4n}_<)$. In order to obtain b), an additional geometrical argument is necessary, see [OS 3, Si 1].

In the following theorem we collect all the properties of the Schwinger functions derived so far and state our main equivalence result.

Proposition I: The Schwinger functions associated to a Wightman theory have the following properties:

Distribution property: For each $n \geqslant 1$

$$\mathfrak{S}_n(\underline{x}) \in \mathcal{S}_0'(\mathbb{R}^{4n}) \quad ; \quad \mathfrak{S}_0 = 1 .$$

(E0) S_n defines an element in $\mathcal{S}'(\mathbb{R}^{4n}_+)$ and is continuous with respect to some $|\cdot|_m'$ -norm.

Euclidean Covariance: For each $n \geqslant 1$ and all $(a, R) \in iSO_4$,

(E1) $\mathfrak{S}_n(\underline{x}) = \mathfrak{S}_n(\underline{R}x + a)$.

Positivity: For all finite sequences $f_0, f_1, \dots f_N$ of test functions $f_n \in \mathcal{S}(\mathbb{R}^{4n}_<)$,

(E2) $\sum_{n,m} \mathfrak{S}_{n+m}(\Theta f_n^* \times f_m) \geqslant 0$,

where $\Theta f_n(\underline{x}) = f_n(\underline{\vartheta x})$.

Symmetry: For all permutations π ,

(E3) $\mathfrak{S}_n(x_1, \dots x_n) = \mathfrak{S}_n(x_{\pi(1)}, \dots x_{\pi(n)})$.

Cluster Property: For all n, m, $f \in \mathcal{S}(\mathbb{R}^{4n}_<)$, $g \in \mathcal{S}(\mathbb{R}^{4m}_<)$, $a = (0, \vec{a}) \in \mathbb{R}^4$,

(E4) $\lim_{\lambda \to \infty} \mathfrak{S}_{n+m}(\Theta f^* \times g_{\lambda a}) = \mathfrak{S}_n(\Theta f^*)\, \mathfrak{S}_m(g)$,

where $g_{\lambda a}$ is defined by $g_{\lambda a}(\underline{x}) = g(\underline{x} + \lambda a)$.

Conversely, Schwinger "functions" obeying (E0) - (E4) are the Schwinger functions associated with a unique Wightman theory.

Proof: The derivation of (E0) - (E4) from the Wightman axioms follows from Lemmas 4, 5, 6, and 10. Because $\int \mathfrak{S}_n(\underline{x}) f(\underline{x})\, dx$, $f \in \mathcal{S}_0(\mathbb{R}^{4n})$, can be defined as an ordinary integral, covariance, positivity, symmetry, and cluster properties hold in the distributional sense if they hold pointwise.

For a proof of the converse statement it suffices to assume instead of (E0) that S_n is in the algebraic dual of $\mathcal{S}(\mathbb{R}_+^{4n})$ and that it is continuous with respect to some $|\cdot|'_m$ -norm. Then the rest of (E0) follows from Lemma 8,(E1) and (E3). By the multivariable version of theorem 9, (E0) implies that there exist distributions $\widetilde{W}(\underline{q}) \in \mathcal{S}'(\mathbb{R}^{4n})$ with $\operatorname{supp}\widetilde{W} \subset \overline{\mathbb{R}}_+^{4n}$, such that (in the distributional sense) for $\underline{\xi} \in \mathbb{R}_+^{4n}$,

(11) $$S_n(\underline{\xi}) = \int \exp\left[-\sum_{k=1}^{n}(\xi_k^0 q_k^0 + i\vec{\xi}_k \vec{q}_k)\right] \widetilde{W}_n(\underline{q})\, d^{4n}q .$$

We define $\widetilde{W}_n(\underline{q})$ to be the Fourier transform of the difference variable Wightman distributions of our theory. From (E1) we conclude that the $\widetilde{W}_n(\underline{q})$ are Lorentz invariant distributions, see Nelson [Ne 2] and hence have support in $\overline{V}_+^n$. Positivity (W2) and the cluster property (W4) follow easily from the corresponding conditions (E2) and (E4). Locality (W3) finally follows from symmetry (E3) and all the other axioms already established, see ref. [Jo 1], p. 83. In the next section, these arguments will be discussed in more detail.

The norms $|\cdot|'_m$ which appear in (E0)-see eq.(6) for the definition - might be difficult to deal with in constructive field theory. We now introduce another distribution property (E0'):

> There is a Schwartz norm $|\cdot|_S$ on $\mathcal{S}(\mathbb{R}_+^4)$ and some $L > 0$, such that for all n and for all $f_k \in \mathcal{S}(\mathbb{R}_+^4)$, k=1, ... n,
>
> (E0') $$|S_n(f_1 \times f_2 \times \dots \times f_n)| \leq (n!)^L \prod_{k=1}^{n} |f_k|_S .$$

Our main result is the following proposition.

Proposition II: Schwinger functions satisfying (E0'), (E1) - (E4) determine a unique Wightman theory (whose Schwinger functions they are).

III. RECONSTRUCTING THE WIGHTMAN THEORY

In this section we start from a set of Schwinger functions $\mathfrak{S}_n$, satisfying (E0'), (E1) - (E4) and reconstruct the Wightman theory belonging to it. This will prove proposition II, but it will also shed some more light on the proof of proposition I.

Let $\underline{\mathcal{S}}_<$ be the vector space consisting of sequences $\underline{f} = (f_0, f_1, \dots)$

where $f_0 \in \mathbb{C}$, $f_n \in \mathcal{S}(\mathbb{R}^{4n}_{<})$, for $1 \leq n \leq N$, and $f_n = 0$ for $n > N$, some finite N. For $\underline{f}, \underline{g} \in \underline{\mathcal{S}}_<$ let

$$\langle \underline{f}, \underline{g} \rangle = \sum_{n,m} \mathfrak{S}_{n+m} (\Theta f_n^* \times g_m)$$

By (E2), $\langle , \rangle$ is a positive semi-definite inner product. Set $\mathcal{N} = \{\underline{f} \mid \underline{f} \in \underline{\mathcal{S}}_<, \|\underline{f}\|^2 = \langle \underline{f}, \underline{f} \rangle = 0\}$. Then the completion of $\underline{\mathcal{S}}_< / \mathcal{N}$ defines a Hilbert space $\mathcal{H}$, which will turn out to be the physical Hilbert space. Denoting by Φ the natural injection of $\underline{\mathcal{S}}_<$ into $\mathcal{H}$, we obtain $(\Phi(\underline{f}), \Phi(\underline{g})) = \langle \underline{f}, \underline{g} \rangle$ for $\underline{f}, \underline{g} \in \underline{\mathcal{S}}_<$. We set $\Omega = \Phi((1, 0, 0 \ldots))$. Let $f \in \mathcal{S}(\mathbb{R}^{4n}_<)$, some n. Then if $\underline{f}$ is of the form $f_n = f$, $f_k = 0$ for $k \neq n$, we write $\Phi(\underline{f}) = \Phi_n(f) = \int \Phi_n(\underline{x}) f(\underline{x}) d^{4n}\underline{x}$. We also define $\Psi_n^E(x_1, \underline{\xi}) = \Phi_n(\underline{x})$, where $\xi_k = x_{k+1} - x_k$, and Ψ_n^E is a vector valued distribution in $\mathcal{S}'(\mathbb{R}^{4n}_+)$. By (E2), positivity, the scalar product of two such vectors is again given by eq. (4) in Lemma 5.

In the following we will analytically extend the "functions" $\Psi_n^E(x_1, \underline{\xi})$ in the zero components $x_1^0, \xi_1^0 \ldots \xi_{n-1}^0$ of their arguments, after smearing them in the remaining variables. We define for $g \in \mathcal{S}(\mathbb{R}^{3n})$, $h \in \mathcal{S}(\mathbb{R}^{3m})$,

(12) $$\Psi_n^E(x_1^0, \underline{\xi}^0 | g) = \int \Psi_n^E(x_1, \underline{\xi})\, g(\vec{x}_1, \underline{\vec{\xi}})\, d\vec{x}_1 d\underline{\vec{\xi}}, \quad \text{and}$$

(13) $$S_{n+m-1}(\overleftarrow{\underline{\xi}}^0, x^0 + x'^0, \underline{\xi}'^0 | gh) = (\Psi_n^E(x^0, \underline{\xi}^0 | g), \Psi_m^E(x'^0, \underline{\xi}'^0 | h))$$

Let $\mathbb{C}_+ = \{z \mid \operatorname{Re} z > 0\}$ and $\mathbb{C}_+^k = (\mathbb{C}_+)^k$

<u>Theorem 11</u>: For fixed gh, the distributions $S_{n+m-1}(\underline{\xi}^0 | gh)$ are restrictions to the product of positive real half axes of functions $S_{n+m-1}(\underline{\zeta}^0 | gh)$, analytic in $\mathbb{C}_+^{n+m-1}$. There are vector valued functions $\Psi_n^E(z^0, \underline{\zeta}^0 | g)$ analytic in $\mathbb{C}_+^n$, such that

$$S_{n+m-1}(\overleftarrow{\underline{\zeta}}^0, \bar{z}^0 + z'^0, \underline{\zeta}'^0 | gh) =$$

(14) $$(\Psi_n^E(z^0, \underline{\zeta}^0 | g), \Psi_m^E(z'^0, \underline{\zeta}'^0 | h)) .$$

Furthermore $S_{n+m-1}(\underline{\zeta}^0 | gh)$ satisfies for $\underline{\zeta}^0 \in \mathbb{C}_+^{n+m-1}$

(15) $$|S_{n+m-1}(\underline{\zeta}^0 | gh)| \leq c |g|_s |h|_s (1 + |\underline{\zeta}^0|)^a (1 + [\min_k \operatorname{Re} \zeta_k^0]^{-1})^b$$

for some Schwartz norms $|\cdot|_s$ and some constants a, b, c depending on n + m - 1.

Remarks: (1) By standard arguments, (15) implies that $S_{n+m-1}(\underline{\xi}^0|gh)$ is the Laplace transform of some distribution in $\mathcal{S}'$ with support in $\overline{\mathbb{R}}_+^{n+m-1}$, hence $S_{n+m-1}(\underline{\xi})$ is of the form (11), and again $\widetilde{W}_n(\underline{q})$ is the Fourier transform of the difference variable Wightman distribution. Note that in the proof of proposition I this fact was a direct consequence of the $|\cdot|'_m$-continuity assumption for the S_n's. Positivity (W2) follows easily from (13).

(2) The proof of Theorem 11 as presented below is due to Osterwalder-Schrader [OS 4]. The construction of the analytic continuation of S_n was also found simultaneously and independently by Glaser [Gl 1,2].

Proof of Theorem 11: A. Constructing the Hamiltonian

For t ⩾ 0 we define $\hat{T}_t : \underline{\mathcal{S}}_< \to \underline{\mathcal{S}}_<$ by

$$(\hat{T}_t \underline{f})_n(\underline{x}) = f_n(x_1^0 - t, \vec{x}_1, \dots x_n^0 - t, \vec{x}_n)$$

By the distribution property of $\mathfrak{S}_n$, for $\underline{f}, \underline{g} \in \underline{\mathcal{S}}_<$, $t \to \langle \underline{f}, \hat{T}_t \underline{g} \rangle$ is continuous and for some c, m, depending on $\underline{f}$,

(16) $$|\langle \underline{f}, \hat{T}_t \underline{f} \rangle| \leq c(1 + t^m)$$

Furthermore for $t \geqslant 0, s \geqslant 0$, $\hat{T}_t \hat{T}_s = \hat{T}_{t+s}$, and $\langle \underline{f}, \hat{T}_t \underline{g} \rangle = \langle \hat{T}_t \underline{f}, \underline{g} \rangle$.

By the Schwarz inequality

(17) $$|\langle \underline{f}, \hat{T}_t \underline{f} \rangle| \leq \|\underline{f}\| \, \|\hat{T}_t \underline{f}\| = \langle \underline{f}, \hat{T}_{2t} \underline{f} \rangle^{1/2}$$

for any $\underline{f} \in \underline{\mathcal{S}}_<$ with $\|\underline{f}\| = 1$. Iterating (17) and substituting (16) we obtain

$$|\langle \underline{f}, \hat{T}_t \underline{f} \rangle| \leq \langle \underline{f}, \hat{T}_{2^n t} \underline{f} \rangle^{2^{-n}} \leq [c(1 + (2^n t)^m)]^{2^{-n}} \longrightarrow 1$$

as $n \to \infty$. As $\hat{T}_t$ leaves the set $\mathcal{N}$ of norm zero vectors invariant, we may define a semigroup T_t in $\mathcal{H}$ by $T_t \Phi(\underline{f}) = \Phi(\hat{T}_t \underline{f})$ and extend it by continuity to all of $\mathcal{H}$. As for all $\Phi, \Phi' \in \mathcal{H}$, $\|T_t \Phi\| \leq \|\Phi\|$ and $(T_t \Phi, \Phi') = (\Phi, T_t \Phi')$, we conclude that T_t, $t \geqslant 0$, is a weakly continuous one parameter semigroup of selfadjoint contractions on $\mathcal{H}$ and $T_t = e^{-tH}$, where H is a selfadjoint positive operator. H is the Hamiltonian.

B. The Analytic Continuation

We will use the holomorphic semigroup $e^{-\tau H}$, $\mathrm{Re}\,\tau > 0$, to construct the analytic continuation in the time variables of the Schwinger functions. For the following arguments the space variables will play no role, so we shall drop them completely and write $S_{n-1}(\underline{\xi})$ and

$\Psi_n^E(x,\underline{\xi})$ instead of $S_{n-1}(\underline{\xi}^\circ | gh)$ and $\Psi_n^E(x^\circ, \underline{\xi}^\circ | h)$ resp., where $\underline{\xi}$ now stands for the n - 1 time variables $\xi_1^\circ, \dots \xi_{n-1}^\circ$.

Sandwiching $e^{-\tau H}$ between two vectors Ψ_n^E and Ψ_m^E we find that for $\tau = t + is$, $t > 0$,

$$(\Psi_n^E(x,\underline{\xi}), \ e^{-\tau H}\Psi_m^E(x',\underline{\xi}')) = S_{n+m-1}(\overleftarrow{\underline{\xi}}, x+x'+t, \underline{\xi}' | s)$$

is a distribution in $\underline{\xi}$, x + x' + t, $\underline{\xi}'$ and s which satisfies the Cauchy-Riemann equations in t and s. It follows (see e.g. Vladimirov [Vl 1], p. 31) that

$$S_{n+m-1}(\overleftarrow{\underline{\xi}}, x+x'+t, \underline{\xi}' | s) = S_{n+m-1}(\overleftarrow{\underline{\xi}}, x+x'+\tau, \underline{\xi}')$$

is a distribution in the $\underline{\xi}$, $\underline{\xi}'$ variables and a function of x + x' + τ= z, analytic in the right half plane $\mathbb{C}_+ = \{z | \operatorname{Re} z > 0\}$, if properly smeared in the other variables. For n + m - 1 = k fixed and m = 0, 1, ... k, the "functions" $S_{n+m-1}(\overleftarrow{\underline{\xi}}, z, \underline{\xi}')$ are all analytic continuations of the same distribution S_k and we are in the situation of an edge of the wedge theorem for flat tubes. More precisely, if we define $\hat{S}_{n+m-1}(\overleftarrow{\underline{u}}, w, \underline{u}') = S_{n+m-1}(\overleftarrow{\underline{\xi}}, z, \underline{\xi}')$,where $\xi_k = e^{u_k}$, $z = e^w = e^{u+iv}$, then the $\hat{S}_{n+m-1}$ are analytic functions in w for $|\operatorname{Im} w| < \pi/2$ and distributions in the u-variables, and for $\operatorname{Im} w \to 0$, they all coincide. The Malgrange-Zerner theorem (see Epstein [Ep 1]) tells us now that there is a function $\hat{S}_k(w_1, \dots w_k)$, analytic in $\{\underline{w} | \sum_{\ell=1}^{k} |\operatorname{Im} w_\ell| < \pi/2\}$, which continues all the $\hat{S}_{n+m-1}(\overleftarrow{\underline{u}}, w, \underline{u}')$. Equivalently we may say that the $S_{n+m-1}(\overleftarrow{\underline{\xi}}, z, \underline{\xi}')$ are all restrictions of a function $S_k(\zeta_1, \dots \zeta_k) = S_k(\underline{\zeta})$, k = n + m - 1, which is analytic in

$$C_k^{(1)} = \{\underline{\zeta} | \zeta_\ell \in \mathbb{C}_+, \ \sum_{\ell=1}^{k} |\arg \zeta_\ell| < \pi/2\}.$$

We claim that the functions $S_k(\underline{\zeta})$ can be interpreted as the scalar products of two vectors in $\mathcal{H}$. Let us define $\mathcal{D}_n^{(1)} \subset \mathbb{C}_+^n$ by

(18) $$\mathcal{D}_n^{(1)} = \{(x, \underline{\zeta}) | x > 0, (\overleftarrow{\underline{\zeta}}, 2x, \underline{\zeta}) \in C_{2n-1}^{(1)}\}.$$

<u>Lemma 12</u>: There are vector valued functions $\Psi_n^E(x,\underline{\zeta}) : \mathcal{D}_n^{(1)} \to \mathcal{H}$, analytic in $\underline{\zeta}$, such that

(19) $$(\Psi_n^E(x,\underline{\zeta}), \Psi_m^E(x',\underline{\zeta}')) = S_{n+m-1}(\overleftarrow{\underline{\zeta}}, x+x', \underline{\zeta}').$$

<u>Proof</u>: For $(\overset{\circ}{x}, \underline{\xi}) \in \mathcal{D}_n^{(1)}$ we choose a "polydisc" $P = \{(\overset{\circ}{x}, \underline{\zeta}) | |\zeta_k - \overset{\circ}{\xi}_k| < r_k, k = 1, \dots n-1\}$, centered at some real point $(\overset{\circ}{x}, \overset{\circ}{\underline{\xi}})$ and containing the

point $(\overset{\circ}{x}, \underline{\zeta})$, with r_k small enough, so that $P \subset \mathcal{D}_n^{(1)}$. (Note that the first variable $\overset{\circ}{x}$ is always kept fixed and real.) Then the Taylor series expansion of $S_{2n-1}(\overleftarrow{\underline{\zeta}}, \overset{\circ}{x}+\overset{\circ}{x}', \underline{\zeta}')$ around the point $(\overleftarrow{\overset{\circ}{\underline{\xi}}}, \overset{\circ}{x}+\overset{\circ}{x}', \overset{\circ}{\underline{\xi}}')$ is convergent for $(\overset{\circ}{x}, \underline{\zeta}) \in P$, $(\overset{\circ}{x}, \underline{\zeta}') \in P'$. Now let $\chi_\nu(x, \underline{\xi})$ be a C_0^∞ approximation to the δ-function $\delta(x - \overset{\circ}{x}) \cdot \prod_{i=1}^{n-1} \delta(\xi_k - \overset{\circ}{\xi}_k)$ such that $\operatorname{supp} \chi_\nu \subset \mathbb{R}_+^n$, $\chi_\nu \geq 0$ and $\int \chi_\nu \, dx \, d\underline{\xi} = 1$.

Furthermore define

$$f_{\nu\mu}(x, \underline{\xi}) = \sum_{|k| \leq \mu} \frac{(\overset{\circ}{\underline{\xi}} - \underline{\zeta})^{\underline{k}}}{\underline{k}!} \frac{\partial^{|k|}}{\partial \underline{\xi}^{\underline{k}}} \chi_\nu(x, \underline{\xi}) ,$$

where $\underline{k} = (k_1, \ldots k_{n-1})$, $\underline{k}! = \prod_i k_i!$, $|k| = \sum_i k_i$, $\underline{x}^{\underline{k}} = \prod_i x_i^{k_i}$,

$$\frac{\partial^{|k|}}{\partial \underline{\xi}^{\underline{k}}} = \prod_i \frac{\partial^{k_i}}{\partial \xi_i^{k_i}}$$

Then

$$\Psi^E_{n,\nu\mu}(\overset{\circ}{x}, \underline{\zeta}) = \int \Psi^E_n(x, \underline{\xi}) \, f_{\nu\mu}(x, \underline{\xi}) \, dx \, d^{n-1}\underline{\xi}$$

are vectors in $\mathcal{H}$, which for fixed ν, μ depend analytically on $\underline{\zeta}$. Using eq. (13) and the fact that $S_{n+m-1}(\overleftarrow{\underline{\xi}}, x+x', \underline{\xi}')$ is a real analytic function we easily check that $\Psi^E_n(\overset{\circ}{x}, \underline{\zeta}) = \lim_{\mu \to \infty} \lim_{\nu \to \infty} \Psi^E_{n,\nu\mu}(\overset{\circ}{x}, \underline{\zeta})$ exist and satisfy equation (19).

Lemma 12 enables us to construct an analytic extension of $S_k(\underline{\zeta})$ to a larger domain $C_k^{(2)}$. For $(x, \underline{\zeta}) \in \mathcal{D}_n^{(1)}$ and $(x', \underline{\zeta}') \in \mathcal{D}_m^{(1)}$ we define for $\tau \in \mathbb{C}_+$

$$(20) \qquad S_{n+m-1}(\overleftarrow{\underline{\zeta}}, x+x'+\tau, \underline{\zeta}') = \left(\Psi^E_n(x, \underline{\zeta}), e^{-\tau H} \Psi^E_m(x', \underline{\zeta}') \right) .$$

With n + m - 1 = k fixed and n = 0, ... k, eq. (20) yields an analytic extension of S_k to the domain

$$(21) \qquad \hat{C}_k^{(2)} = \bigcup_n \{ (\overleftarrow{\underline{\zeta}}, x+x'+\tau, \underline{\zeta}') \mid (x, \underline{\zeta}) \in \mathcal{D}_n^{(1)}, (x', \underline{\zeta}') \in \mathcal{D}_m^{(1)}, \tau \in \mathbb{C}_+, n+m-1 = k \} .$$

In terms of the variables $w_k = u_k + iv_k = \ln \zeta_k$, the domain $\hat{C}_k^{(2)}$ is a tube. By the tube theorem, see e.g. Vladimirov [Vl 1], p. 154, $S_k(\underline{\zeta})$ can be analytically extended to the envelope of holomorphy $C_k^{(2)}$ of $\hat{C}_k^{(2)}$, which is just the convex hull of $\hat{C}_k^{(2)}$ in the variables w_k. As in (18) we define

$$(22) \qquad \mathcal{D}_k^{(2)} = \{ (x, \underline{\zeta}) \mid x > 0, (\overleftarrow{\underline{\zeta}}, 2x, \underline{\zeta}) \in C_{2k-1}^{(2)} \} ,$$

and we prove Lemma 12 with $\mathcal{D}_k^{(2)}$ replacing $\mathcal{D}_k^{(1)}$ as before.

Repeating this procedure N times, we end up with an analytic extension of $S_k(\underline{\zeta})$ to a domain $C_k^{(N)}$, and vector valued functions $\mathcal{U}_n^{\equiv}(x,\underline{\zeta})$ defined on a domain $\mathcal{D}_k^{(N)}$, such that equation (19) holds. The first part of Theorem 11 follows if we can show that $\bigcup_N C_k^{(N)} = \mathbb{C}_+^k$ or equivalently $\bigcup_N \mathcal{D}_n^{(N)} = \mathbb{R}_+ \times \mathbb{C}_+^{n-1}$. We will prove a stronger result:

<u>Lemma 13</u>: For all N = 3, 4, ..., n = 1, 2, ...

(A) $\mathcal{D}_n^{(N)}$ contains the set

$$\{(x,\underline{\zeta}) \mid x \in \mathbb{R}_+ ,\ |\arg \zeta_r| < \tfrac{\pi}{2}\left(1 - 2^{-N}\sum_{t=0}^{r-1}\tbinom{N}{t}\right),\ 1 \le r \le n\}$$

(B) $C_n^{(N)}$ contains the sets

$$\{\underline{\zeta} \mid |\arg \zeta_r| < \tfrac{\pi}{2}\left(1 - 2^{-N}\sum_{t=0}^{r-1}\tbinom{N}{t}\right),\ 1 \le r \le n\}$$, and

$$\{\underline{\zeta} \mid |\arg \zeta_r| < \max\{0, \tfrac{\pi}{2}(1 - 2^{-N/2}\cdot \gamma_n)\},\ 1 \le r \le n\},$$

where $\gamma_n = \max\limits_{N>1} 2^{-N/2}\sum_{t=0}^{n-1}\binom{N}{t}$, and $\sum_{t=0}^{r-1}\binom{N}{t} \equiv 2^N$ for $r > N$.

<u>Proof</u>: By construction the regions $C_k^{(N)}$ and $\mathcal{D}_k^{(N)}$ become tubes after the transformation $\zeta_i = e^{w_i}$. We define $c_k^{(N)}$ and $d_k^{(N)}$ to be the bases of these tubes:

$$c_k^{(N)} = \{\underline{v} \mid v_i = \operatorname{Im} w_i,\ 1 \le i \le k,\ (e^{w_1}, \dots e^{w_k}) \in C_k^{(N)}\}$$
$$d_k^{(N)} = \{(0,\underline{v}) \mid v_i = \operatorname{Im} w_i,\ 1 \le i \le k-1,\ (e^{u_0}, e^{w_1}, \dots e^{w_{k-1}}) \in \mathcal{D}_k^{(N)}\}$$

Note that $c_k^{(N)}$ and $d_k^{(N)}$ are subsets of $[-\pi/2, \pi/2]^n$. From the definitions of $C_k^{(N)}$ and $\mathcal{D}_k^{(N)}$ we find (see (21), (22))

(23) $$c_k^{(N)} = \text{convex hull of } \{\underline{v} \mid \underline{v} = (\overleftarrow{\underline{v}}', v, \underline{v}'') \text{ for } (0,\underline{v}') \in d_n^{(N)},$$
$$(0, v'') \in d_m^{(N)},\ v \in (-\pi/2, \pi/2),\ n+m-1 = k\},$$

(24) $$d_k^{(N)} = \{(0,\underline{v}) \mid (-\overleftarrow{\underline{v}}, 0, \underline{v}) \in c_{2k-1}^{(N)}\}.$$

Equations (23), (24) together with $c_k^{(0)} = (0, \dots 0)$ define the sets $c_k^{(N)}$, $d_k^{(N)}$ for all k and N inductively. Observe that all the sets $c_k^{(N)}$ and $d_k^{(N)}$ are convex. Moreover if $(v_1, \dots v_k) \in c_k^{(N)}$ (or $d_k^{(N)}$), then the whole hyperrectangle with corners $(\pm v_1, \dots \pm v_k)$ is also contained in $c_k^{(N)}$ ($d_k^{(N)}$ resp.).

For a proof of Lemma 13 we first construct a function $h(r,N)$, $r = 1,2,\dots,$ $N = 1, 2, \dots,$ such that for all $t \geq 1$, $s \geq 0$,

(25) $$w(t,s,N) \equiv (\underbrace{0,0,\dots 0}_{t}, h(1,N), \dots h(s,N)) \in d_{t+s}^{(N)}$$

We choose $h(r,0) = 0$ for $r = 1, 2, \dots$. Obviously $w(t,s,0) \in d_{t+s}^{(0)}$.

Suppose now we have already constructed $h(r,N)$ for all $r \geq 1$ and $N = 0, 1, \ldots L$, such that (25) holds. Then by (23) the following points are contained in $C^{(L+1)}_{2(s+t)-1}$, for all $|\alpha| < \pi/2$:

$$(-h(s,L), -h(s-1,L), \ldots -h(2,L), -h(1,L), \underbrace{0,\ldots 0}_{2t-1}, \alpha, h(1,L), \ldots h(s-1,L))$$

and

$$(-h(s-1,L), \ldots\ldots -h(1,L), -\alpha, \underbrace{0,\ldots 0}_{2t-1}, h(1,L), h(2,L), \ldots h(s,L))$$

Because $C^{(L+1)}_{2(s+t)-1}$ is convex, it also contains the point

$$\Big(-\tfrac{1}{2}[h(s,L)+h(s-1,L)], \ldots -\tfrac{1}{2}[h(2,L)+h(1,L)], -\tfrac{1}{2}[h(1,L)+\alpha], \underbrace{0,\ldots 0}_{2t-1}, \tfrac{1}{2}[h(1,L)+\alpha], \ldots \tfrac{1}{2}[h(s,L)+h(s-1,L)]\Big)$$

This means that with

$$\begin{aligned} & h(r,0) = 0, && r = 1,2,\ldots \\ (26)\qquad & h(1,L+1) = \tfrac{1}{2}[h(1,L) + \alpha] && \\ & h(r,L+1) = \tfrac{1}{2}[h(r,L) + h(r-1,L)], && r = 2,3,\ldots \end{aligned}$$

the points $w(t, s, L+1)$, defined by (25), are again in $d^{(L+1)}_{t+s}$. We take (26) as inductive definition of $h(r,N)$. A simple calculation shows that the solution of (26) is

$$h(r,N) = \alpha\Big[1 - 2^{-N}\sum_{t=0}^{r-1}\binom{N}{t}\Big],$$

where we define $\sum_{t=0}^{s}\binom{N}{t} = 2^N$ for $s \geq N$.

Because α can be chosen arbitrarily close to $\pi/2$, the points $w_0(1,n-1,N) = (0, v_1, v_2, \ldots v_{n-1})$ with $|v_r| < \pi/2\Big[1 - 2^{-N}\sum_{t=0}^{r-1}\binom{N}{t}\Big]$ are in $d^{(N)}_k$, which proves part (A) of Lemma 13. For part (B) we just use (A) and the equation $S_n(\underline{\xi}) = (\Omega, \Psi^E_{n+1}(x,\underline{\xi}))$.

C. Estimating $S_k(\underline{\xi})$

We only sketch the main ideas which lead to the estimate (15). All the detailed arguments will be in [OS 4]. In particular we shall continue neglecting the space variables completely, which of course we should not do in a complete derivation of inequality (15).

In a first step we use (E0') to conclude that there are integers α, β, and γ, such that for $\xi_k > 0$, $k = 1, 2, \ldots n$, and all n,

$$(27)\qquad |S_n(\xi_1,\ldots \xi_n)| \leq (\alpha n)^{\beta n}\prod_{i=1}^{n}\big[(1+\xi_i)(1+\xi_i^{-1})\big]^{\gamma}.$$

Now let $\varepsilon > 0$ and define

$$(28)\qquad S_{n,\varepsilon}(\underline{\zeta}) = \prod_{i=1}^{n} [(1+\zeta_i+\varepsilon)(1+\varepsilon^{-1})]^{-\gamma}\, S_n(\underline{\zeta}+\underline{\varepsilon})$$

and

$$\Psi^{E}_{n,\varepsilon}(x,\underline{\zeta}) = [(1+2x+\varepsilon)(1+\varepsilon^{-1})]^{-\gamma/2} \prod_{i=1}^{n-1} [(1+\zeta_i+\varepsilon)(1+\varepsilon^{-1})]^{-\gamma}\, \Psi^{E}_{n}(x+\varepsilon/2, \underline{\zeta}+\underline{\varepsilon}),$$

where $\underline{\zeta}+\underline{\varepsilon}$ is defined by $(\underline{\zeta}+\underline{\varepsilon})_k = \zeta_k+\varepsilon$.
Both $S_{n,\varepsilon}(\underline{\zeta})$ and $\Psi^{E}_{n+1,\varepsilon}(x,\underline{\zeta})$ are defined and analytic for $x \in R_+$, $\underline{\zeta} \in \mathbb{C}^n_+$. By (28) we have for $z = x + iy$,

$$(29)\qquad S_{n+m-1,\varepsilon}(\overleftarrow{\underline{\zeta}}, 2z, \underline{\zeta}') = \left(\frac{1+2x+\varepsilon}{1+2z+\varepsilon}\right)^{\gamma} \left(\Psi^{E}_{n,\varepsilon}(x,\underline{\zeta}),\, e^{-2iyH}\, \Psi^{E}_{m,\varepsilon}(x,\underline{\zeta}')\right).$$

We claim that for $\underline{\zeta} \in C^{(N)}_n$,

$$(30)\qquad |S_{n,\varepsilon}(\underline{\zeta})| \le (\alpha n)^{\beta n}\, 2^{\beta n N}$$

We prove (30) by induction. First we notice that for N = 0, (30) is just ineq. (27). Now assume we have verified ineq. (30) for N = 1, 2, ... L and all n. Then for $(x,\underline{\zeta}) \in \mathcal{D}^{(L)}_n$, $(x,\underline{\zeta}') \in \mathcal{D}^{(L)}_m$, $k = n + m - 1$, by (29),

$$\begin{aligned} |S_{n+m-1,\varepsilon}(\overleftarrow{\underline{\zeta}}, 2(x+iy), \underline{\zeta}')| &\le \|\Psi^{E}_{n,\varepsilon}(x,\underline{\zeta})\| \cdot \|\Psi^{E}_{m,\varepsilon}(x,\underline{\zeta}')\| \\ &= [S_{2n-1,\varepsilon}(\overleftarrow{\underline{\zeta}}, 2x, \underline{\zeta})\; S_{2m-1,\varepsilon}(\overleftarrow{\underline{\zeta}}', 2x, \underline{\zeta}')]^{1/2} \\ &\le [\alpha(2n-1)]^{\beta(n-1/2)}\, [\alpha(2m-1)]^{\beta(m-1/2)}\, 2^{\beta k L} \\ &\le (\alpha k)^{\beta k}\, 2^{\beta k (L+1)} . \end{aligned}$$

As $C^{(L+1)}_k$ is just the envelop of holomorphy of the region

$$\bigcup_{n+m-1=k} \{(\overleftarrow{\underline{\zeta}}, 2(x+iy), \underline{\zeta}') \mid (x,\underline{\zeta}) \in \mathcal{D}^{(L)}_n,\ (x,\underline{\zeta}') \in \mathcal{D}^{(L)}_m\}$$

we can use the maximum principle (see e.g. [Vl 1], p. 178) to conclude that (30) holds for N = L + 1. This proves ineq. (30) for all n and N.

The next step is to eliminate N from the right hand side of ineq. (30). In order to do that we choose and fix $\underline{\zeta} \in \mathbb{C}^n_+$ and determine $N = N(\underline{\zeta})$ so that $\underline{\zeta} \in C^{(N)}_n$. Let N_n be such that $2^{-N_n/2}\gamma_n < 1/2$. Then by Lemma 13(B) $\underline{\zeta} \in C^{(N_n)}_n$ if $|\arg \zeta_r| < \pi/4$ for r = 1, ... n and we may set $N(\underline{\zeta}) = N_n$ for such $\underline{\zeta}$. The dependence on n of N_n will be unimportant. Now take a $\underline{\zeta} \in \mathbb{C}^n_+$ with $\max_{1\le r\le n} |\arg \zeta_r| = |\arg \zeta_s| \ge \pi/4$ for some $1 \le s \le n$.
Then for $\xi = \operatorname{Re}\zeta_s$, $\eta = |\operatorname{Im}\zeta_s|$,

$$|\arg \zeta_s| = \operatorname{arc\,tg}\frac{\eta}{\xi} = \frac{\pi}{2} - \operatorname{arc\,tg}\frac{\xi}{\eta} < \frac{\pi}{2}\left(1 - \frac{\xi}{2\eta}\right) .$$

(use that $\operatorname{arctg} x > \frac{\pi}{4} x$ for $x < 1$.)

We define $N(\underline{\zeta}) = \lceil M \rceil$, where

$$2^{-M/2} \gamma_n = \frac{\xi}{2\eta} \quad , \text{ or } \quad M = c_n + \frac{2}{\ln 2} \ln \eta \xi^{-1} \quad ,$$

for some constant c_n depending on n only. In the following the meaning of c_n might change from line to line, but it will always be an n-dependent constant. With this choice of $N(\underline{\zeta})$ we find that for $1 \le r \le n$

$$|\arg \zeta_r| \le |\arg \zeta_s| \le \frac{\pi}{2}(1 - 2^{-N(\zeta)/2} \gamma_n) \quad ,$$

hence $\underline{\zeta} \in C_n^{(N(\zeta))}$, by Lemma 13 (B). Now we substitute $N(\underline{\zeta})$ in ineq. (30) This gives

(31) $$|S_{n,\varepsilon}(\underline{\zeta})| \le (\alpha n)^{\beta n} 2^{\beta n N_n} = c_n \, , \quad \text{for } \max_r |\arg \zeta_r| < \pi/4 \, ,$$

and

(32) $$|S_{n,\varepsilon}(\underline{\zeta})| \le c_n e^{2\beta n \cdot \ln \eta \xi^{-1}} = c_n (\eta \xi^{-1})^{2\beta n} \quad ,$$

for $\max |\arg \zeta_r| = \arg(\xi + i\eta) \geq \pi/4$. Combining (31) and (32) we finally obtain for some c_n

(33) $$|S_{n,\varepsilon}(\underline{\zeta})| \le c_n \left[(1 + |\max_k \operatorname{Im} \zeta_k|)(1 + (\min_k \operatorname{Re} \zeta_k)^{-1}) \right]^{2\beta n} .$$

Now we combine (33) with (28) and obtain by choosing $\varepsilon = \frac{1}{2} \min_k \operatorname{Re} \zeta_k$

$$\begin{aligned} |S_n(\underline{\zeta})| &= \left| \prod_{i=1}^{n} [(1+\zeta_i)(1+\varepsilon^{-1})]^{\gamma} S_{n,\varepsilon}(\underline{\zeta} - \underline{\varepsilon}) \right| \\ &\le c_n \left| \prod_{i=1}^{n} [(1+\zeta_i)(1+\varepsilon^{-1})]^{\gamma} \right| \cdot \left[(1+|\max_k \operatorname{Im} \zeta_k|)(1+\varepsilon^{-1}) \right]^{2\beta n} \\ &\le c\,(1+|\underline{\zeta}|)^{a} \, (1+(\min_k \operatorname{Re} \zeta_k)^{-1})^{b} \quad , \end{aligned}$$

for some constants a, b, c, depending on n. This proves (15), if we neglect the space variables.

By standard arguments, theorem 11 implies that there are distributions $\widetilde{W}_n(\underline{q})$ in $\mathcal{S}'(\mathbb{R}^{4n})$ with supp $\widetilde{W}_n \subset \overline{\mathbb{R}}_+^{4n}$, such that

$$S_n(\underline{\xi}) = \int \exp\left[-\sum_{k=1}^{n} (\xi_k^0 q_k^0 + i \vec{\xi}_k \vec{q}_k)\right] \widetilde{W}_n(\underline{q}) \, d^{4n}q \quad .$$

We define the Wightman distributions by.

(34) $$\mathcal{W}_{n+1}(\underline{x}) = \int \exp\left[i \sum_k q_k (x_{k+1} - x_k)\right] \widetilde{W}_n(\underline{q}) \, d^{4n}q \quad ,$$

and we have to verify (W0) - (W5).

Verification of the Wightman Axioms

(W0) The distribution property (E0) follows from (34) and

(W1) Invariance of $\mathcal{W}_n$ under translations and space rotations follows immediately from (34) and the corresponding properties of S_n. To prove invariance under Lorentz rotations we need only show that $X_{0j} \widetilde{W}(\underline{q}) = 0$, where $X_{0j} = \sum_{k=1}^{n} (q_k^0 \frac{\partial}{\partial q_k^j} + q_k^j \frac{\partial}{\partial q_k^0})$, j = 1, 2, 3, are the infinitesimal generators of Lorentz rotations, see [Ne 2]. But with $Y_{0j} = \sum_{k=1}^{n} (\xi_k^0 \frac{\partial}{\partial \xi_k^j} - \xi_k^j \frac{\partial}{\partial \xi_k^0})$, the infinitesmal generators of rotations in Euclidean space, we conclude from (E1) that $0 = Y_{0j} S_n(\underline{\xi}) = i \int \exp[-\Sigma(\xi_k^0 q_k^0 + i\vec{\xi}_k \vec{q}_k)]\, X_{0j} \widetilde{W}_n(\underline{q})\, d^{4n}\underline{q}$. Because the kernel of Fourier and Laplace transforms is zero, $X_{0j} \widetilde{W}_n(\underline{q}) = 0$, j = 1, 2, 3.

(W2) Positivity follows simply from Lemma 12 and the fact that the distribution $W_n(\underline{\xi})$ can be obtained as boundary value of the analytic continuation of the Schwinger functions S_n.

(W5) Lorentz covariance of $\widetilde{W}_n(\underline{q})$ and $\operatorname{supp}\widetilde{W}_n \subset \overline{\mathbb{R}}_+^{4n}$ imply that $\operatorname{supp}\widetilde{W}_n \subset \overline{V}_+^n$.

(W4) The cluster property (E4) implies that $\lim_{\lambda\to\infty} (\Phi, U(a,\mathbb{1})\,\Phi') = (\Phi,\Omega)(\Omega,\Phi')$ for any a = (0, $\vec{a}$) and arbitrary vectors $\Phi, \Phi' \in \mathcal{H}$. This immediately gives the cluster property (W4) for the Wightman distributions.

(W3) Locality follows from the symmetry of the Schwinger functions and a theorem in Jost [Jo 1], p. 83.

This ends the proof of proposition II.

Remark: 1) It has become clear from our proof, that condition (E0') was picked somewhat at random; other, similar conditions would do the same job. Work in progress even suggests, that (E0') can be replaced by a similar condition on the Schwinger functions $\mathfrak{S}_n$ rather than on S_n, see [OS 4]. This would probably be the most convenient form for applications.

2) For the Schwinger functions of $P(\varphi)_2$ models with small coupling constant, all the axioms (E0'), (E1) - (E4) follow easily from the cluster expansion estimates of Glimm, Jaffe and Spencer [GJS 1]. In particular (E0') is a consequence of corollary 1.1.9 in [GJS 1], which implies that

$$(35)\qquad |\mathfrak{S}_n(f)| < c^n\, n!\, \|f\|_2$$

for f having support in the product of unit lattice squares $\Delta_{i_1} \times \dots \times \Delta_{i_n} = \Delta_{\underline{i}}$. (We use the notation of [GJS 1].) For, let $h \in \mathcal{S}(\mathbb{R}_+^{2(n-1)})$ and let $\chi_{\underline{j}}(\underline{\xi})$ be the characteristic function of $\Delta_{j_1} \times \dots \times \Delta_{j_{n-1}}$ then

$$|S_{n-1}(h)| \le \sum_{\underline{j}} |S_{n-1}(h\chi_{\underline{j}})| .$$

Furthermore for any $g \in \mathcal{S}(\mathbb{R}^2)$, $\operatorname{supp} g \in \Delta_0$, $\int g(x)\, d^2x = 1$, $\xi_k = x_{k+1} - x_k$,

$$\begin{aligned} S_{n-1}(h\chi_{\underline{j}}) &= \int \mathfrak{S}_n(\underline{x})\, g(x_1)\, h(\underline{\xi})\, \chi_{\underline{j}}(\underline{\xi})\, d^{2n}x \\ (36)\qquad &= \sum_{\underline{i}} \int \mathfrak{S}_n(\underline{x})\, g(x_1)\, h(\underline{\xi})\, \chi_{\underline{j}}(\underline{\xi})\, \chi_{\underline{i}}(\underline{x})\, d^{2n}x . \end{aligned}$$

Observe, that $g(x_1)\chi_{\underline{j}}(\underline{\xi})\chi_{\underline{i}}(\underline{x})$ is different from zero only if $x_1 \in \Delta_0 \cap \Delta_{i_1}$; $x_2 - x_1 \in \Delta_{j_1}$, $x_2 \in \Delta_{i_2}$, ... $x_n - x_{n-1} \in \Delta_{j_{n-1}}$, $x_n \in \Delta_{i_n}$. This implies that for fixed $\underline{j}$, not more than 4^n values of $\underline{i}$ give non-vanishing contributions to $\sum_{\underline{i}}$ in (36). Hence we obtain using (35)

$$\begin{aligned} |S_{n-1}(h\chi_{\underline{j}})| &\le 4^n \max_{\underline{i}} \left| \int \mathfrak{S}_n(\underline{x})\, g(x_1)\, h(\underline{\xi})\, \chi_{\underline{j}}(\underline{\xi})\, \chi_{\underline{i}}(\underline{x})\, d^{2n}x \right| \\ &\le c_1^n\, n!\, \max_{\underline{i}} \left(\int |g(x_1) h(\underline{\xi}) \chi_{\underline{j}}(\underline{\xi}) \chi_{\underline{i}}(\underline{x})|^2\, d^{2n}x \right)^{1/2} \\ &\le c_1^n\, n!\, \|g\|_2\, \|h\chi_{\underline{j}}\|_2 . \end{aligned}$$

In the last step we have used the fact that the substitution $\underline{x} \to (x_1, \underline{\xi})$ has Jacobean 1. Hence for $h = h_1 \times h_2 \times \dots h_{n-1}$

$$\begin{aligned} |S_{n-1}(h)| &\le c_2^n\, n!\, \sum_{\underline{j}} \|h\chi_{\underline{j}}\|_2 \\ &= c_2^n\, n!\, \prod_k \Big(\sum_j \|h_k \chi_j\|_2 \Big) \end{aligned}$$

and this gives immediately (E0').

References

[BEG] BROS, J., EPSTEIN, H., GLASER, V., Commun. Math. Phys. 6, 77 (1967).

[Dy 1] DYSON, F J., Phys. Rev. 75, 1736 (1949).

[Ep 1] EPSTEIN, H., in: Axiomatic Field Theory, Brandeis Lectures 1965, Volume I; ed. M. Chretien and S. Deser, Gordon and Breach, New York, 1966.

[Fe 1] FEYNMAN, R. P., Rev. Mod. Phys. 20, 367 (1948).

[Fe 2] FEYNMAN, R. P., Phys. Rev. 76, 749 (1949).

[Fr 1] FRADKIN, E. S., Dokl. Akad. Nauk. SSSR, 125, 311 (1959); engl. transl. in Sov. Phys. Dokl. 4, 347 (1959).

[Gl 1] GLASER, V., On the equivalence of the Euclidean and Wightman Formulation of Field Theory, CERN Preprint.

[Gl 2] GLASER, V., Problems of Theoretical Physics, "Nauka", Moscow, 1969, p. 69.

[GJS] GLIMM, J., JAFFE, A., SPENCER, T., The Wightman Axioms and Particle Structure in the $P(\varphi)_2$ Quantum Field Model. Preprint.

[Gu 1] GUERRA, F., Phys. Rev. Lett. 28, 1213 (1972).

[Jo 1] JOST, R., The General Theory of Quantized Fields, Amer. Math. Soc. Publ., Providence, R.I., 1965.

[Li 1] LIONS, J. L., J. Analyse Math., 2, 369 (1952).

[Na 1] NAKANO, T., Progr. Theor. Phys. 21, 241 (1959).

[Ne 1] NELSON, E., Quantum Fields and Markoff Fields, Amer. Math. Soc. Summer Institute on PDE, held at Berkeley, 1971.

[Ne 2] NELSON, E., J. Funct. Anal. 12, 97 (1973).

[Ne 3] NELSON, E., J. Funct. Anal. 12, 211 (1973).

[Ne 4] NELSON, E., Erice Lecture Notes (1973).

[OS 1] OSTERWALDER, K., and SCHRADER, R., Phys. Rev. Lett. 29, 1423, (1972).

[OS 2] OSTERWALDER, K., and SCHRADER, R., Euclidean Fermi Fields and a Feynman-Kac Formula for Boson-Fermion Models, to appear in Helv. Phys. Acta.

[OS 3] OSTERWALDER, K., and SCHRADER, R., Commun. Math. Phys. 31, 83 (1973).

[OS 4] OSTERWALDER, K., and SCHRADER, R., Axioms for Euclidean Green's Functions II, in preparation.

[Oz 1] OZKAYNAK, H., Euclidean Fields for Arbitrary Spin, to appear.

[Sc 1] SCHWINGER, J., Proc. Natl. Acad. Sci. U.S. 44, 956 (1958).

[Sc 2] SCHWINGER, J., Phys. Rev. 115, 721 (1959).

[Sz 1] SCHWARTZ, L., Medd. Lunds. Univ. Math. Semin. (Supplementband), 196 (1952).

[Si 1] SIMON, B., Euclidean Formulation of (Relativistic) Quantum Field Theory, Zürich Lectures 1973, to appear in: Princeton Series in Physics.

[SW 1] STREATER, R. F., WIGHTMAN, A. S., PCT, Spin and Statistics and All That, New York: Benjamin, 1964.

[Sy 1] SYMANZIK, K., J. Math. Phys. 7, 510 (1966).

[Sy 2] SYMANZIK, K., Euclidean Quantum Field Theory, in: Proc. of the International School of Physics "ENRICO FERMI", Varenna Course XLV, ed. R. Jost, New York: Academic Press, 1969.

[Vl 1] VLADIMIROV, V. S., Methods of the Theory of Functions of Several Complex Variables, Cambridge and London: MIT Press 1966.

[Wi 1] WICK, G. C., Phys. Rev. 96, 1124 (1954).

[Wg 1] WIGHTMAN, A. S., Phys. Rev. 101, 860 (1956).

PROBABILITY THEORY AND EUCLIDEAN FIELD THEORY

Edward Nelson

Department of Mathematics, Princeton University

I. SECOND QUANTIZATION

Let $\underset{\sim}{H}$ be a real Hilbert space, and let φ be the unit Gaussian process on $\underset{\sim}{H}$ That is, φ is the unique (up to equivalence) real Gaussian process indexed by $\underset{\sim}{H}$ with mean 0 and covariance given by the inner product on $\underset{\sim}{H}$. Thus there is a probability space $(\Omega,\underset{\sim}{S},\mu)$ such that for each f in $\underset{\sim}{H}$, $\varphi(f)$ is a random variable on $(\Omega,\underset{\sim}{S},\mu)$, the σ-algebra $\underset{\sim}{S}$ is generated by the $\varphi(f)$ with f in $\underset{\sim}{H}$, and if $f_1,\ldots,f_n$ are orthonormal in $\underset{\sim}{H}$ and F is a bounded Baire function on $\mathbb{R}^n$ then

$$\int_\Omega F(\varphi(f_1),\ldots,\varphi(f_n))\, d\mu = \frac{1}{(2\pi)^{n/2}} \int_{\mathbb{R}^n} F(x) e^{-\frac{x^2}{2}}\, dx .$$

The choice of $(\Omega,\underset{\sim}{S},\mu)$ is immaterial. Perhaps the simplest concrete construction is to choose an orthonormal basis $\{e_n\}$ of $\underset{\sim}{H}$ and let $(\Omega,\underset{\sim}{S},\mu)$ be the Cartesian product, indexed by n, of copies of $(\mathbb{R}, \underset{\sim}{B}, (2\pi)^{-\frac{1}{2}}e^{-x^2/2}\, dx)$, where $\underset{\sim}{B}$ is the σ-algebra of Borel sets.

If α is a positive or integrable random variable on $(\Omega,\underset{\sim}{S},\mu)$, we denote its expectation by $E\alpha$, so that

$$E\alpha = \int_\Omega \alpha\, d\mu .$$

For the unit Gaussian process φ we have the formulas

(1) $$E\,\varphi(f_1)\cdots\varphi(f_{2n+1}) = 0 ,$$

(2) $$E\,\varphi(f_1)\cdots\varphi(f_{2n}) = \Sigma \langle f_{i_1},f_{j_1}\rangle\cdots\langle f_{i_n},f_{j_n}\rangle ,$$

where the sum is over all pairings of $1,\ldots,2n$ -- that is,

$$i_1 < \ldots < i_n;\ i_1 < j_1,\ \ldots,\ i_n < j_n ,$$

and $(i_1, j_1, \ldots, i_n, j_n)$ is a permutation of $1,\ldots,2n$. The equation (1) follows from the fact that the symmetry $\varphi(f) \longrightarrow \varphi(-f) = -\varphi(f)$ preserves expectations. We need only prove (2) for the case $f_1 = \ldots = f_{2n} = f$, by virtue of the general polarization identity

$$\prod_{i=1}^{m} z_i = \frac{1}{m!} \sum_{r=1}^{m} (-1)^{m-r} \sum_{i_1 < \ldots < i_r} (z_{i_1} + \ldots + z_{i_r})^m \tag{3}$$

which is valid for commuting indeterminates z_i. We may assume that f is a unit vector. Then (2) reduces to

$$E\,\varphi(f)^{2n} = (2n-1)(2n-3)\cdots 3\cdot 1 , \tag{4}$$

since the right hand side of (4) is the number of pairings of $1,\ldots,2n$. But (4) is just the one dimensional integral

$$\frac{1}{\sqrt{2\pi}} \int_{-\infty}^{\infty} x^{2n} e^{-\frac{x^2}{2}}\, dx = (2n-1)(2n-3)\cdots 3\cdot 1 . \tag{5}$$

We denote $\underset{\sim}{L}^p(\Omega, \underset{\sim}{S}, \mu)$, for $1 \leq p \leq \infty$, simply by $\underset{\sim}{L}^p(\underset{\sim}{H})$, and we also use the notation $\Gamma(\underset{\sim}{H})$ for $\underset{\sim}{L}^2(\underset{\sim}{H})$. Let $\Gamma(\underset{\sim}{H})_{\leq n}$ be the closed linear span in $\Gamma(\underset{\sim}{H})$ of all elements of the form $\varphi(f_1)\cdots\varphi(f_m)$ with $m \leq n$, and let $\Gamma(\underset{\sim}{H})_n$ be the orthogonal complement of $\Gamma(\underset{\sim}{H})_{\leq(n-1)}$ in $\Gamma(\underset{\sim}{H})_{\leq n}$. For $f_1,\ldots,f_n$ in $\underset{\sim}{H}$, we define

$$:\varphi(f_1)\cdots\varphi(f_n):$$

to be the orthogonal projection of $\varphi(f_1)\cdots\varphi(f_n)$ into $\Gamma(\underset{\sim}{H})_n$. (If $\underset{\sim}{H}$ is one dimensional, so that $\Gamma(\underset{\sim}{H}) = \underset{\sim}{L}^2(\mathbb{R}, \underset{\sim}{B}, (2\pi)^{-\frac{1}{2}} e^{-x^2/2}\, dx)$, then $\Gamma(\underset{\sim}{H})_n$ is the one dimensional space spanned by the n'th Hermite polynomial, and $:x^n:$ is the n'th

Hermite polynomial normalized so that the leading coefficient is 1.) We have the formula

$$(6) \qquad \langle :\varphi(g_1)\cdots\varphi(g_n):, :\varphi(f_1)\cdots\varphi(f_n): \rangle = \sum_{\pi} \langle g_{\pi(1)}, f_1\rangle \cdots \langle g_{\pi(n)}, f_n\rangle ,$$

where the sum is over all permutations π of $1,\ldots,n$. Since the left hand side of (6) is symmetric, it suffices to prove (6) for $g_1 = f_1, \ldots, g_n = f_n$, and by (3) it suffices to consider the case $f_1 = \ldots = f_n = f$. We may assume that f is a unit vector. Then (6) becomes

$$(7) \qquad \langle :\varphi(f)^n:, :\varphi(f)^n: \rangle = n! ,$$

or

$$(8) \qquad \frac{1}{\sqrt{2\pi}} \int_{-\infty}^{\infty} (:x^n:)^2 e^{-\frac{x^2}{2}} \, dx = n! ,$$

which is the well-known normalization of Hermite polynomials.

Let $\underset{\sim}{H}_1$ be the complexification of $\underset{\sim}{H}$, and let $\underset{\sim}{H}_n$ be the n-fold Hilbert space symmetric tensor product of $\underset{\sim}{H}_1$ with itself. We give $\underset{\sim}{H}_n$ the inner product such that

$$(9) \qquad \langle \mathrm{Sym}\, g_1 \otimes \ldots \otimes g_n, \mathrm{Sym}\, f_1 \otimes \ldots \otimes f_n \rangle = \sum_{\pi} \langle g_{\pi(1)}, f_1\rangle \cdots \langle g_{\pi(n)}, f_n\rangle ,$$

where Sym is the symmetrization operator

$$\mathrm{Sym}\, f_1 \otimes \ldots \otimes f_n = \frac{1}{n!} \sum_{\pi} f_{\pi(1)} \otimes \ldots \otimes f_{\pi(n)} .$$

It is clear, by (6) and (9), that

$$:\varphi(f_1)\cdots\varphi(f_n): \longmapsto \mathrm{Sym}\, f_1 \otimes \ldots \otimes f_n$$

extends uniquely to be unitary from $\Gamma(\underset{\sim}{H})_n$ onto $\underset{\sim}{H}_n$. We shall use this mapping to identify $\Gamma(\underset{\sim}{H})_n$ with $\underset{\sim}{H}_n$. It is well-known that the Hermite polynomials span all of $\underset{\sim}{L}^2(\mathbb{R}, \underset{\sim}{B}, (2\pi)^{-\frac{1}{2}} e^{-x^2/2}\, dx)$, and Segal [13] extended this result to arbitrary real Hilbert spaces, showing that the $\Gamma(\underset{\sim}{H})_n$ span $\Gamma(\underset{\sim}{H})$. Consequently,

$$\Gamma(\underset{\sim}{H}) = \sum_{n=0}^{\infty} \underset{\sim}{H}_n .$$

Thus $\Gamma(\underset{\sim}{H})$ is Fock space.

The space $\Gamma(\underset{\sim}{H})$ is intrinsically attached to the structure of $\underset{\sim}{H}$ as a real Hilbert space. Consequently, if $U: \underset{\sim}{H} \to \underset{\sim}{K}$ is an orthogonal mapping of one real Hilbert space onto another it induces a unitary mapping $\Gamma(U): \Gamma(\underset{\sim}{H}) \to \Gamma(\underset{\sim}{K})$. On $\underset{\sim}{H}_n$, $\Gamma(U)$ is $U \otimes \ldots \otimes U$ (n factors). If $I: \underset{\sim}{H} \to \underset{\sim}{K}$ is an isometric linear imbedding of one real Hilbert space into another it induces an isometric linear imbedding $\Gamma(I): \Gamma(\underset{\sim}{H}) \to \Gamma(\underset{\sim}{K})$, and on $\underset{\sim}{H}_n$, $\Gamma(I)$ is $I \otimes \ldots \otimes I$ (n factors). If $E: \underset{\sim}{H} \to \underset{\sim}{K}$ is the orthogonal projection of a real Hilbert space onto a closed linear subspace, then it induces an orthogonal projection $\Gamma(E): \Gamma(\underset{\sim}{H}) \to \Gamma(\underset{\sim}{K})$, and on $\underset{\sim}{H}_n$, $\Gamma(E)$ is $E \otimes \ldots \otimes E$ (n factors). If $A: \underset{\sim}{H} \to \underset{\sim}{K}$ is any contraction (linear mapping of norm ≤ 1) from one real Hilbert space to another, we define $\Gamma(A)$ to be the direct sum of the $\Gamma(A)_n$, where $\Gamma(A)_n: \underset{\sim}{H}_n \to \underset{\sim}{K}_n$ is given by $A \otimes \ldots \otimes A$ (n factors). Halmos [8] showed that any contraction $A: \underset{\sim}{H} \to \underset{\sim}{K}$ may be represented in the form $A = EUI$ with E, U, and I as above (in fact, $I: \underset{\sim}{H} \longrightarrow \underset{\sim}{H} \oplus \underset{\sim}{K}$ is the obvious imbedding, $E: \underset{\sim}{K} \oplus \underset{\sim}{H} \longrightarrow \underset{\sim}{K}$ is the obvious projection, and the operator $U: \underset{\sim}{H} \oplus \underset{\sim}{K} \longrightarrow \underset{\sim}{K} \oplus \underset{\sim}{H}$ is defined to make the relation $A = EUI$ valid and turns out, by a non-trivial proof, to be orthogonal). Consequently $\Gamma(A) = \Gamma(E)\Gamma(U)\Gamma(I)$. Therefore the operator $\Gamma(A)$ is <u>doubly Markovian</u>, in the sense that

$$\begin{aligned} &\alpha \leq 0 \implies \Gamma(A)\alpha \geq 0 , \\ (10) \qquad &\Gamma(A)1 = 1 , \\ &E\Gamma(A)\alpha = E\alpha , \end{aligned}$$

since $\Gamma(E)$, $\Gamma(U)$, and $\Gamma(I)$ are clearly doubly Markovian.

If Γ is any doubly Markovian operator it is clearly a contraction from $\underset{\sim}{L}^{\infty}$ to $\underset{\sim}{L}^{\infty}$ and from $\underset{\sim}{L}^1$ to $\underset{\sim}{L}^1$. Either by the Riesz-Thorin theorem or by a simple application of Hölder's inequality, it follows that Γ is a contraction from $\underset{\sim}{L}^p$ to $\underset{\sim}{L}^p$ for all $1 \leq p \leq \infty$.

Theorem 1 (Hypercontractivity). Let $A: \underset{\sim}{H} \to \underset{\sim}{K}$ be a contraction from one real Hilbert space to another. Then $\Gamma(A)$ is a contraction from $\underset{\sim}{L}^q(\underset{\sim}{H})$ to $\underset{\sim}{L}^p(\underset{\sim}{K})$, for $1 \leq q \leq p \leq \infty$, provided that

$$\|A\| \leq \sqrt{\frac{q-1}{p-1}}\,. \tag{11}$$

If (11) does not hold, then $\Gamma(A)$ is not a bounded operator from $\underset{\sim}{L}^q(\underset{\sim}{H})$ to $\underset{\sim}{L}^p(\underset{\sim}{K})$.

We sketch here a proof which is different from the rather obscure proof [9] in the literature. If $A = 0$, then $\Gamma(A)$ is just the expectation, which is always a contraction. If $A \neq 0$, then we may write $A = (c^{-1}A)c$, where $c = \|A\|$, so that $\Gamma(A) = \Gamma(c^{-1}A)\Gamma(c)$. But $c^{-1}A$ is a contraction, so that $\Gamma(c^{-1}A)$ is a contraction from $\underset{\sim}{L}^p(\underset{\sim}{H})$ to $\underset{\sim}{L}^p(\underset{\sim}{K})$. Thus, to establish the first part of the theorem, we need only show that $\Gamma(c)$ is a contraction from $\underset{\sim}{L}^q(\underset{\sim}{H})$ to $\underset{\sim}{L}^p(\underset{\sim}{H})$ provided that (11) holds. It suffices to show that

$$(E(\Gamma(c)\alpha)^p)^{\frac{1}{p}} \leq (E\alpha^q)^{\frac{1}{q}} \tag{12}$$

for $\alpha \geq 0$, since for any positivity preserving linear operator Γ we have $|\Gamma\alpha| \leq \Gamma|\alpha|$. (This is easily seen by approximating Γ by an integral operator with positive kernel.)

Suppose that $\underset{\sim}{H} = \underset{\sim}{H}_{(1)} \oplus \underset{\sim}{H}_{(2)}$, so that we may take the probability space $(\Omega,\underset{\sim}{S},\mu)$ for the unit Gaussian process on $\underset{\sim}{H}$ to be the Cartesian product of the probability spaces $(\Omega_1,\underset{\sim}{S}_1,\mu_1)$ and $(\Omega_2,\underset{\sim}{S}_2,\mu_2)$ for the unit Gaussian processes on $\underset{\sim}{H}_{(1)}$ and $\underset{\sim}{H}_{(2)}$. We use the notation E_1 and E_2 for the expectations on these probability spaces, and similarly we use the notations $\Gamma_1(c)$ and $\Gamma_2(c)$. If α is a positive random variable on $(\Omega,\underset{\sim}{S},\mu)$, we claim that

$$\|\Gamma(c)\alpha\|_p = (E(\Gamma(c)\alpha)^p)^{\frac{1}{p}} \leq (E_1(\Gamma_1(c)(E_2(\Gamma_2(c)\alpha)^p)^{\frac{1}{p}})^p)^{\frac{1}{p}}\,. \tag{13}$$

We may write $\Gamma_1(c)$ as an integral operator with kernel $\Gamma_1(\cdot,\cdot)$, and similarly for $\Gamma_2(c)$. If β is a positive random variable and $\frac{1}{p} + \frac{1}{p'} = 1$, we have, by using Hölder's inequality twice,

$$E\ \beta\Gamma(c)\alpha$$

$$= \iiiint \beta(\omega_1,\omega_2)\Gamma_1(\omega_1,\eta_1)\Gamma_2(\omega_2,\eta_2)\alpha(\eta_1,\eta_2)\ d\mu_1(\eta_1)\ d\mu_2(\eta_2)\ d\mu_1(\omega_1)\ d\mu_2(\omega_2)$$

$$\leq \iint \left(\int \beta(\omega_1,\omega_2)^{p'}\ d\mu_2(\omega_2)\right)^{\frac{1}{p'}} \Gamma_1(\omega_1,\eta_1)$$
$$\left(\int \left(\int \Gamma_2(\omega_2,\eta_2)\alpha(\eta_1,\eta_2)\ d\mu_2(\eta_2)\right)^p d\mu_2(\omega_2)\right)^{\frac{1}{p}} d\mu_1(\eta_1)\ d\mu_1(\omega_1)$$

$$\leq \left(\iint \beta(\omega_1,\omega_2)^{p'}\ d\mu_2(\omega_2)\ d\mu_1(\omega_1)\right)^{\frac{1}{p'}}$$
$$\left(\int \Gamma_1(\omega_1,\eta_1)\left(\int \left(\int \Gamma_2(\omega_2,\eta_2)\alpha(\eta_1,\eta_2)\ d\mu_2(\eta_2)\right)^p d\mu_2(\omega_2)\right)^{\frac{1}{p}} d\mu_1(\eta_1)\right)^p d\mu_1(\omega_1)\Big)^{\frac{1}{p}},$$

which is just $\|\beta\|_{p'}$ times the right hand side of (13). Therefore (13) holds.

It follows at once from (13) that if the first part of the theorem holds for $\underset{\sim}{H}_{(1)}$ and $\underset{\sim}{H}_{(2)}$, it holds for $\underset{\sim}{H}_{(1)} \oplus \underset{\sim}{H}_{(2)}$. Therefore we need only prove it for a one dimensional real Hilbert space, for it then follows by induction for all finite dimensional real Hilbert spaces and, by approximation, for all real Hilbert spaces.

Let F be a function on $\mathbb{R}$ which is bounded below by a strictly positive constant and which is bounded together with its derivatives up to third order. It suffices to prove (12) for $\alpha = F(\varphi)$, where φ is a Gaussian random variable of mean 0 on a probability space $(\Omega_1,\underset{\sim}{S}_1,\mu_1)$. Let φ_h be a Gaussian random variable of mean 0 and variance h on a probability space $(\Omega_2,\underset{\sim}{S}_2,\mu_2)$, so that φ and φ_h are independent random variables on the product space $(\Omega,\underset{\sim}{S},\mu)$. Notice that $\varphi + \varphi_h$ is a Gaussian random variable of mean 0 whose expectation is h plus the variance of φ. We use the notation E_1, E_2, $\Gamma_1(c)$, $\Gamma_2(c)$ as above. Then

$$(14)\quad (E_2(\Gamma_2(c)\dot{F}(\varphi + \varphi_h))^p)^{\frac{1}{p}}$$

$$= (E_2(\Gamma_2(c)(F(\varphi) + F'(\varphi)\varphi_h + \tfrac{1}{2}F''(\varphi)h + o(h)))^p)^{\frac{1}{p}}$$

$$= (E_2(F(\varphi) + cF'(\varphi)\varphi_h + \tfrac{1}{2}F''(\varphi)h + o(h))^p)^{\frac{1}{p}}$$

$$= (E_2(F(\varphi)^p + F(\varphi)^{p-1}(pcF'(\varphi)\varphi_h + p\tfrac{1}{2}F''(\varphi)h) + F(\varphi)^{p-2}\frac{p(p-1)}{2}c^2F'(\varphi)^2h + o(h)))^{\frac{1}{p}}$$

$$= (F(\varphi)^p + F(\varphi)^{p-1}p\tfrac{1}{2}F''(\varphi)h + F(\varphi)^{p-2}\frac{p(p-1)}{2}c^2F'(\varphi)^2h + o(h))^{\frac{1}{p}}$$

$$= F(\varphi) + (\tfrac{1}{2}F''(\varphi) + \frac{p-1}{2}c^2\frac{F'(\varphi)^2}{F(\varphi)})h + o(h) .$$

With $c = 1$ and $p = q$ this gives

$$(15)\qquad (E_2F(\varphi + \varphi_h)^q)^{\frac{1}{q}} = F(\varphi) + (\tfrac{1}{2}F''(\varphi) + \frac{q-1}{2}\frac{F'(\varphi)^2}{F(\varphi)})h + o(h)$$

which, by (11), is greater than (14), up to terms which are $o(h)$.

By (13), we have that

$$(16)\qquad \|\Gamma(c)F(\varphi + \varphi_h)\|_p \le \|F(\varphi + \varphi_h)\|_q + o(h) ,$$

provided that we have already established that

$$(17)\qquad \|\Gamma(c)G(\varphi)\|_p \le \|G(\varphi)\|_q ,$$

where

$$G(x) = F(x) + \frac{1}{2}F''(x) + \frac{q-1}{2}\frac{F'(x)^2}{F(x)} .$$

Let $H(x,t)$ be the solution of the non-linear equation

$$(18)\qquad \frac{\partial H}{\partial t} = \frac{1}{2}\frac{\partial^2 H}{\partial x^2} + \frac{q-1}{2}\frac{(\frac{\partial H}{\partial x})^2}{H} ,$$

with initial value $H(x,0) = F(x)$. Notice that (18) is equivalent to the heat equation

$$\frac{\partial H^q}{\partial t} = \frac{1}{2}\frac{\partial^2 H^q}{\partial x^2}$$

for H^q, so that (18) has a unique solution which for each t as a function of x has the same properties as those assumed above for the initial value F. It follows from the above argument that if φ is a Gaussian random variable of mean 0 and variance t, then

$$\|\Gamma(c)F(\varphi)\|_p \leq H(0,t) = \|F(\varphi)\|_q , \tag{19}$$

which establishes (12).

Now suppose that (11) does not hold, so that $A: \underset{\sim}{H} \to \underset{\sim}{K}$ is a contraction with

$$\|A\| > \sqrt{\frac{q-1}{p-1}} . \tag{20}$$

By restricting A to the orthogonal complement of its null space, it is enough to consider the case that A has null space zero. Then A has a polar decomposition $A = UP$ where $U: \underset{\sim}{H} \to \underset{\sim}{K}$ is orthogonal and $P: \underset{\sim}{H} \to \underset{\sim}{H}$ is positive. Since $\Gamma(U)^{-1}$ is a contraction, it is enough to consider the case that $A = P: \underset{\sim}{H} \to \underset{\sim}{H}$ and is positive. By the spectral theorem, there is a non-zero subspace of $\underset{\sim}{H}$ and a constant c with $c > \sqrt{\frac{q-1}{p-1}}$ such that $A \geq c$ on that subspace, so again it suffices to consider the case that A is a scalar operator c. But if φ is a Gaussian random variable of mean 0 and variance 1, we can establish [9] that

$$\Gamma(c)e^{a\varphi} = e^{ca\varphi}e^{\frac{1}{2}(1-c^2)a^2} ,$$

$$\|\Gamma(c)e^{a\varphi}\|_p = e^{\frac{1}{2}[(p-1)c^2+1]a^2} ,$$

$$\|e^{a\varphi}\|_q = e^{\frac{1}{2}qa^2} ,$$

so that $\|\Gamma(c)e^{a\varphi}\|_p/\|e^{a\varphi}\|_q$ is arbitrarily large if a is sufficiently large.

II. EUCLIDEAN FIELDS

Let $\mathbb{E}^d$ be d-dimensional Euclidean space; that is, $\mathbb{E}^d$ is the real d-dimensional space $\mathbb{R}^d$ with the inner product $x\cdot y = x^1y^1 + \ldots + x^dy^d$. By $\underset{\sim}{S}(\mathbb{E}^d)$ we mean the Schwartz space. If $\underset{\sim}{X}$ is a topological vector space, a linear process over $\underset{\sim}{X}$ is a stochastic process φ indexed by $\underset{\sim}{X}$ which is linear and such that if $f_\alpha \to f$ in $\underset{\sim}{X}$ then $\varphi(f_\alpha) \to \varphi(f)$ in measure. Thus the $\varphi(f)$, for f in $\underset{\sim}{X}$, are random variables on some probability space $(\Omega,\underset{\sim}{S},\mu)$, and for convenience we assume that the σ-algebra $\underset{\sim}{S}$ is generated by the $\varphi(f)$ with f in $\underset{\sim}{X}$.

Let φ be a linear process over $\underset{\sim}{S}(\mathbb{E}^d)$. If Λ is an open set in $\mathbb{E}^d$, we let $\mathfrak{O}(\Lambda)$ be the σ-algebra generated by the $\varphi(f)$ with $\operatorname{supp} f \subseteq \Lambda$, and if Λ is an arbitrary subset of $\mathbb{E}^d$ we let

(1)
$$\mathfrak{O}(\Lambda) = \bigcap_{\Lambda' \supset \Lambda} \mathfrak{O}(\Lambda') ,$$

where the intersection is over all open sets Λ' containing Λ. We also use the notation $\mathfrak{O}(\Lambda)$ for the set of all random variables which are measurable with respect to the σ-algebra $\mathfrak{O}(\Lambda)$. Conditional expectations with respect to $\mathfrak{O}(\Lambda)$ are denoted by $E\{\cdot|\mathfrak{O}(\Lambda)\}$. We let Λ^c denote the complement of Λ and $\partial\Lambda$ the boundary of Λ. A Markoff field on $\mathbb{E}^d$ is a linear process φ over $\underset{\sim}{S}(\mathbb{E}^d)$ such that whenever α is a positive or integrable random variable in $\mathfrak{O}(\Lambda)$, where Λ is an open set in $\mathbb{E}^d$, then

(2)
$$E\{\alpha|\mathfrak{O}(\Lambda^c)\} = E\{\alpha|\mathfrak{O}(\partial\Lambda)\} .$$

We call this the Markoff property.

Let $IO(d)$ be the Euclidean group of $\mathbb{E}^d$, consisting of all isometries of $\mathbb{E}^d$. By a representation T of $IO(d)$ on a probability space $(\Omega,\underset{\sim}{S},\mu)$ we mean a homomorphism $\eta \mapsto T(\eta)$ of $IO(d)$ into the group of automorphisms of the measure algebra. We note that if S is an automorphism of the measure algebra then S acts in a natural way on the random variables. A <u>Euclidean field</u> is a Markoff field φ over $\underset{\sim}{S}(\mathbb{E}^d)$ together with a representation T of $IO(d)$ on the underlying probability space $(\Omega,\underset{\sim}{S},\mu)$ of φ such that for all f in $\underset{\sim}{S}(\mathbb{E}^d)$ and η in $IO(d)$,

$$T(\eta)\varphi(f) = \varphi(f \circ \eta^{-1}) , \tag{3}$$

and which has the following property: If ρ is the reflection in the hyperplane $\mathbb{E}^{d-1}$ then

$$T(\rho)\alpha = \alpha , \qquad \alpha \in \mathcal{O}(\mathbb{E}^{d-1}) . \tag{4}$$

Relation (3) is called <u>Euclidean covariance</u> and (4) is called the <u>reflection property</u>. Notice that by Euclidean covariance, if the reflection property holds for one hyperplane it holds for all hyperplanes, so that no special choice is involved.

We need an assumption which guarantees that certain expectations of products of fields exist. The following assumption is convenient, although stronger than necessary.

(B) *For all f in $\underset{\sim}{S}(\mathbb{E}^d)$, $\varphi(f)$ is in $\underset{\sim}{L}^p$ for $1 \le p < \infty$. The mapping $\underset{\sim}{S}(\mathbb{E}^d)^n \mapsto \mathbb{C}$ given by*

$$S_n(f_1,\ldots,f_n) = E\,\varphi(f_1)\cdots\varphi(f_n)$$

is continuous.

By The Schwartz kernel theorem, there is a tempered distribution S_n on $\mathbb{E}^{dn}$ such that

$$S_n(f_1,\ldots,f_n) = S_n(f_1 \otimes \ldots \otimes f_n) .$$

<u>Theorem</u> 2. *Let φ be a Euclidean field satisfying assumption* (B). *Then the sequence of distributions S_n satisfies the axioms* E0, E1, E2, *and* E3 *of*

Osterwalder and Schrader [11].

Proof. We use the notation of [11]. Axiom E0 asserts that $S_0 = 1$ (which is true since the empty product is by convention 1, and $E1 = 1$) and that S_n is in $\underset{\sim}{S}'(\mathbb{E}^{dn})$. Since $\underset{\sim}{S}(\mathbb{E}^{dn})$ is a closed subspace of $\underset{\sim}{S}(\mathbb{E}^{dn})$ and since S_n is in $\underset{\sim}{S}'(\mathbb{E}^{dn})$ it defines, by restriction, an element of $\underset{\sim}{S}'(\mathbb{E}^{dn})$.

Axiom E1 asserts Euclidean invariance:

$$S_n(f) = S_n(f_{(a,R)}) , \quad R \in SO(d) , \quad a \in \mathbb{E}^d , \quad f \in \underset{\sim}{S}(\mathbb{E}^{dn}) .$$

We need only show this for f a tensor product, $f = f_1 \otimes \ldots \otimes f_n$, in which case it reduces to

$$E\, \varphi(f_1)\cdots\varphi(f_n) = E\, T(\eta)(\varphi(f_1)\cdots\varphi(f_n))$$

for $\eta = (a,R)$, which is true by Euclidean covariance.

Axiom E2 we prefer to call positive definiteness (to distinguish it from the positivity condition of Symanzik [15] which results from the fact that the expectation of a positive random variable is positive). It asserts that

$$(5) \qquad \sum_{n,m} S_{n+m}(\Theta f_n^* \times f_m) \geq 0 , \qquad \underline{f} \in \underline{S}_+ .$$

It is enough to prove (5) for the case that each f_n is a tensor product, $f_n = f_{n1} \otimes \ldots \otimes f_{nn}$. To say that $\underline{f}$ is in $\underline{S}_+$ means that

$$f_n(x_1,\ldots,x_n) = f_{n1}(x_1)\cdots f_{nn}(x_n) = 0$$

unless

$$0 < x_1{}^d < \ldots < x_n{}^d .$$

Let

$$\alpha = \sum_n \varphi(f_{n1})\cdots\varphi(f_{nn}) ;$$

this is a finite sum. In our notation, (5) becomes

(6) $$E[(T(\rho)\bar{\alpha})\alpha] \geq 0 \; .$$

Let Λ be the half-space $x^d > 0$, so that $\partial\Lambda$ is the hyperplane $\mathbb{E}^{d-1}$ and Λ^c is the half-space $x^d \leq 0$. Then α is in $\mathcal{O}(\Lambda)$ and $T(\rho)\bar{\alpha}$ is in $\mathcal{O}(\Lambda^c)$. We have

$$E[(T(\rho)\bar{\alpha})\alpha]$$

$$= E[(T(\rho)\bar{\alpha})E\{\alpha|\mathcal{O}(\Lambda^c)\}]$$

$$= E[(T(\rho)\bar{\alpha})E\{\alpha|\mathcal{O}(\mathbb{E}^{d-1})\}]$$

(by the Markoff property)

$$= E[E\{T(\rho)\bar{\alpha}|\mathcal{O}(\mathbb{E}^{d-1})\}E\{\alpha|\mathcal{O}(\mathbb{E}^{d-1})\}]$$

$$= E[(T(\rho)E\{\bar{\alpha}|\mathcal{O}(\mathbb{E}^{d-1})\})E\{\bar{\alpha}|\mathcal{O}(\mathbb{E}^{d-1})\}]$$

(by Euclidean covariance)

$$= E[E\{\bar{\alpha}|\mathcal{O}(\mathbb{E}^{d-1})\}E\{\alpha|\mathcal{O}(\mathbb{E}^{d-1})\}]$$

(by the reflection property)

$$= E|E\{\alpha|\mathcal{O}(\mathbb{E}^{d-1})\}|^2 \geq 0 \; .$$

Axiom E3, symmetry, follows immediately from the fact that random variables commute, so that

$$S_n(f_1,\dots,f_n) = E\,\varphi(f_1)\cdots\varphi(f_n) = S_n(f_{\pi(1)},\dots,f_{\pi(n)})$$

for all permutations π. This concludes the proof.

The Osterwalder-Schrader theorem [11] asserts that there is a set of tempered distributions $\underset{\sim}{W}_n$ on $\mathbb{M}^{dn}$, where $\mathbb{M}^d$ is d-dimensional Minkowski space, which are the vacuum expectation values of a quantum field satisfying all of the Wightman axioms except uniqueness of the vacuum (cluster decomposition property of $\underset{\sim}{W}_n$), and such that the $\underset{\sim}{W}_n$ are boundary values of holomorphic functions which agree with $S_n(x_1,\ldots,x_n)$ whenever the x_j are distinct. At the time of writing these notes, there is said to be a gap in the proof, but one can confidently expect that this will soon be remedied. For another approach to the problem of obtaining quantum fields on Minkowski space from Euclidean fields, not using the Osterwalder-Schrader axioms, see [10]. The axiom E4 of Osterwalder and Schrader, which implies uniqueness of the vacuum, does not follow without additional assumptions on the Euclidean field, and in fact it is a question of great interest whether it always holds in the $P(\varphi)_2$ models.

III. THE FREE EUCLIDEAN FIELD

Let $\underset{\sim}{S}_{\mathbb{R}}(\mathbb{E}^d)$ be the real Schwartz space on $\mathbb{E}^d$, let m be a strictly positive constant or merely a positive constant if $d \geq 3$, and let $\underset{\sim}{H}$ be the real Hilbert space completion of $\underset{\sim}{S}_{\mathbb{R}}(\mathbb{E}^d)$ with respect to the inner product

$$\langle g\,,\,(-\Delta + m^2)^{-1} f\rangle \ ,$$

where Δ is the Laplace operator. Let φ be the unit Gaussian process on $\underset{\sim}{H}$ and extend it by linearity to the complexification of $\underset{\sim}{H}$. Restricted to the Schwartz space, φ is a linear process over $\underset{\sim}{S}(\mathbb{E}^d)$.

Theorem 3. *Let* φ *be as above. Then* φ *is a Euclidean field satisfying assumption* (B).

Proof. Let Λ be an open set in $\mathbb{E}^d$, and let

$$\underset{\sim}{U} = \{f \in \underset{\sim}{H}:\ \operatorname{supp} f \subset \Lambda\}\ ,$$

$$\underset{\sim}{M} = \{f \in \underset{\sim}{H}:\ \operatorname{supp} f \subset \Lambda^c\}\ ,$$

$$\underset{\sim}{N} = \{f \in \underset{\sim}{H}:\ \operatorname{supp} f \subset \partial\Lambda\}\ ,$$

$$\underset{\sim}{L} = \underset{\sim}{M} \cap \underset{\sim}{N}^{\perp}\ .$$

Let f be in $\underset{\sim}{U}$ and let h be the orthogonal projection of f onto $\underset{\sim}{M}$. We claim that h is in $\underset{\sim}{N}$. To see this, observe that

$$<g, (-\Delta + m^2)^{-1}h> = <g, (-\Delta + m^2)^{-1}f>$$

for all g in $\underset{\sim}{M}$, and in particular for all C^∞ functions g with compact support in the interior Λ^{co} of the complement of Λ. That is, $(-\Delta + m^2)^{-1}h = (-\Delta + m^2)^{-1}f$ as distributions on Λ^{co}. Since $-\Delta + m^2$ is a local operator, $h = f$ as distributions on Λ^{co}, but $f = 0$ on Λ^{co}. Therefore $\operatorname{supp} h \subset \Lambda^c - \Lambda^{co} = \partial\Lambda$, so that h is indeed in $\underset{\sim}{N}$.

If $\underset{\sim}{K}$ is a closed linear subspace of $\underset{\sim}{H}$, let $\tilde{\underset{\sim}{K}}$ be the σ-algebra generated by the $\varphi(f)$ with f in $\underset{\sim}{K}$. If $\underset{\sim}{K}_n \downarrow \underset{\sim}{K}$, it is easily seen that $\tilde{\underset{\sim}{K}}_n \downarrow \tilde{\underset{\sim}{K}}$. Let Λ_n be a sequence of open sets with $\Lambda_n \downarrow \Lambda^c$ and let $\underset{\sim}{K}_n = \{f \in \underset{\sim}{H}: \operatorname{supp} f \subset \Lambda_n\}$. Then $\underset{\sim}{K}_n \downarrow \underset{\sim}{M}$, so that $\sigma(\Lambda_n) = \tilde{\underset{\sim}{K}}_n \downarrow \tilde{\underset{\sim}{M}}$, and consequently $\sigma(\Lambda^c) = \tilde{\underset{\sim}{M}}$. Similarly, $\sigma(\partial\Lambda) = \tilde{\underset{\sim}{N}}$, and of course $\sigma(\Lambda) = \tilde{\underset{\sim}{U}}$.

We showed above that $\underset{\sim}{U} \perp \underset{\sim}{L}$, so that the σ-algebras $\tilde{\underset{\sim}{U}}$ and $\tilde{\underset{\sim}{L}}$ are independent. Now $\underset{\sim}{M} = \underset{\sim}{N} \oplus \underset{\sim}{L}$, so that $\tilde{\underset{\sim}{M}}$ is the σ-algebra generated by $\tilde{\underset{\sim}{N}}$ and $\tilde{\underset{\sim}{L}}$. Therefore if α is a positive or integrable random variable in $\tilde{\underset{\sim}{U}}$, $E\{\alpha|\tilde{\underset{\sim}{M}}\} = E\{\alpha|\tilde{\underset{\sim}{N}}\}$. Thus φ is a Markoff field.

If η is in $IO(d)$, we define the orthogonal operator $U(\eta)$ on $\underset{\sim}{H}$ by

$$U(\eta)f = f \circ \eta^{-1} .$$

Then $\eta \mapsto U(\eta)$ is an orthogonal representation of $IO(d)$ on $\underset{\sim}{H}$, and consequently $\eta \mapsto T(\eta) = \Gamma(U(\eta))$ is a representation of $IO(d)$ on the underlying probability space $(\Omega, \underset{\sim}{S}, \mu)$ of φ. Clearly,

$$T(\eta)\varphi(f) = \varphi(f \circ \eta^{-1}) ,$$

so the Euclidean covariance holds.

Let

$$\underset{\sim}{H}_o = \{f \in \underset{\sim}{H}: \operatorname{supp} f \subset \mathbf{E}^{d-1}\}$$

and let f be in $\underset{\sim}{H}_o$. Since f is in $\underset{\sim}{H}$, its Fourier transform $\tilde{f}$ is in $\underset{\sim}{L}^2$ with respect to the measure

(1) $$\frac{dk}{k^2 + m^2},$$

and since $\text{supp } f \subset \mathbb{E}^{d-1}$, we have

$$\tilde{f}(k) = \sum_i f_{oi}(\vec{k}) p_i(k^d)$$

where f_{oi} is some function of $\vec{k} = (k^1, \ldots, k^{d-1})$ and p_i is a polynomial. But since $\tilde{f}$ is square integrable with respect to the measure (1), the polynomial p must be a constant. Consequently, if ρ is the reflection in $\mathbb{E}^{d-1}$,

$$U(\rho)f = f .$$

Therefore

$$T(\rho)\varphi(f) = \varphi(f) ,$$

and since $\sigma(\mathbb{E}^{d-1}) = \underset{\sim}{\tilde{H}}_o$, if α is in $\sigma(\mathbb{E}^{d-1})$ then

$$T(\rho)\alpha = \alpha .$$

That is, the reflection property holds. Thus φ is a Euclidean field.

Assumption (B) is obviously true for the Gaussian process φ.

We call φ the free Euclidean field of mass m on $\mathbb{E}^d$.

IV. MULTIPLICATIVE FUNCTIONALS

Let φ be a Markoff field over $\underset{\sim}{S}(\mathbb{E}^d)$ with the underlying probability space $(\Omega, \underset{\sim}{S}, \mu)$. We say that a random variable α is additive in case for every finite open cover $\{\Lambda_i\}$ of $\mathbb{E}^d$ there exist real α_i in $\sigma(\Lambda_i)$ with $\alpha = \sum \alpha_i$. Similarly, we say that a random variable β is multiplicative in case for every finite open cover $\{\Lambda_i\}$ of $\mathbb{E}^d$ there exist strictly positive β_i in $\sigma(\Lambda_i)$ with $\beta = \prod \beta_i$. Thus α is additive if and only if $\beta = e^\alpha$ is multiplicative.

Theorem 4. *Let* φ *be a Markoff field over* $\underset{\sim}{S}(E^d)$ *with the underlying probability space* $(\Omega,\underset{\sim}{S},d\mu)$ *and let* β *be a multiplicative random variable with* $E\beta = 1$. *Then* φ *is a Markoff field over* $\underset{\sim}{S}(E^d)$ *on the probability space* $(\Omega,\underset{\sim}{S},\beta\, d\mu)$.

To prove this, we need a pair of lemmas. If $\underset{\sim}{A}$ and $\underset{\sim}{B}$ are complete σ-algebras of measurable sets on a probability space, we let $\underset{\sim}{A} \cup \underset{\sim}{B}$ be the smallest complete σ-algebra containing $\underset{\sim}{A}$ and $\underset{\sim}{B}$.

Lemma 1. *Let* $\underset{\sim}{A}$ *and* $\underset{\sim}{B}_n$ *be complete* σ-*algebras of measurable sets on a probability space, with* $\underset{\sim}{B}_n$ *decreasing. Then*

$$\bigcap_n (\underset{\sim}{A} \cup \underset{\sim}{B}_n) = \underset{\sim}{A} \cup \bigcap_n \underset{\sim}{B}_n .$$

Proof. If $\underset{\sim}{A}$ is a σ-algebra of measurable sets on a probability space, let $\bar{\underset{\sim}{A}}$ denote the von Neumann algebra of multiplication operators on $\underset{\sim}{L}^2$ by bounded random variables which are measurable with respect to $\underset{\sim}{A}$. Then we need only show that

$$\bigcap_n (\bar{\underset{\sim}{A}} \cup \bar{\underset{\sim}{B}}_n) = \bar{\underset{\sim}{A}} \cup \bigcap_n \bar{\underset{\sim}{B}}_n . \tag{1}$$

But using ' to denote the commutant, we see with a little thought that

$$\begin{aligned}\Big(\bigcap_n (\bar{\underset{\sim}{A}} \cup \bar{\underset{\sim}{B}}_n)\Big)' &= \bigcup_n (\bar{\underset{\sim}{A}} \cup \bar{\underset{\sim}{B}}_n)' = \bigcup_n (\bar{\underset{\sim}{A}}' \cap \bar{\underset{\sim}{B}}_n{}') \\ &= \bar{\underset{\sim}{A}}' \cap \bigcup_n \bar{\underset{\sim}{B}}_n{}' = \bar{\underset{\sim}{A}}' \cap \Big(\bigcap_n \bar{\underset{\sim}{B}}_n\Big)' = \Big(\bar{\underset{\sim}{A}} \cup \bigcap_n \bar{\underset{\sim}{B}}_n\Big)' .\end{aligned}$$

By the double commutant theorem applied to the first and last terms, (1) holds.

Lemma 2. *Let* φ *be a Markoff field over* $\underset{\sim}{S}(E^d)$, *let* Λ *be an open set in* E^d, *and let* Λ' *be a closed subset of* Λ^c *containing* $\partial\Lambda$. *Then*

$$E\{\sigma(\Lambda \cup \Lambda') | \sigma(\Lambda^c)\} = \sigma(\Lambda') .$$

Proof. Notice that $\Lambda \cup \Lambda'$ is a closed set, since $\Lambda' \supset \partial\Lambda$. Let Λ_n be a sequence of open sets decreasing to Λ'. Then $\Lambda \cup \Lambda_n$ is a sequence of open sets decreasing to $\Lambda \cup \Lambda'$. Since $\underset{\sim}{S}(\mathbb{E}^d)$ admits partitions of unity, $\sigma(\Lambda \cup \Lambda_n) = \sigma(\Lambda) \cup \sigma(\Lambda_n)$. Therefore, by Lemma 1,

$$\sigma(\Lambda \cup \Lambda') = \bigcap_n \sigma(\Lambda \cup \Lambda_n) = \bigcap_n (\sigma(\Lambda) \cup \sigma(\Lambda_n))$$

$$= \sigma(\Lambda) \cup \bigcap_n \sigma(\Lambda_n) = \sigma(\Lambda) \cup \sigma(\Lambda') .$$

If α_i and β_i are bounded random variables in $\sigma(\Lambda)$ and $\sigma(\Lambda')$ respectively, then, since $\Lambda' \subset \Lambda^c$,

$$E\{\Sigma\, \alpha_i\beta_i | \sigma(\Lambda^c)\} = \Sigma\; E\{\alpha_i | \sigma(\Lambda^c)\}\beta_i$$

$$= \Sigma\; E\{\alpha_i | \sigma(\partial\Lambda)\}\beta_i$$

by the Markoff property, so that

$$E\{\sigma(\Lambda \cup \Lambda') | \sigma(\Lambda^c)\} \subset \sigma(\Lambda') .$$

The reverse inclusion is obvious.

Now we prove Theorem 4. We denote expectations and conditional expectations with respect to $d\mu$ by E and $E\{\cdot|\cdot\}$, and we denote expectations and conditional expectations with respect to $\beta\, d\mu$ by E_β and $E_\beta\{\cdot|\cdot\}$.

Let Λ be open in $\mathbb{E}^d$, and let α be a positive random variable on $\sigma(\Lambda)$. We need to show that

$$E_\beta\{\alpha | \sigma(\Lambda^c)\} = E_\beta\{\alpha | \sigma(\partial\Lambda)\} .$$

That is, by the Radon-Nikodym theorem there is a unique positive random variable $\tilde{\alpha} = E_\beta\{\alpha | \sigma(\Lambda^c)\}$ in $\sigma(\Lambda^c)$ such that for all positive random variables γ in $\sigma(\Lambda^c)$,

$$E_\beta\, \alpha\gamma = E_\beta\, \tilde{\alpha}\gamma ,$$

or equivalently

$$E\, \alpha\gamma\beta = E\, \tilde{\alpha}\gamma\beta , \tag{2}$$

and we need to show that $\tilde{\alpha}$ is in $\sigma(\partial\Lambda)$. Let Λ_o be any open set containing $\partial\Lambda$. Now $\{\Lambda, \Lambda_o, \Lambda^{co}\}$, where Λ^{co} is the interior of Λ^c, is an open cover of $\mathbb{E}^d$. Therefore there exist strictly positive random variables β_1 in $\sigma(\Lambda)$, β_2 in $\sigma(\Lambda_o)$, and β_3 in $\sigma(\Lambda^{co})$ with $\beta = \beta_1\beta_2\beta_3$. We know, by (2), that

$$E\, \alpha\gamma\beta_1\beta_2\beta_3 = E\, \tilde{\alpha}\gamma\beta_1\beta_2\beta_3$$

for all positive random variables γ in $\sigma(\Lambda^c)$. But β_3^{-1} is a positive random variable in $\sigma(\Lambda^c)$, so that

$$E\, \alpha\gamma\beta_1\beta_2 = E\, \tilde{\alpha}\gamma\beta_1\beta_2 \tag{3}$$

for all positive random variables γ in $\sigma(\Lambda^c)$. Notice that (3) is also equal to

$$E[\tilde{\alpha}\gamma\; E\{\beta_1\beta_2 | \sigma(\Lambda^c)\}]\ .$$

Since γ is arbitrary, this means that

$$E\{\alpha\beta_1\beta_2 | \sigma(\Lambda^c)\} = \tilde{\alpha}\; E\{\beta_1\beta_2 | \sigma(\Lambda^c)\}\ ,$$

and since $\beta_1\beta_2$ is strictly positive this may be written as

$$\frac{E\{\alpha\beta_1\beta_2 | \sigma(\Lambda^c)\}}{E\{\beta_1\beta_2 | \sigma(\Lambda^c)\}} = \tilde{\alpha}\ . \tag{4}$$

By Lemma 2 applied to Λ and $\Lambda' = \bar{\Lambda}_o \cap \Lambda^c$, both the numerator and the demoninator of (4) are in $\sigma(\Lambda')$, which is contained in $\sigma(\bar{\Lambda}_o)$. Therefore $\tilde{\alpha}$ is in $\sigma(\bar{\Lambda}_o)$, and since Λ_o is an arbitrary open set containing $\partial\Lambda$, this shows that $\tilde{\alpha}$ is in $\sigma(\partial\Lambda)$. This concludes the proof.

The proof shows that if α is a positive random variable in $\sigma(\Lambda)$ then

$$E_\beta\{\alpha | \sigma(\Lambda^c)\} = \frac{E\{\alpha\beta | \sigma(\Lambda^c)\}}{E\{\beta | \sigma(\Lambda^c)\}}\ , \tag{5}$$

since we may insert β_3 inside the conditional expectations of numerator and denominator in (4).

If φ is the free Euclidean field, then $\varphi(f)$ for f in $\underset{\sim}{S}_{\mathbb{R}}(\mathbb{E}^d)$ is additive. We shall now construct a more interesting example in dimension $d = 2$.

Let φ be the free Euclidean field of mass $m > 0$ on $\mathbb{E}^2$. We define its Fourier transform $\tilde{\varphi}$ by

$$\tilde{\varphi}(\tilde{f}) = \varphi(f)$$

where

$$\tilde{f}(k) = \int e^{-ix\cdot k} f(x)\, dx$$

is the Fourier transform of f, and we use the notation

$$\tilde{\varphi}(\tilde{f}) = \int \tilde{f}(k)\tilde{\varphi}(dk) .$$

Let

$$\varphi_\kappa(x) = \int_{|k| \leq \kappa} e^{ix\cdot k}\tilde{\varphi}(dk) .$$

Then $\varphi_\kappa(x)$, for x in $\mathbb{E}^2$, is a well-defined Gaussian random variable of mean 0 and variance c_κ^2, where

$$c_\kappa^2 = \frac{1}{(2\pi)^2} \int_{|k| \leq \kappa} \frac{dk}{k^2 + m^2} = \frac{1}{4\pi} \log \frac{\kappa^2 + m^2}{m^2} = O(\log \kappa) . \tag{6}$$

Hence $:\varphi_\kappa(x)^n:$ is well-defined. For g in $\underset{\sim}{L}^1 \cap \underset{\sim}{L}^\infty(\mathbb{E}^2)$, let

$$:\varphi_\kappa^n:(g) = \int :\varphi_\kappa(x)^n: g(x)\, dx .$$

If $\lambda \geq \kappa$ then we claim that

$$\| :\varphi_\lambda^n:(g) - :\varphi_\kappa^n:(g) \|_2^2 = \langle g, G_\lambda^n * g \rangle - \langle g, G_\kappa^n * g \rangle \tag{7}$$

where

$$G_\kappa(x) = \frac{1}{(2\pi)^2} \int_{|k| \leq \kappa} e^{ix\cdot k} \frac{dk}{k^2 + m^2} .$$

This follows from (6,I). As a consequence, $:\varphi_\kappa{}^n:(g)$ converges in $\underset{\sim}{L}^2(\Omega,\underset{\sim}{S},\mu)$ as $\kappa \to \infty$. We denote the limit by $:\varphi^n:(g) = \int g(x):\varphi(x)^n:\, dx$.

Next we assert that if Λ is open in $\mathbb{E}^2$ and g is in $\underset{\sim}{L}^1 \cap \underset{\sim}{L}^\infty(\mathbb{E}^2)$ with $\operatorname{supp} g \subset \Lambda$, then

$$(8) \qquad :\varphi^n:(g) \in \mathcal{O}(\Lambda) .$$

To see this, let $\underset{\sim}{M}_\kappa(\Lambda)$ be the closed linear subspace of $\underset{\sim}{L}^2(\Omega,\underset{\sim}{S},\mu)$ spanned by the $\varphi_\lambda(x)$ with x in Λ and $\lambda \geq \kappa$. This is the same as the closed linear subspace spanned by the $\varphi_\lambda(f)$ with f in $\underset{\sim}{L}^1 \cap \underset{\sim}{L}^\infty(\mathbb{E}^2)$, $\operatorname{supp} f \subset \Lambda$, and $\lambda \geq \kappa$. Let $\mathcal{O}_\kappa(\Lambda)$ be the σ-algebra generated by $\underset{\sim}{M}_\kappa(\Lambda)$. Then $:\varphi_\kappa{}^n:(g)$ is in $\mathcal{O}_\kappa(\Lambda)$. Let $\underset{\sim}{M}(\Lambda)$ be the closed linear subspace spanned by the $\varphi(f)$ with f in $\underset{\sim}{L}^1 \cap \underset{\sim}{L}^\infty(\mathbb{E}^2)$, $\operatorname{supp} f \subset \Lambda$. Then

$$\bigcap_\kappa \underset{\sim}{M}_\kappa(\Lambda) = \underset{\sim}{M}(\Lambda)$$

and consequently

$$\bigcap_\kappa \mathcal{O}_\kappa(\Lambda) = \mathcal{O}(\Lambda) .$$

Since $:\varphi_\kappa{}^n:(g) \to :\varphi^n:(g)$, we have

$$:\varphi^n:(g) \in \bigcap_\kappa \mathcal{O}_\kappa(\Lambda) = \mathcal{O}(\Lambda) ,$$

which proves (8).

Since $\underset{\sim}{L}^1 \cap \underset{\sim}{L}^\infty(\mathbb{E}^2)$ admits partitions of unity, it follows from (8) that $:\varphi^n:(g)$ is additive. In fact, it has the stronger property that if $\{\Lambda_i\}$ is a finite partition of $\mathbb{E}^2$ into measurable sets then there exist random variables α_i in $\mathcal{O}(\Lambda_i)$ with $:\varphi^n:(g) = \Sigma\, \alpha_i$, for we may take $\alpha_i = :\varphi^n:(g\chi_{\Lambda_i})$. This stronger property makes the analogue of Theorem 4 trivial to prove, without Lemmas 1 or 2. It is not clear whether Theorem 4 will prove useful.

From (7) it follows that

$$(9) \qquad \|{:}\varphi^n{:}(g) - {:}\varphi_\kappa{}^n{:}(g)\|_2^2 = \langle g, G^n * g\rangle - \langle g, G_\kappa{}^n * g\rangle$$

where

$$G(x) = \frac{1}{(2\pi)^2} \int e^{ix\cdot k} \frac{dk}{k^2 + m^2} .$$

We claim that for some $\epsilon > 0$, (9) is $O(\kappa^{-\epsilon})$. To prove this, it suffices to show that

$$(10) \qquad \|G^n - G_\kappa{}^n\|_r = O(\kappa^{-\epsilon})$$

for some r with, say, $2 \leq r < \infty$, and by the Hausdorff-Young theorem it suffices to show that

$$(11) \qquad \|\tilde{G}^{*n} - \tilde{G}_\kappa^{*n}\|_s = O(\kappa^{-\epsilon})$$

for some s with $1 < s \leq 2$. But

$$\tilde{G}^{*n} - \tilde{G}_\kappa^{*n} = (\tilde{G} - \tilde{G}_\kappa) * \tilde{G}_\kappa * \ldots * \tilde{G}_\kappa + \tilde{G} * (\tilde{G} - \tilde{G}_\kappa) * \tilde{G}_\kappa * \ldots * \tilde{G}_\kappa$$
$$+ \ldots + \tilde{G} * \ldots * \tilde{G} * (\tilde{G} - \tilde{G}_\kappa) .$$

Now if $q > 1$, $\|\tilde{G}_\kappa\|_q$ is uniformly bounded in κ, and if $q - 1 < \frac{1}{n}$ then by Young's inequality we have that (11) is smaller than a constant times $\|\tilde{G} - \tilde{G}_\kappa\|_q$, so that we need only show that

$$(12) \qquad \|\tilde{G} - \tilde{G}_\kappa\|_q = O(\kappa^{-\epsilon}) .$$

But a simple computation shows that (12) is true.

By Theorem 1,

$$(13) \qquad \left\|\Gamma\left(\frac{1}{\sqrt{p-1}}\right)({:}\varphi^n{:}(g) - {:}\varphi_\kappa{}^n{:}(g))\right\|_p \leq \|{:}\varphi^n{:}(g) - {:}\varphi_\kappa{}^n{:}(g)\|_2 ,$$

which is $O(\kappa^{-\epsilon})$ for some $\epsilon > 0$. But by definition of Γ, the left hand side of (13) is

$$\left(\frac{1}{\sqrt{p-1}}\right)^{n} \| :\varphi^{n}:(g) - :\varphi_{\kappa}{}^{n}:(g) \|_{p} .$$

Consequently, there is an $\epsilon > 0$ and a $C < \infty$ such that for all κ and all p,

$$\| :\varphi^{n}:(g) - :\varphi_{\kappa}{}^{n}:(g) \|_{p} \leq (p-1)^{\frac{n}{2}} C \kappa^{-\epsilon} . \tag{14}$$

If P is any polynomial in one variable,

$$P(\xi) = a_{n}\xi^{n} + \ldots + a_{1}\xi + a_{0} , \tag{15}$$

we define $:P(\varphi):(g) = \int g(x):P\varphi(x):\,dx$ for g in $\underset{\sim}{L}^{1} \cap \underset{\sim}{L}^{\infty}(\mathbb{E}^{2})$ to be

$$a_{n}:\varphi^{n}:(g) + \ldots + a_{1}\varphi(g) + a_{0}$$

and similarly for $:P(\varphi_{\kappa}):(g)$. Again, there is an $\epsilon > 0$ and a $C < \infty$ such that for all κ and all p,

$$\| :P(\varphi):(g) - :P(\varphi_{\kappa}):(g) \|_{p} \leq (p-1)^{\frac{n}{2}} C \kappa^{-\epsilon} . \tag{16}$$

Now suppose that P is bounded below; that is, P is real and in (15) n is even and $a_{n} > 0$. If ψ is Gaussian of mean 0 and variance 1 then $:P(\psi):$ is bounded below, since for a certain polynomial Q with the same leading term as P, we have $:P(\psi): = Q(\psi)$. It follows that for any Gaussian random variable ψ of mean 0 we have that $:P(\psi):$ is bounded below by a constant times the variance of ψ to the power $n/2$. Consequently, if $g \geq 0$ is in $\underset{\sim}{L}^{1} \cap \underset{\sim}{L}^{\infty}(\mathbb{E}^{2})$ then by (6) there is an $a > 0$ such that

$$:P(\varphi_{\kappa}):(g) \geq -a(\log \kappa)^{n} + 1 \tag{17}$$

for all κ (greater than 2, say), where we have put in the term 1 on the right hand side of (17) for later convenience. By (16),

$$\mu\{\omega: \ |:P(\varphi):(g) - :P(\varphi_\kappa):(g)| \geq 1\}$$

$$\leq \|:P(\varphi):(g) - :P(\varphi_\kappa):(g)\|_p^p \leq (p-1)^{\frac{n}{2}p} C^p \kappa^{-\epsilon p} ,$$

so that $:P(\varphi):(g) \geq -a(\log \kappa)^n$ except on a set of measure at most

$$(p^{\frac{n}{2}} C\kappa^{-\epsilon})^p . \tag{18}$$

If we choose the value of p which minimizes (18) we find that for some $b > 0$ and $\epsilon > 0$ it is less than

$$e^{-b\kappa^\epsilon} . \tag{19}$$

Consequently,

$$\mu\{\omega: \ e^{-:P(\varphi):(g)} \geq e^{a(\log \kappa)^n}\} = \mu\{\omega: \ :P(\varphi):(g) \leq -a(\log \kappa)^n\} \leq e^{-b\kappa^\epsilon} .$$

Let $\lambda = e^{a(\log \kappa)^n}$, so that $\kappa = e^{\left(\frac{\log \lambda}{a}\right)^{1/n}}$. Then

$$\mu\{\omega: \ e^{-:P(\varphi):(g)} \geq \lambda\} \leq e^{-be^{\left(\frac{\log \lambda}{a}\right)^{\epsilon/n}}} . \tag{20}$$

By (20), $\gamma = e^{-:P(\varphi):(g)}$ is in $\underset{\sim}{L}^1(\Omega,\underset{\sim}{S},\mu)$, so that $\beta = \gamma/E\gamma$ is a multiplicative random variable.

To summarize, let $d = 2$, let $g \geq 0$ be in $\underset{\sim}{L}^1 \cap \underset{\sim}{L}^\infty(\underset{\sim}{E}^2)$, let P be bounded below. Then φ is a Markoff field on the probability space $(\Omega,\underset{\sim}{S},\beta\, d\mu)$ where

$$\beta = \frac{e^{-\int g(x):P(\varphi(x)):\,dx}}{E\, e^{-\int g(x):P(\varphi(x)):\,dx}} . \tag{21}$$

V. LATTICE FIELDS

Let Λ be a subset of the integral lattice points Z^d. If f is a function defined on Λ, let

(1) $$\Delta_\Lambda f(x) = -2d\, f(x) + \sum_{\substack{|y-x|=1 \\ y \in \Lambda}} f(y) , \qquad x \in \Lambda .$$

For $x,y \in \Lambda$ let

(2) $$A_\Lambda(x,y) = \begin{cases} 2d + m^2 , & x = y \\ -1 , & |x - y| = 1 \\ 0 , & \text{otherwise} . \end{cases}$$

Then

(3) $$(-\Delta_\Lambda + m^2)f(x) = \sum_{y \in \Lambda} A_\Lambda(x,y)f(y) , \qquad x \in \Lambda .$$

If $m > 0$ (or if $d \geq 3$ and $m \geq 0$ or if Λ is finite and $m \geq 0$) then $(-\Delta_\Lambda + m^2)$ has an inverse on $\ell^2(\Lambda)$. Let G_Λ on $\Lambda \times \Lambda$ be such that

(4) $$(-\Delta_\Lambda + m^2)^{-1}f(x) = \sum_{y \in \Lambda} G_\Lambda(x,y)f(y) , \qquad x \in \Lambda .$$

Since $(-\Delta_\Lambda + m^2)^{-1}$ is a positive operator, the function G_Λ is of positive type on $\Lambda \times \Lambda$. We let φ_Λ be the Gaussian stochastic process indexed by Λ with mean 0 and covariance G_Λ. We call it the <u>free lattice field of mass</u> m <u>on</u> Λ.

If Λ is finite then the $\varphi_\Lambda(x)$ are just the coordinate functions u_x on $R^{\#\Lambda}$, where $\#\Lambda$ is the cardinality of Λ, with respect to the Gaussian measure

(5) $$(\det 2\pi G_\Lambda)^{-\frac{1}{2}} e^{-\frac{1}{2} \sum_{x,y \in \Lambda} u_x A_\Lambda(x,y) u_y} \prod_{x \in \Lambda} du_x .$$

Two properties of this measure are evident. First, only <u>nearest neighbors</u>

are coupled. Second, it is ferromagnetic in the sense that the off-diagonal terms in the exponent are all positive.

VI. THE INFINITE VOLUME LIMIT

Let P be an even real polynomial which is bounded below. That is,

$$P(\xi) = a_n \xi^n + a_{n-2}\xi^{n-2} + \dots + a_2 \xi^2 + a_0$$

with n even and $a_n > 0$. Notice that if φ is the free lattice field of mass m on $\mathbb{Z}^d$ then $:P(\varphi(x)): = Q(\varphi(x))$ where Q is also an even polynomial which is bounded below, so that for the purposes of a general discussion we may ignore Wick ordering.

Now let Λ be a finite subset of $\mathbb{Z}^d$ and consider the measure

$$N_\Lambda e^{-\frac{1}{2}\sum_{x,y\in\Lambda} u_x A_\Lambda(x,y) u_y - \sum_{x\in\Lambda} P(u_x)} \prod_{x\in\Lambda} du_x \tag{1}$$

where A_Λ is given by (5,V) and N_Λ is the normalizing constant which makes this a probability measure.

We have an instance of the situation studied by Ginibre [2 , Model 2, p. 321] for the formulation of Griffiths inequalities. At each site x in Λ we have the even measure

$$e^{-\frac{1}{2}(2d + m^2)u_x{}^2 - P(u_x)} du_x \, . \tag{2}$$

We let σ be the product of these measures on $\mathbb{R}^{\#\Lambda}$.

We let S_x be all functions of the form $f(|u_x|)$ or $\operatorname{sgn} u_x f(|u_x|)$, where f is positive, continuous, and increasing on $[0,\infty)$, we let S be all products of such functions for the various sites x in Λ, and we let $Q(S)$ be the set of all limits of polynomials with positive coefficients of such functions.

Notice that $Q(S)$ contains $-h$, where

(3)
$$-h = \frac{1}{2} \sum_{\substack{|x-y|=1 \\ x,y\in\Lambda}} u_x u_y \, .$$

We let

(4)
$$Z_h = \int e^{-h} \, d\sigma \, ,$$

so that $N_\Lambda = Z_h^{-1}$. If f is a function on Λ, we use Ef or $<f>_h$ to denote its expectation with respect to the measure (1), so that

(5)
$$Ef = <f>_h = \frac{\int f e^{-h} \, d\sigma}{\int e^{-h} \, d\sigma} \, .$$

As Ginibre shows [2], we have Griffiths' second inequality: If f and g are in $Q(S)$ then

(6)
$$Efg \geq Ef\, Eg \, .$$

Actually, Ginibre gives the proof for the case that the functions in S_x are bounded, and so extend to be continuous on $[0,\infty]$, and in Ginibre's notation this interval is replaced by $[0,1]$. However, the general statement made above follows readily from this case.

A familiar application of Griffiths' second inequality is the existence of the infinite volume limit. In fact, it follows readily from (6) that if $x_1,\ldots,x_n \in \Lambda \subset \Lambda'$ then

(7)
$$E\, \varphi_\Lambda(x_1)\cdots\varphi_\Lambda(x_n) \leq E\, \varphi_{\Lambda'}(x_1)\cdots\varphi_{\Lambda'}(x_n) \, ,$$

and consequently as Λ increases to $\mathbb{Z}^d$, the left hand side of (7) increases to a limit

(8)
$$S(x_1,\ldots,x_n) \, .$$

The same result holds if we replace the lattice $\mathbb{Z}^d$ by the lattice $\epsilon\mathbb{Z}^d$ with spacing ϵ between nearest neighbors and include a factor ϵ^{-2} in the definition (1,V) of Δ_Λ. For smooth functions f on $\mathbb{R}^d$, we then have that the limit as $\epsilon \to 0$ of the difference operator (1,V) applied to f is the Laplacian Δf. In dimension $d = 2$ it can be shown (see [7]) that the limit as $\epsilon \to 0$ of the expectation values $E\,\varphi_r(x_1)\cdots\varphi_r(x_n)$ exists, and is the expectation value of what Guerra, Rosen, and Simon call the half-Dirichlet theory. The inequality (7) then carries over to this case, yielding the existence of the infinite volume limit. Guerra, Rosen, and Simon [7] show how to establish boundedness of the infinite volume limit, and that the limit is the Schwinger functions of a theory satisfying all of the Wightman axioms except possibly uniqueness of the vacuum.

We conclude by showing that uniqueness of the vacuum need not hold in the lattice case. Specifically, we consider the case $d = 2$ and

$$P(\xi) = \xi^4 - a\xi^2 . \tag{9}$$

At each site x in Λ the measure (2) is then

$$e^{E(u_x)}\,du_x \tag{10}$$

where

$$E(u_x) = -u_x^4 - (2 + \tfrac{1}{2}m^2 - a)u_x^2 . \tag{11}$$

The function E is an increasing function on the interval [0,b], where

$$b = \sqrt{\frac{1}{2}a - 1 - \frac{m^2}{4}} \tag{12}$$

and by choosing a large enough we may make b as large as we please. Let

$$R(u_x) = \begin{cases} E(u_x) , & |u_x| \le b \\ \infty , & |u_x| > b . \end{cases} \tag{13}$$

Then R is a limit of functions in S_x. Consequently, it follows from (6) that expectations of products of field operators decrease if we replace the measure (10) by

$$e^{E(u_x) - R(u_x)} du_x \ . \tag{13}$$

But (13) is just

$$\chi_{[-b,b]}(u_x)\, du_x \tag{14}$$

and we have the continuous spin Ising model. Griffiths shows [4] that, for b large enough, this model possesses long range order, and the cluster decomposition property fails.

NOTES

Thanks to the work of Glimm and Jaffe, constructive quantum field theory is now a large and vigorously growing subject. We shall not review here the origins or principal applications of the techniques discussed in these lectures, but shall confine ourselves in these notes to some comments on matters of detail.

Lecture I: A reference for Hermite series is [16]. The fact that the $\Gamma(A)$ can actually be contractions into $\underset{\sim}{L}^p$ is due to Glimm [3] - this was an essential step in passing from box quantization to field quantization. After these lectures were written, a preprint by L. Gross [6] appeared which contains a beautiful, clean proof of Theorem 1. Gross differentiates with respect to p, and establishes a logarithmic Sobolev inequality as equivalent to the best possible hypercontractivity result. He then establishes the result for Bosons by first proving it for the one degree of freedom Fermion case and applying the central limit theorem.

Lecture II: The fact that the Markoff and reflection properties lead to the Osterwalder-Schrader axioms was noted in [7]. I am grateful to Jay Rosen for the proof which is given here. Dobrushin and Minlos [1] have announced that the uniqueness of the vacuum fails in some $P(\varphi)_2$ models.

Lecture IV: A reference for von Neumann algebras and the double commutant theorem is [12]. A reference for the Hausdorff-Young theorem and for Young's inequality is [14]. The locality of $:\varphi^n:$, formula (8), would have been easier to establish if we had not used a sharp momentum cutoff.

Lecture VI: This material is a comment on the work of Guerra, Rosen, and Simon [7]. For a discussion of the uses of Griffiths' inequalities, e.g. in proving monotonicity, see [5]. The proof given for the failure of the cluster property relies heavily on the fact that we have a fixed lattice spacing. It would be interesting to know whether the argument can be refined to give the Dobrushin-Minlos result [1].

REFERENCES

[1] R. L. Dobrushin and R. A. Minlos, Construction of a one-dimensional quantum field via a continuous Markoff field, submitted to Functional Analysis and its Applications.

[2] J. Ginibre, General formulation of Griffiths' inequalities, Communications in Math. Phys. 16 (1970), 310-328.

[3] J. Glimm, Boson fields with nonlinear self-interaction in two dimensions, Communications in Math. Phys. 8 (1968), 12-25.

[4] Robert B. Griffiths, Rigorous results for Ising ferromagnets of arbitrary spin, J. of Mathematical Physics 10 (1969), 1559-1565.

[5] Robert B. Griffiths, Phase transitions, in Statistical Mechanics and Quantum Field Theory (Les Houches 1970) ed. C. DeWitt and R. Stora, Gordon and Breach, New York (1971), 241-279.

[6] Leonard Gross, Logarithmic Sobolev Inequalities, Cornell University preprint (1973).

[7] F. Guerra, L. Rosen, and B. Simon, The $P(\varphi)_2$ Euclidean quantum field theory as classical statistical mechanics, to appear in Annals of Mathematics.

[8] P. R. Halmos, Normal dilations and extensions of operators, Summa Brasiliensis Math. 2 (1950), 125-134.

[9] Edward Nelson, The free Markoff field, J. Functional Anal. 12 (1973), 211-227.

[10] Edward Nelson, Construction of quantum fields from Markoff fields, J. Functional Anal. 12 (1973), 97-112.

[11] K. Osterwalder and R. Schrader, Axioms for Euclidean Green's functions, Communications in Math. Phys. 31, 83 (1973).

[12] Shôichirô Sakai, C*-Algebras and W*-Algebras, Ergebnisse der Math. und ihrer Grenzgebiete, Band 60, Springer-Verlag, New York, 1971.

[13] I. E. Segal, Tensor algebras over Hilbert spaces, Trans. Amer. Math. Soc. 81 (1956), 106-134.

[14] Elias M. Stein and Guido Weiss, Introduction to Fourier Analysis on Euclidean Spaces, Princeton University Press, Princeton (1971).

[15] K. Symanzik, Euclidean quantum field theory, Rend. Scuola Int. Fis. E. Fermi, XLV Corso.

[16] G. Szegö, Orthogonal Polynomials, Amer. Math. Soc. Coll. Publ. XXIII, New York (1939).

THE GLIMM-JAFFE ϕ-BOUND : A MARKOV PROOF

Barry Simon*,†
Departments of Mathematics and Physics
Princeton University

§1. Introduction

One of the most useful estimates in the control of the thermodynamic limit for $P(\phi)_2$ is the ϕ-bound of Glimm-Jaffe (1972) [henceforth GJ]:

$$\pm \phi(h) < |||h||| (\hat{H}_\ell + 1) \tag{1}$$

for suitable h and a suitable norm, $|||\ \ |||$. Here $\hat{H}_\ell$ is defined by:

$$H_\ell = H_0 + \int_{-\ell/2}^{\ell/2} :P(\phi(x)):dx \tag{2}$$

$$E(A) = \inf \operatorname{spec}(A) \tag{3a}$$

$$\hat{A} = A - E(A) \tag{3b}$$

Shortly after the appearance of GJ, Guerra, Rosen, and Simon (1972) [henceforth GRS] provided an abbreviated proof of bounds of the form (1). The GRS bounds were weaker than the GJ bounds in the types of functions, h , allowed and in the norm, $|||\ \ |||$, used. In particular, GJ allow $|||\ \ |||$ to be the L^1 norm and GRS do not. For the original applications, this distinction did not matter but recently Fröhlich (1973) exploited the L^1-bound to prove the existence of equal time VEV's in the infinite volume limit. One of our goals is the extension of the GRS proof to cover these L^1-bounds.

It is possible to merely modify one step in the GRS proof. However, we wish to rephrase the GRS proof in a way that we think makes the mechanism of proof more transparent. To explain our point, we recall the GRS proof: one rewrites the bound (1) as a set of bounds on matrix elements of the semigroup $\exp[-t(H_\ell \pm \phi(h)]$ and then uses Nelson's symmetry. In this new form, one bounds the matrix element as the norm of an operator times the product of the norms of two vectors. Nelson's symmetry is then applied to each of the vector norms. Our improved proof can be phrased as applying Nelson's symmetry also to the operator norm. But then we have exploited Nelson's symmetry twice which suggests that the two uses of the symmetry "cancel" and that somehow the symmetry is not needed.

In a narrow sense, this is the case: what we wish to demonstrate is that what is really critical is the Markov property for constant space planes which

---

* A. Sloan Foundation Fellow

† Research partially supported by USAFOSR under Contract F44620-71-C-0108 and USNSF under Grant GP 39048

provides a sort of decoupling of spatial regions. GRS (or at least a subset of them!) did not really understand the Markov property and so used Nelson's symmetry to reduce to the semigroup property in time-like directions. While we will emphasize the ϕ bound, our remark applies equally well to the other material in GRS. Of course in a deeper sense, "Nelson's symmetry" is involved as the critical element in a Euclidean invariant path integral.

What we will prove (in the third section) is the following result which is a large part of Spencer's (1973) $N_{\tau,loc}$-bounds [which generalize (1)]:

Theorem 1 Let F be any function of the (time zero) fields smeared in $[a,a+1]$. Then for any ℓ. with $[a,a+1] \subset [-\ell/2,\ell/2]$:

$$- F \leq \hat{H}_\ell + c_1 - c_2 \, E(H_0 + c_2^{-1} F) \tag{4}$$

for suitable constants c_1, c_2 independent of ℓ and F.

Corollary 2 Let $||f||_{-1} = (\int |\hat{f}(k)|^2 (k^2+m^2)^{-1} dk)^{\frac{1}{2}}$. Then for a suitable constant c and any f with $\mathrm{supp}\, f \subset [a,a+1] \subset [-\ell/2,\ell/2]$

$$\pm \phi(f) \leq \hat{H}_\ell + c(||f||_{-1}^2 + 1) \tag{5}$$

In particular

$$\pm \phi(f) \leq c' ||f||_{L^1} (\hat{H}_\ell + 1) \tag{6}$$

We prove (5) for Theorem 1 in the next section. (6) follows from (5) as in GJ. We also have:

Corollary 3 Let $||f||_{-\frac{1}{2}} = (\int |\hat{f}(k)|^2 (k^2+m^2)^{-\frac{1}{2}} dk)^{\frac{1}{2}}$. Then for a suitable constant c

$$\phi(f)^2 \leq c ||f||_{-\frac{1}{2}}^2 (\hat{H}_\ell + 1) \tag{5'}$$

In particular

$$\phi(f)^2 \leq c' ||f||_{L^1}^2 (\hat{H}_\ell + 1) \tag{6'}$$

§2. Nelson's Bound

Our notation for the free Euclidean field follows Simon (1974); see also Guerra et al. (1973). If $\Lambda \subset \mathbb{R}^2$, we say F, a function of the Euclidean fields, is Λ-measurable if F is measurable with respect to the σ-field generated by $\{\phi(f) | f \in N;\ \mathrm{supp}\, f \in \Lambda\}$. If $\Lambda \subset \mathbb{R}$, we use $\Lambda \times \mathbb{R}$ (resp. $\mathbb{R} \times \Lambda$) to denote $\{(x,s) | x \in \Lambda\}$ (resp. $\{(x,s) | s \in \Lambda\}$). Later when we deal with the time zero fields and $\Lambda \subset \mathbb{R}$, Λ-measurable will denote measurable with respect to the σ-field generated by $\{\phi_F(f) | f \in F,\ \mathrm{supp}\, f \subset \Lambda\}$. Finally J_a (resp. $\tilde{J}_a$) will denote the isometry of $\mathcal{F}$ into $\mathcal{N}$ induced by the map $j_a : F \to N$ (resp. $\tilde{j}_a$) given by $j_a f(x,s) = f(x)\delta(s-a)$ (resp. $(\tilde{j}_a f)(x,s) = \delta(x-a)f(s)$).

A basic role is played by:

Theorem 4 (Nelson's Bound) Let m be the mass of the free Euclidean field. Let $p = 2/1 - \exp[-m(b-a)]$. Then as a map from $\mathcal{F}$ to $\mathcal{F}$

$$||J_a^* V J_b|| \leq ||V||_p \qquad (\text{resp. } ||\tilde{J}_a^* V \tilde{J}_b|| \leq ||V||_p) \tag{7}$$

if V is $\mathbb{R} \times [a,b]$-measurable (resp. $[a,b] \times \mathbb{R}$-measurable).

Proof This is just an expression of hypercontractivity and Hölder's inequality. The basic idea is Nelson's (1973a) although we have used a result from the later Nelson (1973b). For details see Guerra et al. (1973) or Simon (1974). ∎

Proof of Corollary 2 By Nelson's bound and the FKN formula:

$$\begin{aligned} e^{-E(H_0+\phi(f))} &\leq ||\exp(-\textstyle\int_0^1 ds\ \phi(f,s))||_p \\ &= [\textstyle\int d\mu_0 \exp(-p\phi(f \otimes \chi_{(0,1)}))]^{1/p} \\ &\leq \exp(c||f \otimes \chi_{(0,1)}||_N^2) \leq \exp(c'||f||_{-1}^2) \end{aligned}$$

From this and (4) we immediately conclude (5). Thus for all f with $||f||_{-1} = 1$ and $\operatorname{supp} f \subset [a,a+1]$,

$$\pm\ \phi(f) \leq \hat{H}_\ell + d \leq d(\hat{H}_\ell + 1)$$

for some fixed $d \geq 1$. By homogenity,

$$\pm\ \phi(f) \leq d||f||_{-1}(\hat{H}_\ell + 1)$$

(6) follows from this. ∎

Remark By simply modifying the above one shows that for any $\varepsilon > 0$, there is a $d(\varepsilon)$ with

$$\pm\ \phi(f) \leq ||f||_{L^1}(\varepsilon\, \hat{H}_\ell + d(\varepsilon))$$

Proof of Corollary 3 This is similar to that of Corollary 2. We use that fact that

$$\begin{aligned} e^{-E(H_0-\phi(f)^2)} &\leq ||\exp(+\textstyle\int_0^1 ds(J_\sigma\phi(f))^2 ds)||_p \\ &\leq ||\exp(+\ \phi(f)^2)||_p \\ &\leq \quad \text{const} \end{aligned}$$

so long as $||f||_F \leq d$ for d sufficiently small. ∎

§3. The Proof of Theorem 1

Theorem 1 depends on the following result of some independent interest:

Theorem 5 Let V_1, V_2, V_3 be functions of the time zero fields which are respectively $(-\infty, a], [a, a+1]$ and $[a+1, \infty)$ measurable. Let $\tilde{V}_1$ (resp. $\tilde{V}_3$) be the reflection of V_1 (resp. V_3) in the point $x = a$ (resp. $x = a+1$). Let $\lambda_0 = 2/1 - \exp(-m)$. Then

$$-E(H_0+V_1+V_2+V_3) \leqslant -1/2\ E(H_0+V_1+\tilde{V}_1) - 1/2\ E(H_0+V_3+\tilde{V}_3) - 1/\lambda_0\ E(H_0+\lambda_0 V_2) \tag{8}$$

Proof We need only show that

$$<\Omega_0, e^{-t(H_0+V_1+V_2+V_3)}\Omega_0> \leqslant <\Omega_0 e^{-t(H_0+V_1+\tilde{V}_1)}\Omega_0>^{\frac{1}{2}} \cdots \tag{9}$$

By the FKN formula and the Markov property:

$$<\Omega_0, e^{-t(H_0+V_1+V_2+V_3)}\Omega_0> = \int F_1F_2F_3\, d\mu_0 = \int (\tilde{J}_a^* F_1)(J_a^* F_2 J_{a+1})(J_{a+1}^* F_3) d\mu_0$$

with $F_i = \exp(-\int_0^t (J_s V_i) ds)$. On account of Nelson's bound, this last quantity is bounded by:

$$(\int (\tilde{J}_a^* \bar{F}_1)^2 d\mu_0)^{\frac{1}{2}} (\int F_2^{\lambda_0})^{1/\lambda_0} (\int (\tilde{J}_{a+1}^* F_3)^2 d\mu_0)^{\frac{1}{2}}$$

and by the Markov property again

$$\int (\tilde{J}_a^* F_1)^2 d\mu_0 = \int \tilde{F}_1 F_1\ d\mu_0$$

so (8) follows. ▨

Remark Among other things, (8) implies the linear lower bound of Glimm-Jaffe (1970) that $-E_\ell \leqslant d\ell$ for ℓ large. For (8) implies that (with $V_2 = \int_{-\frac{1}{2}}^{\frac{1}{2}} P\, dx$ and $V_1+V_2+V_3 = \int_{-\ell-\frac{1}{2}}^{\ell+\frac{1}{2}} P\, dx$) $- E(2\ell+1) \leqslant - 1/2\ E(2\ell) - 1/2\ E(2\ell) + c = - E(2\ell) + \text{const.}$

Proof of Theorem 1 Let $V_1 = \int_{-\ell}^{a} {}_2{:}P(\phi(x)){:}dx$; $V_3 = \int_{a+1}^{\frac{1}{2}} {:}P(\phi(x)){:}dx$ and let $V_2 = F + \int_a^{a+1} {:}P(\phi(x)){:}dx$. Then, by (8):

$$-E(H_\ell+F) \leqslant - 1/2\ E(H_{\ell+2a}) - 1/2\ E(H_{\ell-2a-1}) - 1/\mu_0\ E(H_0+\mu_0 V_2) \tag{10}$$

Now, by bounds of GRS:

$$\alpha_\infty \ell + \beta_\infty \leqslant -E(H_\ell) \leqslant \alpha_\infty \ell.$$

so that:

$$- \frac{1}{2} E(H_{\ell+2a}) - 1/2\ E(H_{\ell-2a-1}) \leqslant - E(H_\ell) - \beta_\infty - \alpha_\infty$$

Moreover, by the convexity of the $-E(A)$ in A :

$$-E(H_0+\mu_0 F+\mu_0 \int_a^{a+1} :P:) \leq -1/2\, E(H_0+2\mu_0 F) - 1/2\, E(H_0+2\mu_0 \int_0^1 :P)$$

so that (10) becomes:

$$-E(H_\ell+F) \leq -E(H_\ell) - 1/2\mu_0\, E(H_0+2\mu_0 F) + C$$

which implies (4). ▨

§4. Fröhlich's Bounds

Fröhlich (1973) remarked a very convenient form of the ϕ-bound which provides the neatest form of the bounds needed to complete the proof of Nelson's convergence theorem for the half-Dirichlet Schwinger functions (see Nelson's and Rosen's lectures). Since Rosen uses these bounds in his lecture, we sketch their proof:

Theorem 6 (Fröhlich(1973)) Let $d\nu_\Lambda$ denote the spatially cutoff Markov measure for a fixed $P(\phi)_2$ model. Then for any bounded region D , there exists ℓ-independent constants c_1 and c_2 so that for any f in $L^2(D)$, real-valued:

$$\lim_{t\to\infty} \int e^{\phi(f)} d\nu_{[-\ell/2,\ell/2]\times[-t/2,t/2]} \leq c_1 \exp(c_2||f||_2^2) \tag{11}$$

so long as $D \subset [-\ell/2,\ell/2] \times \mathbb{R}$.

Proof If Ω_ℓ is the vacuum for $\hat{H}_\ell$, the right side of (11) is

$$\langle\Omega_\ell, \exp(-\int_{-\infty}^{\infty} \hat{H}_\ell - \phi(f_t))\Omega_\ell\rangle \tag{12}$$

Without loss of generality, take $D = [-1/2,1/2] \times [-1/2,1/2]$. Then (12) is bounded by

$$\exp\left(-\int_{-1/2}^{1/2} E(\hat{H}_\ell-\phi(f_t))\right) \leq \exp(c \int_{-1/2}^{1/2}(||f_t||_{-1}^2+1))$$
$$\leq \exp(c \int_{-1/2}^{1/2}(||f_t||_{L^2}^2+1))$$

which is (11). ▨

Theorem 7 (Fröhlich (1973)) Let $d\nu_\Lambda^{HD}$ deonote the spatially cutoff half-Dirichlet theory for a fixed $P(\phi)_2$ model with $P(X) = Q(X) - \mu X$ with Q even. Then for each bounded region D , there are constants c_1 and c_2 so that for all $\Lambda \supset D$ with Λ bounded and $f \in L^2(D)$, complex valued:

$$|\int e^{\phi(f)} d\nu_\Lambda^{HD}| \leq c_1 \exp(c_2||f||_2^2) \tag{13}$$

Proof Clearly $|\int e^{\phi(f)} d\nu_\Lambda^{HD}| \leq \int e^{\phi(\text{Re}f)} d\nu_\Lambda^{HD}$. By the first GKS inequality. $\int e^{\phi(\text{Re}f)} d\nu_\Lambda^{HD} \leq \int e^{\phi(|\text{Re}f|)} d\nu_\Lambda^{HD} \leq \int e^{\phi(|f|)} d\nu_\Lambda^{HD}$. Let $\Lambda \subset [-\ell/2,\ell/2] \times [-t/2,t/2]$. Then by Nelson's Monotonicity theorem and the bound $S_\Lambda^{HD} \leq S_\Lambda^{Free}$:

$$\int e^{\phi(|f|)} d\nu_\Lambda^{HD} \leq \lim_{t\to\infty} \int e^{\phi(|f|)} d\nu^{HD}_{[-\ell/2,\ell/2]\times[-t/2,t/2]}$$

$\leq \lim_{t\to\infty} \int e^{\phi(|f|)} d\nu_{[-\ell/2,\ell/2]\times[-t/2,t/2]}$. (11) now implies (13) . ▨

§5. The Lower Bound on the Wave Function Renormalization

We have reported on one simplification of the paper GRS. There is a new bound which helps illuminate another of their results, namely that $\ln(\Omega_\ell,\Omega_0) \geq -c\ell$.

Theorem 8 For any vector Ω in $L^2(Q_F,d\mu_0)$ with $||\Omega||_2 = 1$

$$||\Omega||_1 \geq \exp[-(\Omega,N\Omega)] \qquad (14)$$

Remarks

1. This bound in a disguised form was first shown me by C. Newman who proved it by the method GRS use to prove $\ln(\Omega_\ell,\Omega_0) \geq -c\ell$.

2. Since
$$\begin{aligned}(\Omega,N\Omega) &\leq \frac{1}{m_0}(\Omega,H_0\Omega)\\ &= \frac{1}{m_0}(\Omega,2(H_0+V_\ell)\Omega) - \frac{1}{m_0}(\Omega,(H_0+2V_\ell)\Omega)\\ &\leq \frac{1}{m_0}[2\alpha_\ell(1) - \alpha_\ell(2)]\ell\end{aligned}$$

(14) implies the $\ln(\Omega_\ell,\Omega_0) \geq -c\ell$ bound.

Proof By the infinitesimal form of hypercontractivity of Gross (1973):

$$-(\Omega,N\Omega) \leq -\int |\Omega|^2 \ln|\Omega|\, d\mu_0. \qquad (15)$$

Since $|\Omega|^2 d\mu_0$ is a probability measure,

$$\begin{aligned}\exp(-\int |\Omega|^2 \ln|\Omega|\, d\mu_0) &\leq \int \exp(-\ln|\Omega|)|\Omega|^2 d\mu_0\\ &= \int |\Omega|\, d\mu_0\end{aligned} \qquad (16)$$

(14) follows from (15) and (16) . ▨

References

FRÖHLICH, J. (1973): Schwinger Functions and Their Generating Functionals, Univ. of Geneva Preprint, in preparation.

GLIMM, J., JAFFE, A. (1970): Acta. Math. 125 203-261.

GLIMM, J., JAFFE, A. (1972): J. Math. Phys. 13 1568-1584.

GROSS, L. (1973): Logarithmic Sobolev Inequalities, Cornell Preprint.

GUERRA, F., ROSEN, L., SIMON, B. (1972): Commun. Math. Phys. 27 10-22.

GUERRA, F., ROSEN, L., SIMON, B. (1973): The $P(\phi)_2$ Euclidean Quantum Field Theory as Classical Statistical Mechanics, Ann. Math., to appear.

NELSON, E. (1973a): in Proc. Summer Institute of P.D.E., Berkeley, 1971.

NELSON, E. (1973b): J. Func. Anal. 12, 211-227.

SIMON, B. (1974): The $P(\phi)_2$ Euclidean (Quantum) Field Theory, to appear Princeton Series in Physics.

SPENCER, T. (1973: J. Math. Phys., to appear.

THE PARTICLE STRUCTURE OF THE WEAKLY COUPLED $P(\varphi)_2$ MODEL AND OTHER APPLICATIONS OF HIGH TEMPERATURE EXPANSIONS, PART I:

PHYSICS OF QUANTUM FIELD MODELS

James Glimm*
Courant Institute
New York, N.Y.

Arthur Jaffe†
Harvard University
Cambridge, Mass.

Thomas Spencer‡
Courant Institute
New York, N.Y.

1. FIVE YEARS OF MODELS

1.1. An Overview

Constructive quantum field theory has moved rapidly in the past few years. A continuation of this progress clearly requires the introduction of new formal ideas, methods and/or points of view. In these lectures, we will review past progress, present some new results and point out possible directions for future work. In this Part I, we emphasize the basic ideas and concepts, while in Part II we present in full detail some core results from our program. In particular, we present the vacuum cluster expansion and estimates to establish its convergence.

For the two dimensional $P(\varphi)$ model, a fairly detailed structure is known, and this structure is in qualitative agreement with basic ideas of physics. From the Haag-Ruelle axioms (verified, for instance, for small coupling), we know that fields describe discrete mass particles. An isometric scattering operator expresses the interactions between these particles. We know that the vacuum state is unique for certain coupling constants (e.g., small coupling, or a large - or in some cases, a nonzero - external field). Dobrushin and Minlos have announced that there are multiple phase solutions for even $P(\varphi) + m_0^2 \varphi^2$ models with $\lambda >> m_0^2$. Symmetry breaking plays a key role in current theories of weak interactions, hence the interest in this phenomenon. There is no direct experimental evidence for or against occurence of broken symmetries in elementary particle physics, since the interparticle coupling constants cannot be varied experimentally (in distinction to the case of statistical mechanics where we can, for example, turn off a magnetic field). Consequently the definitive argument in favor of broken symmetries may come from constructive quantum field theory.

The Yukawa$_2$ (Y_2) and φ_3^4 models are less highly developed. Yet many of the formal ideas developed for $P(\varphi)_2$ models appear to apply to superrenormalizable models in general. Clearly then, one set of problems is to develop stronger techniques, to make these ideas applicable to Y_2, φ_3^4 and Y_3. We propose, in fact, four groups of problems.

I. Physical Properties. One important direction for future work is to develop further the physics of existing quantum field models. The particle structure program, bound states, resonances and scattering present interesting problems. Likewise, the long distance and infrared behavior of our models contains much physics. The general particle structure program is: Which interaction polynomials and which coupling constants give rise to which particles, bound states and resonances? How do the masses and half lives depend on the coupling constants? How do cross sections behave asymptotically? We discuss these problems further in Section 1.5 and Chapter 3.

The long distance behavior of our models pertains to the existence of multiple phases, to the existence of a critical point and to the scaling behavior of the models at a critical point. We ask: Does the φ_2^4 model have a critical point? Does it admit scaling properties with anomalous dimensions? What parameters describe the critical point? We discuss these questions further below and in Section 1.5.

II. Four Dimensions (Renormalizable Models). A second important direction is the question of four space-time dimensions, or in other words how to deal with renormalizable interactions, since there are no super-renormalizable interactions in four dimensions. Clearly this is our most challenging goal, to prove the existence of, for example, φ_4^4. Our present methods have been tied to superrenormalizability ($4 - \epsilon$ dimensions) and for $\epsilon = 0$ new ideas are required. We ask: Can an understanding of the renormalization group be an aid to removing the $\epsilon = 0$ ultraviolet cutoff? Do the ideas in the lectures of Symanzik yield insight into charge renormalization? We discuss these questions further in Section 1.5.

III. Simplification. Aside from these two major directions, there is the question of simplifying the present methods. Clearly the major need for simplification concerns problems with ultraviolet divergences, and a major goal of such a program would be to improve the techniques and isolate their essential elements in order to make tractable more complicated superrenormalizable models, such as Y_3, or even Y_2.

IV. Esthetic Questions. Furthermore, there are esthetic or foundational questions. For example, the Schrödinger representation $\mathcal{H} = \mathcal{L}_2(dq)$ exists for $\mathcal{P}(\varphi)_2$ models ; what is the fermion representation corresponding to this non-Gaussian boson measure on $\mathcal{S}'$? What are the properties of the path space measures in models with interaction? Related are interesting, but purely mathematical questions motivated by field theory, which we do not pursue here.

In this connection, we remark that the drive toward simplicity and elegance is important and also has been quite successful in the $\mathcal{P}(\varphi)_2$ model. However, we emphasize here those methods that admit (or appear to admit) generalization to other more singular interactions. The reason for this emphasis is two-fold. First, we believe that, in the long run, our ability to handle more singular problems will determine the extent to which the model program has succeeded. Second, we believe that a premature emphasis on the simplicity and elegance of the details can divert energy away from central issues, and thereby delay or obstruct progress.

1.2. Survey of Results

To begin, we review the status of the φ_2^4, Y_2 and φ_3^4 models. We give a chronological summary in Figure 1, plotting models of increasing complexity versus results of increasing complexity. In this chart, we enter the years in which these results were proved.

In Figure 2 we give details and references for various φ_2^4 models. The results quoted for $\lambda/m_0^2 << 1$, also hold for $\lambda P(\varphi)_2$ models with $\lambda/m_0^2 << 1$. We make several comments: The Wightman axioms require a unique ground state (vacuum), namely the existence of a single vector, invariant under inhomogeneous Lorentz transformations. Alternatively, we consider the C^* algebra vacuum state of a finite volume theory, and its infinite volume limits. Each infinite volume state yields a representation, and a Hilbert space vacuum vector. Uniqueness of the vacuum, as required for the Wightman axioms, refers to vectors in this Hilbert space, and is equivalent to irreducibility of the representation. The infinite volume vacuum state is determined as a limit of finite volume states. The latter are determined by parameters in the energy density $\mathcal{H}(x)$ and the boundary conditions. If the parameters in $\mathcal{H}(x)$ alone are sufficient to specify a unique vacuum, independent of the boundary conditions, then there is said to be a unique phase, and otherwise there are multiple phases. Convergent cluster expansions [Gl Ja Sp 1, 2] yield for certain couplings both a unique vacuum and a single phase.

In a $P(\varphi)_2$ theory satisfying the Wightman axioms, except for the uniqueness of the vacuum, the decomposition theorem of Bratteli [Br 1] allows us to decompose the observables and recover a unique vacuum. The local perturbation estimate [Gl Ja IV] and a result of Streater [St] ensure the spectral condition for the decomposed theory. In this manner we arrive at a Wightman theory for an arbitrary $\lambda\varphi^4 + \frac{1}{2}m_0^2\varphi^2 - \mu\varphi$ interaction. In the case $\mu \neq 0$, the Lee-Yang theorem shows that the decomposition is unnecessary [Gr Si, Si II].

In Figure 3, we have details and references for Y_2, and clearly much work needs to be done to bring it to the level of φ_2^4.

	$P(\varphi)_2$	Y_2	φ_3^4	φ_4^3	Y_3	
Critical Point						3 = 1973
Asymptotic Completeness						0 = 1970
Resonances						9 = 1969
Bound States	3					4 ≤ 1964
Broken Symmetries	3					
Analyticity in Coupling	3					
Perturbation Theory Asymptotic	3					
Single Particle States	3					
Mass Gap	2					
Wightman Axioms, $V \to$ Convergence	2					
Euclidean Formulation	1	2				
Wightman Functions	1					
Physical Representation	9	1				
Haag-Kastler Axioms	9					
Equations of Motion	9	1				
Space Time Covariance	8	0				
$H_V = H_V^*$	8	9				
$0 \le H_V$	5	8	2			
Local Observables	4	7	1			
H_V	4	7	8	0		

Figure 1. Main Results and Year Established

Renormalized Charge, Anomalous Dimensions, Renormalization Group				
Critical Points				
S-Unitary (Asymptotic Completeness)				
Bound States Resonances		Chapter 3		
Multiple Phases (Broken Symmetries)	No: Gl Ja Sp 1	Yes: $\lambda \gg m_0^2$ (Dob Min)	No: Gr Si, Si II	→
Analyticity of the Schwinger Fns.	Re $\lambda > 0$, $0 < \lvert\lambda\rvert < \lambda_0$, $\lvert\mu\rvert < \mu_0$ Gl Ja Sp 2		Re $\lambda > 0$ Sp 2	→
One Particle States and S-Matrix	Gl Ja Sp 1			
Measure dq Schrödinger Repn.	New 1, Fr 2 Fr 2	Fr 2	→	→ Fr 2
Perturbation Theory is Asymptotic	Di 3			Di 3
m	$m > 0$ Gl Ja Sp 1	monotonic in m_0 Gu Ro Si 3, Si I	monotonic in $\lvert\mu\rvert$ Gr Si	$m > 0$ Sp 2
Verify Wightman Axioms	Gl Ja Sp 1	Nel 5, Gu Ro Si 3, Gl Ja IV, Os Sch 3, Br 1, St	→ Gr Si, Si II	→ → Sp 2
Formulate Euclidean Axioms	Sy 2, Nel 3, Os Sch 3	→	→	→
Physical Representation	Gl Ja III-IV, 5	→	→	→
Haag-Kastler Axions	Gl Ja I-II, Ca Ja Ro 1-3	→	→	→
Preliminary	Ja 1, 2, Nel 1, Gl 1 Se 1	→	→	→
	$\frac{\lambda}{m_0^2} \ll 1$, $\frac{\lvert\mu\rvert}{m_0^2} \ll 1$. Also $\lambda P(\varphi)_2$.	$\mu = 0$	$\mu \neq 0$	$\lvert\mu\rvert$ large

Figure 2. Details for $\left(\lambda\varphi^4 + \frac{1}{2} m_0^2 \varphi^2 - \mu\varphi\right)_2$

Bound States	
Particle Structure	
Schrödinger representation	
Wightman axioms	
Euclidean Fields and Feynman-Kac formula	Os Sch 4
Properties of Currents	McB 1
Equations of Motion	Di 1
Physical Representation Exists	Sch 2
Space - Time Covariance	Gl Ja 10
$0 \leq H_V = H_V^*$	Gl 2,3,Gl Ja 9, 11

Figure 3. Details for Y_2

We conjecture that the Euclidean fields given by Osterwalder and Schrader [Os Sch 4], and by Ozkaynak [Oz] for higher spin, can be used as a starting point for a cluster expansion, yielding for small coupling: the $V \to \infty$ limit, the Wightman axioms, a unique vacuum, and single particle states. We also hope that Euclidean gauge fields will lead to an understanding of function space integrals for higher spin. In particular, we mention as problems the factorization of the integration over the symmetry (gauge) group, and the renormalization scheme of Faddeev and Popov [Fa Po, Sa St].

The φ_3^4 interaction is at an even more primitive stage. The φ_3^4 renormalized Hamiltonian is bounded from below [Gl Ja 8]. Similar estimates [Fe 2] show that in a finite space time volume, approximate Schwinger functions have $\varkappa \to \infty$ limits, see Chapter 4. We hope that the Wightman axioms will soon be verified for φ_3^4 .

There are several recent developments which do not fit into the above description and we mention in particular: Federbush's study of the generalized Yukawa$_2$ model, $P(\varphi) + \bar{\psi}\psi Q(\varphi)$, see [Fed] the work of Albeverio and Hoegh-Krohn on bounded nonpolynomial interactions [AHK 1 - 4], and recent work on the polaron model by Gross[Gro 1] and by Fröhlich [Fr 1]. Furthermore Fröhlich's recent results on $P(\varphi)_2$ models [Fr 2] yield the Markov property, DLR equations and cyclicity of the vacuum (for time zero fields) when $\lambda << m_0^2$ or $|\mu| >> m_0^2, \lambda$.

There are three main unifying techniques, yielding the results charted above:

1. Euclidean Formulation. The use of imaginary time in constructive quantum field theory dates to Symanzik's Euclidean field formulation and to Nelson's positivity proof for the $\varphi_{2,V}^4$ Hamiltonian. This approach was used in most of the key technical estimates of the $P(\varphi)_2$ and φ_3^4 boson models, and in proving finite propagation speed for the Y_2 model. The covariant formulation in terms of Euclidean fields or Green's functions is important as a conceptual advance yielding Euclidean axioms, the relativistic Feynman-Kac formula and the Markov property [Nel 4, Os Sch 3, Sy 1, Nel 3, Fel 1]. The covariant formulation is important as a technical advance in proving estimates [Gu 1,Gl Ja 8], in simplifying estimates [Gu Ro Si 1, Si], and in reducing the number of properties which must be verified in order to construct a Wightman-Haag-Ruelle theory [Nel 3, Os Sch 2]. Furthermore, the Euclidean framework makes clear the connection of field theory with statistical mechanics, a central tool in the heuristic approach of Fisher and Wilson and in the work of Guerra, Rosen and Simon. The lattice approximation [Ko Wi, Gu Ro Si 3] then becomes natural, and it relates boson interactions to an Ising model with continuous spin and nearest neighbor interaction.

2. Correlation Inequalities [Gu Ro Si 3, Nel 5, Gr Si, SiII]. The convergence of the lattice approximation and the existence of the thermodynamic limit leads to the generalization of correlation inequalities and the Lee-Yang theorem from statistical mechanics to quantum field theory. These methods appear especially suited to boson interactions and are discussed in the lectures of Guerra, Rosen, Simon, Nelson and in Chapter 3 of these lectures.

3. Cluster Expansions [Gl Ja Sp 1, 2]. These high temperature expansions provide (within their region of convergence - see Figure 6) the most detailed information about $P(\varphi)_2$ models. In combination with low temperature expansions (which we believe to exist), the expansions should converge for all interactions

sufficiently far from a critical point. In fact the inductive expansions for $P(\varphi)_2$ yield an analysis of an arbitrary interval $[0, a]$ of the spectrum of the Hamiltonian H. Related are the inductive bounds for $P(\varphi)_2$ [Gl Ja IV] and for φ_3^4 [Gl Ja 8]. The bounds for φ_3^4 have not yet been refined to yield such detailed results as are known for $P(\varphi)_2$. In contrast to the expansions, the bounds hold without restriction on the magnitude of the coupling constants.

1.3. The Goldstone Structure of $P(\varphi)_2$

We now return to the $P(\varphi)_2$ model, and give a more detailed discussion. There is a very simple picture, partly due to Goldstone, and now part of the physics folklore, which accounts for the fact that some interaction polynomials P produce degenerate vacuums, while others do not. See [Go, Go Sa We, Jo L], and for recent discussions see [Co 1, Wi]. Classically, the (Euclidean) ground state of the $P(\varphi)$ model is the function $\Phi \in \mathcal{S}'(R^2)$ which minimizes the classical (Euclidean) energy

$$\int [\nabla \Phi(x)^2 + P(\Phi)]dx \quad .$$

This function Φ is a constant, equal to the value ξ which minimizes $P(\xi)$. Quantum mechanically, the (Euclidean) ground state is a probability distribution on $\mathcal{S}'(R^2)$, centered (in some sense) about the classical minimum. The quantum mechanical picture requires corrections, including Wick ordering and the higher order mass counterterms, see Coleman and Weinberg [Co We]. In $P(\Phi)_2$ models, Hepp obtained the classical limit $\hbar \to 0$ explicitly, see [He 3]. The Goldstone picture also states that (i) the physical mass m is the curvature of P at its minimum and (ii) in the case of a double (or multiple) minimum, multiple phases exist. In our discussion, we include in P the mass in the free Hamiltonian H_0, but because we ignore quantum corrections beyond (bare) Wick order, our discussion is approximate. Let $\frac{1}{2}m_0^2$ be the coefficient of Φ^2 in P. We consider the polynomials

(a) $$P(\xi) = \sum_{j=1}^{2n} c_j \xi^j \ ; \ c_{2n} > 0 \ , \ \tfrac{1}{2}m_0^2 = c_2 \gg c_j \quad , \ j \neq 2 \quad .$$

(b) $$P(\xi) = \lambda\xi^4 + \tfrac{1}{2}m_0^2\xi^2 - \mu\xi \ , \ \lambda > 0 \quad , \ \mu \neq 0 \quad .$$

(c) $$P(\xi) = P_0(\xi) - \mu\xi \ ; \ P_0 \text{ bounded below, } \mu \gg 1 \quad .$$

(d) $$P(\xi) = \lambda P_0(\xi) + \tfrac{1}{2}m_0^2\xi^2 \ ; \ \lambda > 0 \ , \ m_0^2 < 0 \ , \ P_0 \text{ even.}$$

(e) $$P(\xi) = \lambda\xi^4 \ ; \quad \lambda > 0 \quad .$$

See Figure 4.

In case (b) a cubic term may be added through the trivial transformation $\xi \to \xi + \text{const.}$ In (c), ξ may be replaced by any odd power ξ^{2j+1}, $2j+1 < \deg \mathcal{P}_0$. In (d), a polynomial $\xi^6 + \cdots$ of higher degree may have more than two minima.

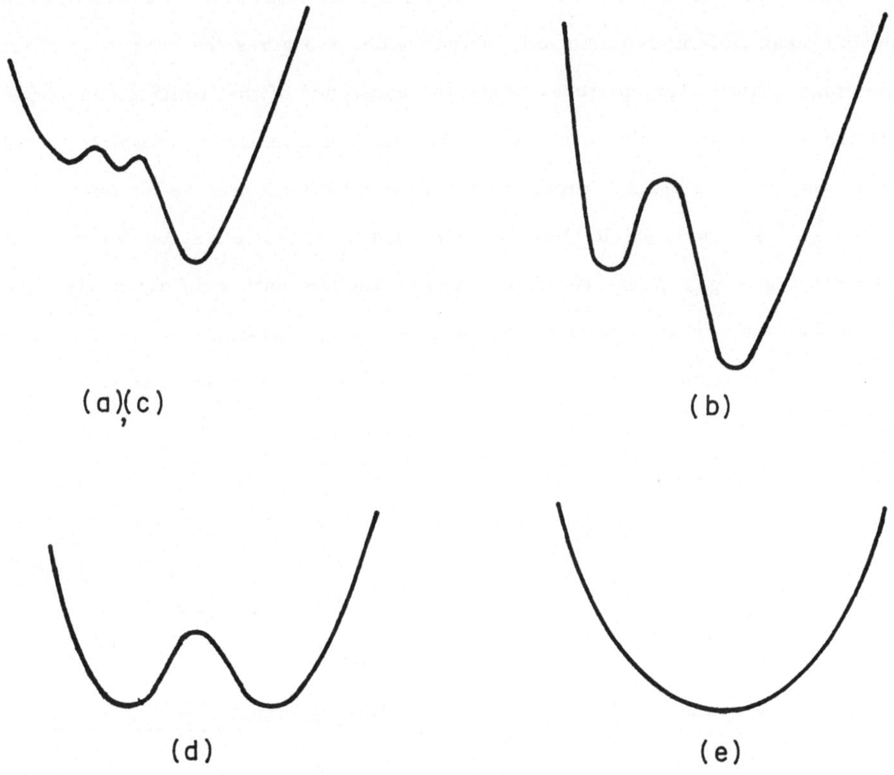

Figure 4. Various interaction polynomials $\mathcal{P}(\xi)$.

In cases (a) - (c) the polynomial $P(\xi)$ has a unique minimum with positive curvature. This suggests the existence of a single phase, and the existence of massive particles. Case (d) has a double minimum, corresponding to the symmetry $\varphi \to -\varphi$, with positive curvature at each minimum. Thus we expect two pure phases, with positive mass particles in each pure phase. The two vacuum states for the two pure phases are localized approximately at minima $\xi = \pm a$. Thus we expect in the vacuum states for these two phases that the expectation value of the field, $\langle\varphi\rangle$, approximately equals $+a$ or $-a$. Case (e) is the limiting case between Case (d) and a particular Case (a). Here $P(\xi)$ has a unique minimum at $\xi = 0$, with curvature zero. The Goldstone picture suggests the presence of zero mass particles, and is the Goldstone picture of a critical point. These figures yield the mean field approximation, which we believe gives the correct qualitative, rather than quantitative, picture. Thus the existence of the limiting case (e) suggests the existence of critical values of the coupling constants for which $m = 0$. In this connection, Figure 5 represents the Goldstone conjecture for even $\lambda P_0(\Phi) + \frac{1}{2} m_0^2 \Phi^2$ models. R. Baumel [Ba] remarked that changing the definition of Wick ordering (e.g., Wick ordering in a bare vacuum with a different bare mass) permits the same H to be written with different bare parameters m_0, P. Thus the location of the critical point, when written in these parameters, can be changed, so the sign of m_0^2 at the critical point is not significant.

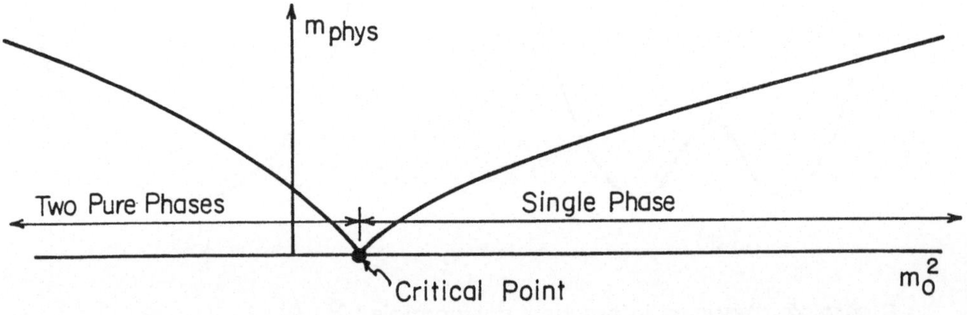

Figure 5. Conjectured symmetry breaking in even $\lambda P_0(\phi) + \frac{1}{2} m_0^2 \phi^2$

Results. We give the rigorous results proved in cases (a) - (e). In case (a), convergent cluster expansions show that the mass operator $M = (H^2 - P^2)^{1/2}$ has isolated eigenvalues at 0 and $m > 0$, [Gl Ja Sp 1]. These results have been generalized to case (c) [Sp 2]. A single phase exists, the vacuum is unique and the mass $M = m$ eigenvalue is nondegenerate in the sense that the Lorentz group acts irreducibly on the one particle space. Any additional mass spectrum in the interval $(m, 2m)$ should be discrete and would describe bound states.

In case (b), Griffiths and Simon [Gr Si] extend the Lee-Yang theorem to φ_2^4 and use this result to show [Si II] that the ground state is unique. For λ large and μ small, the structure of the mass spectrum above zero is not known.

In case (d), Dobrushin and Minlos [Dob Min] have announced the existence of at least two phases.

Continuous Symmetry Breaking. Another form of the classical picture [Go Sa We] concerns a continuous symmetry group, rather than the discrete reflection group $\varphi \rightarrow -\varphi$ that we discussed above. In the continuous case, the minimum of $P(\xi)$ occurs on a manifold of dimension greater than zero, and translation along this manifold leaves $P(\xi)$ constant. The Goldstone picture now states that in the case of broken symmetry ($\delta \neq 0$ as defined below) particles of mass zero occur. Thus these particles have a mass given by the minimum curvature of $P(\xi)$ at its minimum.

In the physics literature, this broken symmetry is defined in terms of the conserved current, $\partial^\mu j_\mu = 0$, associated with the symmetry group. For classical field theories, a standard variational argument establishes the existence of the conserved current. The generator of the symmetry group is $Q = \int j_0 d\vec{x}$, and the Goldstone picture concerns the vacuum expectation value

$$\delta = \langle [iQ, \varphi] \rangle$$

If $\delta \neq 0$, discrete zero mass particles (Goldstone bosons) are believed to exist. In case that the symmetry group of automorphisms can not be unitarily implemented,

Kastler, Robinson and Swieca have shown that the mass spectrum extends down to zero [Ka Ro Sw], and Swieca [Sw] has shown that zero mass particles exist.

Two simple examples of such conserved currents are (1) a zero mass free field, with energy density $\pi^2 + (\nabla\varphi)^2$, invariant under translations $\varphi \to \varphi + \text{const.}$, and (2) a multicomponent field invariant under orthogonal transformations in the space of components.

In case (1), $j_\mu = \partial_\mu \varphi$, the parameter $\delta = 1$ and the Goldstone bosons are just the zero mass free particles. (We note this argument is not applicable in two dimensions where zero mass free scalar fields do not exist.)

Case (2) corresponds to Figure 4(d) with a rotational symmetry about the $\xi = 0$ axis. If our field has two components φ_i, then $\delta_i = \pm \langle \varphi_j \rangle$ is the vacuum expectation value of the field components. The Goldstone bosons occur when the vector, formed of the components δ_i, does not vanish i.e., $\vec{\delta} \neq 0$. In terms of $\vec{\varphi}$, the broken symmetry condition requires that the vacuum expectation $\langle \vec{\varphi} \rangle \neq 0$. Let us suppose that $\langle \vec{\varphi} \rangle \neq 0$, and let $\vec{n}$ be the unit vector in the $\langle \vec{\varphi} \rangle$ direction. We decompose $\vec{\varphi}$ into longitudinal and transverse parts, $\vec{\varphi} = \vec{\varphi}_L + \vec{\varphi}_T$, where

$$\vec{\varphi}_L = (\vec{\varphi} \cdot \vec{n})\vec{n} \quad , \quad \vec{\varphi}_T \cdot \vec{n} = 0 \quad .$$

The conventional picture is that along its trough, $P(\cdot)$ has zero curvature and displays zero mass excitations, while in the orthogonal direction the curvature of $P(\cdot)$ is positive and one expects massive particles. Thus one expects exponential clustering for $\vec{\varphi}_L$,

$$\langle \vec{\varphi}_L(\vec{x}) \vec{\varphi}_L(\vec{y}) \rangle \leq O(e^{-md}) \quad , \quad m > 0 \; ,$$

but polynomial clustering for $\vec{\varphi}_T$.

As we remarked above, two dimensions displays especially singular infra red behavior. In fact, Coleman [Co 2] has proved that for $d = 2$, a continuous symmetry as above always yields $\delta = 0$. Thus the usual Goldstone picture never ensures the existence of zero mass particles. The basic ingredient of Coleman's proof is the distribution character of the two point function. This precludes zero mass particles appearing in a scalar two point function and forces $\delta = 0$. His argument does not, however, prevent the mass spectrum from going down to zero. Therefore we ask: Does the $\lambda(\vec{\varphi}^{\,2})^2$ model with $\lambda >> m_0^2$ have a mass gap? We conjecture that the mass gap for small λ vanishes as λ is increased to the critical point λ_c , and that the mass gap is zero for $\lambda \geq \lambda_c$. In other words, for $\lambda \geq \lambda_c$ we expect that neither Goldstone bosons nor a mass gap occur.

1.4 Field Theory and Statistical Mechanics

The equivalence of relativistic quantum field theory with statistical mechanics has a long history. Older work includes both the Landau-Ginzberg theory and Symanzik's program to construct Euclidean models. Recent work includes that of Fisher, Wilson, Griffiths and a number of lecturers at this conference. We mention here some selected aspects of this correspondence for boson quantum fields. Ideas of this nature in models with fermions have not yet proved fruitful.

The Partition Function. Let dq denote the Euclidean measure for a boson quantum field model. The partition function

$$Z\{J\} = \int e^{\Phi(J)}\, dq$$

is the generating function for Schwinger functions, and has been studied in $\mathcal{P}(\varphi)_2$ models by Fröhlich [Fr 2]. As mentioned above, the Euclidean field model has a natural approximation by a continuous spin ferromagnetic Ising lattice with nearest neighbor interaction, see for instance [Ko Wi]. The convergence of the lattice approximation [Gr Ro Si 3] and the approximation of φ^4 by spin 1/2 Ising models [Gr Si] sharpens this correspondence, see also [New 2]. The one point Schwinger function $\int \phi(x)\, dq$, which parameterizes symmetry breaking in the Goldstone picture above, corresponds to spontaneous magnetization in the Landau-Ginzberg theory. The coupling constant λ/m_0^2, which measures the deviation from a free theory, corresponds to the inverse temperature $\beta = (kT)^{-1}$. In this way the Goldstone picture of the vacuum corresponds to a picture of phase transitions in many body systems. The existence of a gap in the spectrum of H, and exponential clustering, corresponds to a finite correlation length $\xi = m^{-1}$ in the many body system.

One Particle Structure. We define G{J} by

$$G\{J\} = \ln Z\{J\} - \int \Phi(J)\, dq \quad ,$$

and then $G\{J\}$ is the generating function for the connected (truncated) Euclidean Green's functions. The one particle structure is displayed by an entropy principle (Legendre transformation)

$$\Gamma\{A\} = \inf_J \left[-J \cdot A + G\{J\} \right] ,$$

or in differential form,

$$\Gamma\{A\} = -J \cdot A + G\{J\} ,$$

where J is determined by $A(x) = \delta G\{J\} / \delta J(x)$. This transformation was introduced in statistical mechanics by De Dominicis and Martin [De Ma], in quantum field theory by Jona-Lasinio [Jo L], and was developed by Symanzik [Sy 4]. The analysis of $\Gamma\{A\}$ in quantum field models [Gl Ja 13] may complement our study of the spectrum of the Hamiltonian by expansions described below. The functional $\Gamma\{A\}$ generates the (amputated, one particle irreducible) vertex functions. These functions are directly related to the magnitude of interparticle forces, i.e., the physical charge.

Bound States. In Chapter 3 we study the presence and absence of bound states in certain quantum field models. Our results in Section 3.3 about the absence of bound states in pure φ_2^4 models depend on methods both from field theory and from statistical mechanics. We use high temperature expansions from field theory (see below). We also use an idea of Lebowitz from statistical mechanics to obtain two-particle clustering for the four point vertex function.

Conversely, in Section 3.4 we sketch a proof that bound states occur in φ_2^4 models in a strong external field. We remark that in statistical mechanics, bound state excitations appear in the transfer matrix for large values of chemical potential μ.

High Temperature Expansions. The high temperature expansions in statistical mechanics yield the existence of the thermodynamic limit and high temperature analyticity, i.e., the absence of phase transitions. These Kirkwood-Salsburg or

Mayer-Montroll expansions converge for T/T_c sufficiently large, where T_c is the critical temperature. Related to these expansions are the virial expansions which converge for large values of the chemical potential μ, and which also yield analyticity (absence of phase transitions), see [Ru]. In field theory, the cluster expansions play an analagous role. They converge for large inverse coupling m_0^2/λ [Gl Ja Sp 1, 2] (large T) and for large external field [Sp 2] (large μ). As a result, the cluster expansions establish the existence of the infinite volume limit in field theory, and the existence of a single phase with a unique vacuum vector. These high temperature expansions do not, in general, arise from Kirkwood-Salsburg (or other) integral equations, but have a wider range of validity. We have, however, obtained Kirkwood-Salsburg equations for the partition function Z, see Chapter 6 of Part II. These integral equations are a useful tool in our proof of analyticity of the Schwinger functions.

In addition to yielding information about the vacuum, the high temperature expansions give us detailed information about the spectrum of the Hamiltonian H, e.g., the particle structure and the presence or absence of bound states, see Chapters 2, 3 of these lectures. We remark that these more detailed field theory techniques may yield insights into statistical mechanics.

Low Temperature Expansions. The Peierls argument [Pe] is the basic proof of the existence of phase transitions at low temperatures. The proof considers the energy associated with boundaries (contours) separating up spins from down spins. For temperatures T sufficiently below T_c, it is energetically favorable to have spins all up or all down. Griffiths, Dobrushin and others have modified and extended these results, see for example [Dob 1-3, Gi, Gr 1, ML, Min Sin 1-2]. In particular, the contour methods yield exponential clustering in pure phases of low temperature spin systems. Some continuous spin systems have been studied [Bo Gr].

We regard these methods as convergent low temperature expansions. We believe that such low temperature contour expansions exist in quantum field models.

They should converge sufficiently far from the critical point. (Such an expansion may have been used in the proof of the announced result [Dob Min].) We believe that low temperature expansions exist independent of whether multiple phases exist. In a pure phase, we believe that they exhibit exponential clustering and thus are useful to investigate particle structure.

In Figure 6 we show our conjectured region of convergence of the high temperature (cluster) expansion and presumed low temperature (contour) expansion in the φ^4 model. For models such as $\lambda\varphi^4 - \mu\varphi$, $\mu \gg \lambda$, in which symmetry breaking does not occur, the regions of convergence of the high and low temperature expansions may overlap.

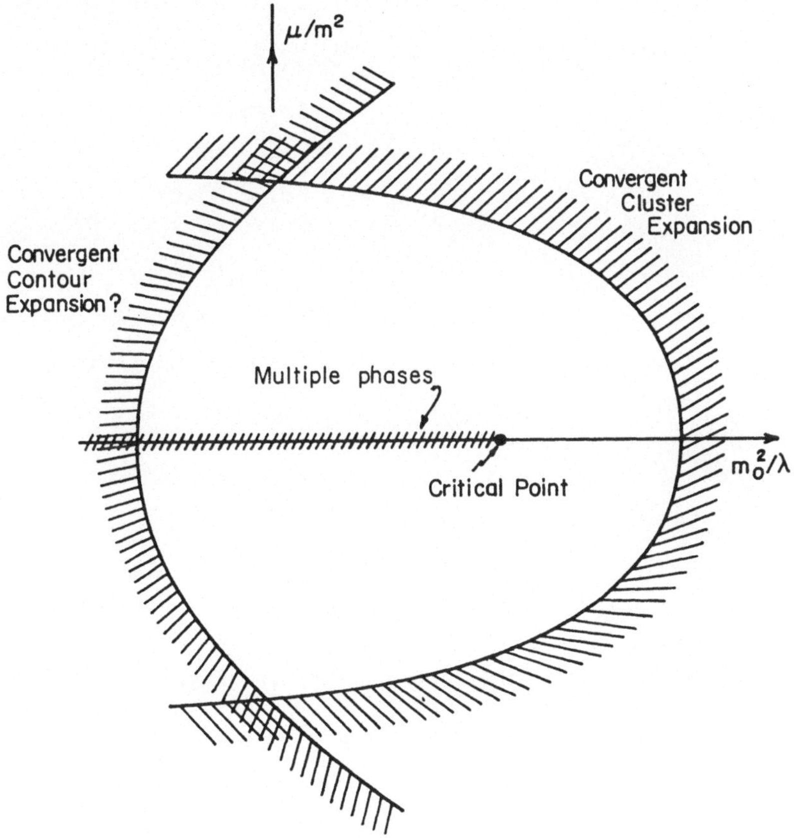

Figure 6. Presumed convergence of cluster and contour expansions.

Correlation Inequalities and the Lee-Yang Theorem. These methods yield the convergence of the Schwinger functions for even $P(\varphi)_2$ models, [Gu Ro Si 3, Nel 5], and a unique phase for $(\varphi^4 + m_0^2 \varphi^2 - \mu\varphi)_2$ models, $\mu \neq 0$, [GrSi, Si II]. These methods and related developments are included in the lectures of Guerra, Nelson, Rosen, Simon, to which we refer the reader for further discussion.

1.5. Some Problems

We discuss several open problems for $P(\varphi)_2$. In addition, problems closely related to other sections are mentioned throughout the lectures.

Asymptotic completeness. In a pure φ^4 model with small coupling, is the S-matrix unitary? Can this be proved (see Chapter 3)? In models with bound states, does the inclusion of the bound state in and out fields for mass spectrum below $2m$ yield asymptotic completeness? Related to these questions is the possibility of performing a cluster expansion in asymptotic fields, as suggested formally by the LSZ expansion of the scattering matrix or the Yang-Feldman equations.

Asymptotic Perturbation Theory. It is known that the Euclidean Green's functions are asymptotic to all orders in the coupling, in the region of convergent cluster expansions, [Di 3]. We conjecture that the S-matrix is asymptotic to its Feynman perturbation series, $S = I + \lambda S_1 + \cdots$. Since $S_1 \neq 0$, this would yield $S \neq I$. We conjecture that the physical mass m is asymptotic in the coupling constant expansion,

$$m = m_0 + \lambda^2 m_2 + \lambda^3 m_3 + \cdots + \lambda^n m_n + O(\lambda^{n+1}) \quad .$$

Cluster Expansions. The high temperature expansions in Part II are based on expanding the Gibbs factor e^{-V_0} in the Gaussian measure $d\Phi$ into $e^{-V_0} = 1 + (e^{-V_0} - 1)$. They yield Kirkwood-Salsburg equations for Z, and related expansions for Z times the Schwinger functions, $ZS(x_1, \cdots, x_n)$. Do these expansions generalize in a natural way to yield particle structure? What is the optimal convergence domain for these expansions?

Contour Expansions. We believe that a low temperature contour expansion exists and converges, independent of whether a $P(\varphi)_2$ model has an internal symmetry. What is this expansion? Does it yield the existence of the infinite volume limit, of particles and of other properties of the mass spectrum.

Analyticity. In Part II, we show that the $\lambda P(\varphi)_2$ Schwinger functions are analytic in a half circle $0 < |\lambda| < \lambda_0$, Re $\lambda > 0$. What is the complex domain into which the Schwinger functions can be continued? In statistical mechanics, the Lee-Yang theorem is used to extend the analyticity domain of high temperature (small λ/m_0^2 , large m_0^2/λ) expansions and of virial (large μ) expansions. Are the Schwinger functions for $\lambda\varphi^4 - \mu\varphi$ real analytic in λ , μ except for a cut from λ_{crit} to ∞ ? In other words, are the Schwinger functions real analytic in all of Figure 6, except for a cut along the line of multiple phases? What is the domain of complex analyticity? For $\lambda > 0$ and Re $\mu \neq 0$, the pressure is analytic [Sp 2].

Haag-Doplicher-Roberts axioms. Is duality, the missing HDR axiom, valid for $P(\varphi)_2$ models? The HDR analysis of superselection sectors applies only in three and four dimensions, but duality is still presumably true for $P(\varphi)_2$.

Critical Points. If a critical point exists (See Figure 5) how do m , $\langle\varphi\rangle$, etc., behave in a neighborhood of it? Do the mass, spontaneous magnetization, etc., vary with power laws (given by critical exponents)? For $\lambda < \lambda_{crit}$, m is monotone in m_0^2 [Gu Ro Si 3]. Is the mass monotone above the critical point? Since Coleman has shown that $\delta = 0$ for $\lambda(\vec{\varphi}^{\,2})^2_2$ models, do multiple phases exist for this model? Is there more than one phase at the critical point for $\lambda\varphi^4_2$? Do zero mass particles occur in φ^4_2 at the critical point? (We remark that zero mass particles do not occur in the two point function, since it is a tempered distribution [Gl Ja IV].) What is the locus of multiple phases for a φ^6 or φ^8 model, etc? Do critical manifolds exist for these models?

Structure Analysis. With our control over the particle spectrum, we have the ingredients to carry out the particle structure analysis of Green's functions, as proposed by Symanzik [Sy 1]. It is also of interest to perform a structure analysis of models in statistical mechanics. As a first step, one can prove the existence and analyticity of the generating functional for one particle irreducible (IPI) Green's functions, as given in [Gl Ja 13]. These vertex parts are important in the study of symmetry breaking and of the renormalization group. In the former direction, Jona-Lasinio has an effective potential which one believes gives corrections to the mean field Goldstone picture of Section 1.3. Such potentials have by studied heuristically in [Co We]. In what sense is the mean field or the effective potential model a limit of quantum field theory?

Anomalous Dimensions. An extremely interesting circle of problems concerns the more refined aspects of $P(\varphi)_2$ models at the critical point. These ideas also make close contact with ideas of high energy theorists. The short distance behavior of $P(\varphi)_2$ and $\varphi_3^4(g)$ models is canonical, and a rigorous proof should follow from the local perturbation estimates [Gl Ja IV, Fel 2]. Since these estimates hold for all λ, they hold in particular at a critical point for $P(\varphi)_2$, giving a logarithmic singularity. On the other hand, the long distance behavior at the critical point for $P(\varphi)_2$ models is not canonical, since $\langle\varphi(x)\varphi(y)\rangle \to$ const. as $|x - y| \to \infty$. Consequently, we do not expect that any $P(\varphi)_2$ model we have constructed is scale invariant. In fact, a scale invariant vacuum would ensure that scale transformations are unitarily implemented. This would ensure in turn that the long and short distance scaling properties were the same.

Let us assume that a critical point exists. Then we conclude that the theory at the critical point must contain a fundamental length. This length characterizes the distance at which the small distance asymptotic behavior is replaced by the long distance asymptotic behavior. Scale transformations change this length, so if a critical point exists, there are continuously many zero mass theories related to one another by scaling. One can attempt to force scale invariance by performing an infinite scale transformation. Do such limits exist? Some of the problems raised here are unresolved for the three dimensional Ising model, and a serious effort might start with this case.

The Renormalization Group. Above we parameterized zero mass $P(\varphi)_2$ theories by a fundamental length. An alternative description is based on the renormalization group, which itself has intrinsic interest. Can the Callen-Symanzik equations be used to investigate the long distance behavior of $P(\varphi)_2$ models?

2. FROM ESTIMATES TO PHYSICS

How do we obtain physical properties of particles from our expansions and bounds? In this lecture we show how properties of the one particle states follow from known cluster expansions. These basic estimates for quantum field models exhibit the decoupling $\exp(-d/\xi)$ of disjoint regions in Euclidean phase space. In two space-time dimensions $(d = 2)$, cluster expansions yield space-time decoupling, as in Part II. For $d = 3$, related bounds yield phase space decoupling and the positivity of φ_3^4.

We recall that the theory of a single type of particle with mass m has the energy-momentum spectrum

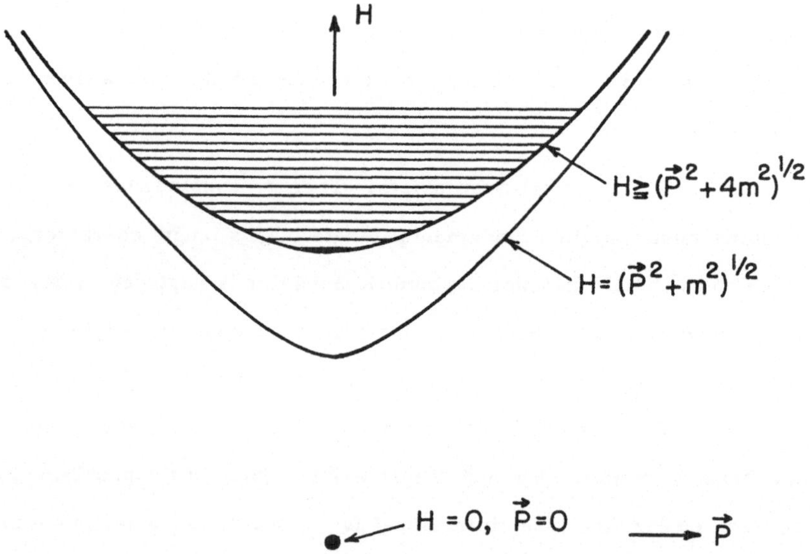

divided into three disjoint parts, the vacuum $\vec{P} = 0$, $H = 0$, the one particle hyperboloid $H^2 - \vec{P}^2 = m^2$ and the continuum $H^2 - \vec{P}^2 \geq (2m)^2$. The two particle states with momentum $\vec{p}_1, \vec{p}_2$ are conveniently parameterized by the relative momentum $p_R = \vec{p}_1 - \vec{p}_2$ and the total momentum $\vec{p}_T = \vec{p}_1 + \vec{p}_2$. The invariant mass for the two particle states is $2^{1/2}(\mu_1\mu_2 - \vec{p}_1 \cdot \vec{p}_2 + m^2)^{1/2}$, which for $\vec{p}_R = 0$ equals $2m$.

The mass operator $M = (H^2 - \vec{P}^2)^{1/2}$ has the corresponding spectrum

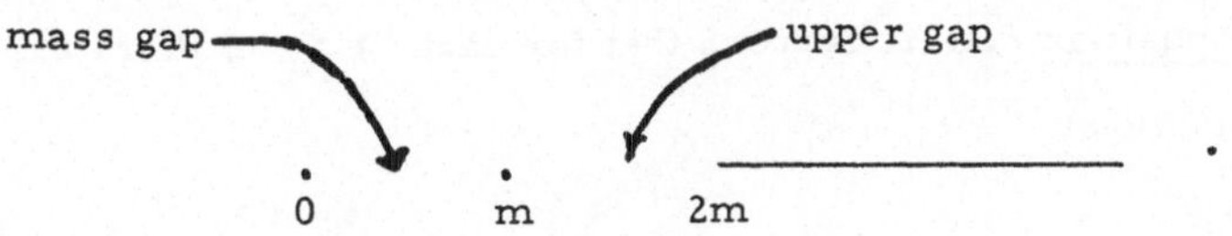

The eigenspace of 0 is the vacuum, and the eigenspace of m is defined to be the one particle space.

In order to establish spectral properties of H and M we use estimates proved by cluster expansions:

(1) Uniform vacuum cluster estimates yield convergence as the volume $\Lambda \to R^2$, and cluster estimates carry over to the infinite volume limit.

(2) The limiting Schwinger functions (for real coupling constants) satisfy the Osterwalder-Schrader axioms, and hence yield a Wightman theory. The cluster property of the vacuum (asymptotic factorization) yields uniqueness of the vacuum vector.

(3) The vacuum cluster expansion bounds the exponential decay to a factorizing vacuum and determines the mass gap. From the one particle cluster expansion, we obtain the upper mass gap and an isolated eigenvalue $M = m$.

In Section 2.1 we give some simple functional analysis. We apply these results in Section 2.2 to establish (1) - (3) above.

2.1. Functional Analysis

Let $0 \le H = H^*$ and let E_a be the spectral projection for $[0, a]$. Let $\mathcal{D}$ be a dense subset of $\mathcal{H}$, and let $\mathcal{D}_0 \subset \mathcal{D}$ be given.

Proposition 2.1.1. Suppose that for each $\theta \in \mathcal{D}$, there exists $\chi \in \mathcal{D}_0$ and $\epsilon > 0$ such that

$$\langle \theta - \chi , e^{-tH}(\theta - \chi)\rangle \le M_\theta e^{-(a+\epsilon)t} \quad . \tag{2.1.1}$$

Then $E_a \mathcal{D}_0$ is dense in $E_a \mathcal{H}$ and

$$\langle \theta - \chi , e^{-tH}(\theta - \chi)\rangle \le \|\theta - \chi\|^2 e^{-(a+\epsilon)t} \quad .$$

Proof. For $\theta \in \mathcal{D}$,

$$\begin{aligned} \|E_a(\theta - \chi)\| &= \|E_a e^{tH} e^{-tH}(\theta - \chi)\| \\ &\le e^{at}\langle \theta - \chi , e^{-2tH}(\theta - \chi)\rangle^{1/2} \\ &\le M_\theta^{1/2} e^{-\epsilon t} \longrightarrow 0 \quad . \end{aligned}$$

Thus $E_a \mathcal{D}_0 = E_a \mathcal{D}$, which is dense in $E_a \mathcal{H}$. By applying the Schwarz inequality n times,

$$\begin{aligned} \langle \theta - \chi , e^{-tH}(\theta - \chi)\rangle &\le \|\theta - \chi\| \, \|e^{-tH}(\theta - \chi)\| \\ &\le \|\theta - \chi\|^{2-2^{-n}} . e^{-(a+\epsilon)t} M_\theta^{2^{-n}} \\ &\longrightarrow \|\theta - \chi\|^2 e^{-(a+\epsilon)t} \quad . \end{aligned}$$

We now let $\mathcal{H}$ be a Hilbert space carrying a unitary representation $U(a, \Lambda)$ of the inhomogeneous Lorentz group. Let $\mathcal{H}_0 \subset \mathcal{H}$ be a subspace of bounded energy and momentum, $P_0 \leq a$, $|P| \leq b$. Let $U(a)\mathcal{H}_0 \subset \mathcal{H}_0$ and $(\cup U(a, \Lambda)\mathcal{H}_0)^- = \mathcal{H}$.

Proposition 2.1.2. If $\mathcal{H}_0 \neq \{0\}$ contains a cyclic vector for the space translation subgroup $U(\vec{a})$, then the spectrum of $M \restriction \mathcal{H}$ contains exactly one point and $U(a, \Lambda)$ is irreducible on $\mathcal{H}$.

Proof. The family $U(\vec{a})$ is maximal abelian on $\mathcal{H}_0$, so any commuting operator is a function of $\vec{P}$. In particular the energy momentum spectrum is a set of the form $\{H(\vec{P}), \vec{P}\}$, and by Lorentz invariance $H = (\vec{P}^2 + \bar{m}^2)^{1/2}$ for some $\bar{m}$. (Here we assume the nontriviality of $\mathcal{H}_0$.) Thus $M = \bar{m}$ on $\mathcal{H}_0$ and by Lorentz $M = \bar{m}$ on $\mathcal{H}$. Since reducibility would be accompanied by multiplicity in the mass spectrum, the representation $U(a, \Lambda)$ is irreducible.

2.2. Relevance to Physics

The Schwinger functions with a space time cutoff h are given by

(2.2.1) $$S_h(x_1, \cdots, x_n) = \int \Phi(x_1) \cdots \Phi(x_n) dq_h$$

where dq_h is the measure

(2.2.2) $$dq_h = \frac{e^{-V(h)} d\Phi}{\int e^{-V(h)} d\Phi} ,$$

$d\Phi$ is the Gaussian measure with mean zero and covariance $(-\Delta + m_0^2)^{-1} = C$, and

$$V(h) = \lambda \int : P(\Phi(x)) : h(x) dx$$

is the $P(\varphi)_2$ Euclidean action. If $h(x)$ is the characteristic function for a set $\Lambda \subset R^2$ with area $|\Lambda|$, then $V(h)$ is the action for Λ . We denote the corresponding Schwinger functions S_Λ .

We state the vacuum cluster expansion, which bounds the rate of asymptotic factorization of the vacuum state. Let A be a function of Euclidean fields,

(2.2.3) $$A = \prod_{i=1}^{n} \int : \Phi(x)^{n_i} : f_i(x) dx$$

where $f_i(x)$ is either an $L_2(R^2)$ function, or else $\delta_s(t) f_i(\vec{x})$, where $f_i(\vec{x})$ is $L_2(R)$. Let $\{\Delta\}$ be a cover of R^2 by unit lattice squares Δ , and define suppt. A as the smallest union of Δ's containing suppt. $f_1 \cup \cdots \cup$ suppt. f_n . In the following, A , B have the form (2.2.3).

<u>Theorem</u> 2.2.1 (<u>Vacuum Cluster Expansion</u> [Gl Ja Sp 1, 2].) Let $\gamma = m_0 - \epsilon$, with $\epsilon > 0$. Consider $\lambda P(\varphi)_2$ models with $\lambda < \lambda(\epsilon, P, m_0)$. Let d = dist. {suppt. A , suppt. B} . Then there exists a constant $M_{A,B}$ such that

(2.2.4) $$\left|\int AB\,dq_h - \int A\,dq_h \int B\,dq_h\right| \le M_{A,B}\,e^{-\gamma d} \quad ,$$

uniformly in h .

[The constant $M_{A,B}$ can be bounded explicitly in terms of the f_i . We suppose each f_i is supported in a single Δ_i (an arbitrary A is a sum of such localized A's). We let $N(\Delta)$ be the sum of n_i's , for suppt. $f_i \in \Delta$. Let K_1 be given and let

$$\eta = \prod_{\Delta} (K_1 N(\Delta)!) \quad .$$

We define for a localized A ,

(2.2.5) $$\|A\| = \eta \prod_{i=1}^{n} \|f_i\| \quad .$$

Here $\|f\| = \|f\|_{L_2(R^2)}$ for $f \in L_2(R^2)$, or if $f(x) = \delta_s(t) f(\vec{x})$, $\|f\| = \|f(\vec{x})\|_{L_2(R)}$. Let us assume $n_i \le \bar{n}$. Then for K_1 sufficiently large,

(2.2.6) $$M_{A,B} \le \|A\|\ \|B\| \quad .$$

If it is not the case that $n_i \le \bar{n}$, we obtain (2.2.6) with η^3 replacing η in (2.2.5). Also

(2.2.7) $$\int A\,dq_h \le \|A\|$$

uniformly in h .]

Theorem 2.2.2. The Schwinger functions $S_\Lambda(x_1, \cdots, x_n)$ converge in $\mathcal{S}'(R^{2n})$ as $\Lambda \to R^2$, to limits $S(x_1, \cdots, x_n)$ obeying the Osterwalder-Schrader axioms.

Proof. As explained in the lectures of Osterwalder and Nelson, it is sufficient to prove convergence as $h \to 1$, and a simple φ bound that follows from the vacuum cluster expansion. Here we establish convergence. Let $0 \le h_0 \le h_1 \le 1$ be two space-time cutoffs, with $h_1 - h_0$ supported on a bounded set Γ. Let $A = \Phi(f_1) \cdots \Phi(f_n)$, let suppt. $A \subset$ suppt. h_0 and let $d = \text{dist}(\Gamma, \text{suppt. } A)$. Let $\bar{\Gamma}$ be the set of lattice squares intersecting Γ.

Define the function

$$F(\alpha) = \int A dq_\alpha$$

where $dq_\alpha = dq_{h_\alpha}$ and $h_\alpha = \alpha h_1 + (1-\alpha) h_0$. Then by differentiating (2.2.2) we obtain

$$\left| \int A dq_1 - \int A dq_0 \right| = \left| \int_0^1 F'(\alpha) d\alpha \right|$$

$$= \int_0^1 d\alpha \sum_{\Delta \subset \bar{\Gamma}} \left[\int A V(\Delta) dq_\alpha - \int A dq_\alpha \int V(\Delta) dq_\alpha \right] .$$

By Theorem 2.2.1 and (2.2.6), the sum above is

$$\sup_{\Delta} M_{A, V(\Delta)} e^{-\gamma' d} \le O(1) e^{-\gamma' d} ,$$

for $\gamma' = m_0 - 2\epsilon$. We let h_0, h_1 be characteristic functions of sets $\Lambda_0, \Lambda_1 \subset R^2$. Then $d \to \infty$ as $\Lambda_0, \Lambda_1 \to R^2$ and we obtain the desired convergence.

The λ dependence of the cluster expansion shows immediately that $S(f_1, \cdots, f_n)$ are continuous functions of λ. In fact, in Part II we use the cluster expansion to establish analyticity in λ in the half circle $0 < |\lambda| < \lambda_0$, $\text{Re}\,\lambda > 0$.

<u>Theorem</u> 2.2.3. (<u>Mass Gap</u>). For $\lambda P(\varphi)_2$ models with small coupling λ, the vacuum Ω spans the subspace of energy less than $\gamma = m_0(1-\epsilon)$.

<u>Proof.</u> Let $A = \Phi(f_1) \cdots \Phi(f_n)$, suppt. A compact, and let θ_A be the vector in the relativistic Hilbert space $\mathcal{H}$ associated with the Euclidean function A. The plan is to apply Proposition 2.1.1 with $\mathcal{D}$ the dense subset of $\mathcal{H}$ spanned by such

θ_A , and with $\mathcal{D}_0$ the subspace of $\mathcal{D}$ spanned by Ω .

Since the cluster estimates are uniform in h , they carry over to the infinite volume limit h = 1 . We choose A_d to be the translate of A in the (Euclidean) time direction. Thus by the vacuum cluster expansion, Theorem 2.2.1, with $\gamma = m_0 - \epsilon$,

$$\langle \theta_A , e^{-dH} \theta_A \rangle = \int A^- A_d dq$$

$$= \int A^- dq \int A dq + O(M_A e^{-\gamma d})$$

$$= |\langle \theta_A , \Omega \rangle|^2 + O(M_A e^{-\gamma d}) \quad .$$

In other words, if $\theta_A^\perp = \theta_A - \langle \Omega , \theta_A \rangle \Omega$ is the component of θ_A orthogonal to Ω ,

$$\langle \theta_A^\perp , e^{-dH} \theta_A^\perp \rangle \leq O(M_A e^{-\gamma d}) \quad .$$

The theorem now follows as planned.

We have established stronger cluster properties, which provide an analysis of arbitrary intervals of the energy spectrum [Gl Ja Sp 1]. These expansions are defined inductively, rather than in closed form or in the form of Kirkwood-Salsburg equations. We now state the n = 1 expansion, or one particle cluster expansion. Let $\gamma = 2(m_0 - \epsilon)$ with $\epsilon > 0$.

Theorem 2.2.4. (One Particle Cluster Expansion). Given $\epsilon > 0$ consider $\lambda P(\varphi)_2$ models with $\lambda < \lambda(\epsilon , m_0 , P)$. Then given θ_A as above, there exists an $L_2(R)$ function h such that for $\chi = \langle \Omega , \theta_A \rangle \Omega + \varphi(h)\Omega$, we have

$$\langle \theta_A - \chi , e^{-tH}(\theta_A - \chi) \rangle \leq M_A e^{-\gamma t}$$

We apply Proposition 2.1.1 once again. We choose $\mathcal{D}$ as in the previous example, and $\mathcal{D}_0$ the span of $\{\Omega, \varphi(h)\Omega\}$, where $h \in L_2(R)$.

Corollary 2.2.5. The vectors $E_{2m_0-\epsilon}\mathcal{D}_0$ span states of energy $< 2m_0-\epsilon$.

Theorem 2.2.6. (Upper Mass Gap). For $\lambda P(\varphi)_2$ models with small coupling, the mass operator M has eigenvalues 0, m and no other spectrum in $[0, 2m_0-\epsilon]$.

Proof. Let $E = E_{2m_0-\epsilon}(I-E_0)$, let $\mathcal{K}_0 = E\mathcal{H}$ and let $\mathcal{K}$ equal the union of the Lorentz translates of $\mathcal{K}_0$. Below we obtain a cyclic vector χ for the space translation subgroup on $\mathcal{K}_0$. By Proposition 2.1.2, the spectrum of M on $\mathcal{K}$ contains exactly one point (unless $\mathcal{K}_0 = \{0\}$). We show $\mathcal{K}_0 \neq 0$: The two point function converges in $\mathcal{S}'$ to the free two point function as $\lambda \to 0$, using the λ dependence of the cluster bounds. Since the free theory has one particle states with mass m_0, the interacting theory must have spectrum in a neighborhood of m_0, for λ sufficiently small. Thus $\mathcal{K}_0 \neq 0$. M has the eigenvalues 0 and $m \in [m_0 - \epsilon, m_0 + \epsilon]$, and no other spectrum in $[0, 2m_0 - \epsilon]$.

To complete the proof we construct χ. Let $h_1 \in \mathcal{S}(R)$, $\tilde{h}_1 > 0$. We show that $E\varphi(h_1)\Omega$ is cyclic on $\mathcal{K}_0$. Let $h_a(x) = h(x-a)$. Then

$$U(\vec{a})\varphi(h)\Omega = \varphi(h_a)\Omega$$

and

$$\varphi(h_1 * h_2)\Omega = \varphi\left(\int h_1(\cdot - a)\, h_2(a)\, da\right)\Omega$$

$$= \int da\, h_2(a)\varphi(h_{1a})\Omega$$

$$= \int da\, h_2(a)\, U(\vec{a})\varphi(h_1)\Omega \quad .$$

Since E and $U(\vec{a})$ commute,

$$E\varphi(h_1 * h_2)\Omega = \int da\, h_2(a)\, e^{-iPa}\, E\varphi(h_1)\Omega$$

lies in the span of translates of $E\varphi(h_1)\Omega$. Since $(h_1 * h_2)^{\sim} = \tilde{h}_1 \tilde{h}_2$ are dense in C_0^∞ as h_2 ranges over C_0^∞, $\chi = E\varphi(h_1)\Omega$ is cyclic for $U(\vec{a})$ on $\mathcal{K}_0$. Here we have also used Corollary 2.2.5 to identigy $\mathcal{K}_0$ with the span of $E\varphi(f)\Omega$.

3. BOUND STATES AND RESONANCES

3.1 Introduction

An important problem in physics is how particles form composites, namely bound states and resonances. In atomic physics, familiar consequences of Coulomb forces and the Schrödinger Hamiltonian are atoms and molecules: their existence and their scattering. The spectrum of atomic and molecular Hamiltonians has been the subject of extensive mathematical analysis.

The realm of nuclear and elementary particle structure includes qualitatively similar ideas, but without detailed justification. Thus a crucial physical question is whether a particular quantum field model does or does not have bound states. For instance: Do mesons bind nucleons to form stable nuclei? Are nucleons bound states of quarks? Are the ρ and the η mesons really π meson resonances?

Little is known about such important questions in quantum field theory. In fact, no quantum field models are known to have bound states, and heuristic calculations based on perturbation theory and the Bethe-Salpeter equation are inconclusive.

In this lecture we give a physical picture of when to expect or not to expect bound states in $\mathcal{P}(\varphi)_2$ models with weak coupling or a strong external field. We prove the absence of two particle bound states in weakly coupled, pure φ^4 models. We outline an argument to prove the presence of bound states in the presence of a strong external field, and certain other models.

Bound states are eigenvalues of the mass operator M, introduced in Chapter 2. Two particle bound states lie below the two particle continuum; otherwise no energetic reason would prevent their decay into free particles. (The decay of bound states in the mass continuum may, however, be forbidden by additional selection rules included in the interaction.) On the other hand, there is no physical interpretation of continuous mass spectrum in the spectral interval $[0, 2m)$. Hence none is believed to exist, and two particle bound states may occur in the "bound state interval" $(m, 2m)$ of the mass spectrum, as illustrated in Figure 7.

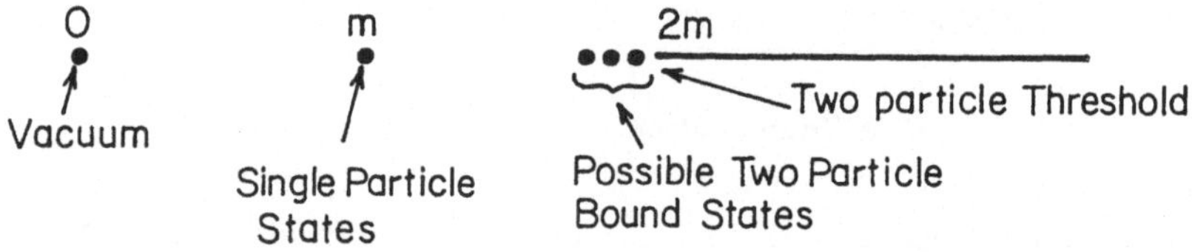

Figure 7. Spectrum of the mass operator M

In an even theory, e.g., φ^4, we can decompose the Hilbert space according to whether states are even or odd under the symmetry $\varphi \to -\varphi$. States with an even number of particles lie in the even subspace. Restricted to the odd subspace, M has the spectrum of Figure 8.

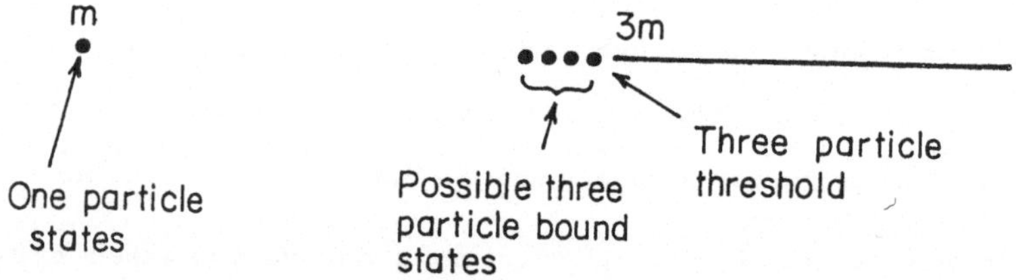

Figure 8. Mass Spectrum on the Odd Subspace of an Even Theory.

The resolvent $(M - z)^{-1} = R(z)$ of the mass operator is an analytic function of z, for $\operatorname{Im} z \neq 0$. It has a pole at each eigenvalue of M (particles and bound states) and presumably a cut starts at each n-particle threshold. The question of resonances concerns the analytic properties of $R(z)$ (or suitable matrix elements) after continuation across a threshold cut. A complex pole, close to the cut, is called a resonance. Such a pole appears in the scattering of particles as a peak in the cross section. Another interpretation of a resonance is an unstable particle. The real part of the position of the pole determines the mass of the resonance, while the distance to the real axis determines the lifetime. It is a challenging question to make a detailed investigation of resonances, and to determine: Are there coupling constants for which $P(\varphi)_2$ models have resonances?

The presence or absence of composite particle states depends on whether the interparticle forces are attractive or repulsive. We pose the related questions: Does the mutual interaction of two particles raise or lower their energy, compared with the state in which they are asymptotically far apart? If the energy is raised, binding does not occur. If the energy is lowered below the continuum, we expect a bound state. In Section 3. 2 we motivate our point of view on this question by perturbation theory. In Section 3. 3, we use cluster estimates and correlation inequalities to study the same question. In Section 3. 4, we show how binding occurs.

Our picture of a two particle bound state is best understood in terms of the relative momentum $\vec{p}_R$. We describe three kinds of forces: <u>attractive</u>, <u>repulsive</u> and <u>dispersive.</u> The attractive and repulsive forces are self explanatory. The dispersive effect arises from the curvature of the mass hyperboloid. A state of two free particles, with $\vec{p}_T = 0$, has a total energy $(4m^2 + \vec{p}_R^{\,2})^{1/2}$, and in general, for small momentum, a two particle state has energy $2m + O(\vec{p}_R^{\,2} + \vec{p}_T^{\,2})$. This raising of the energy away from zero momentum is what we call the dispersive force. For bound states to occur, the attractive force must dominate the repulsive and dispersive forces.

We introduce a parameter δ to measure the spread of the bound state wave packet. For a momentum space distribution concentrated in $|\vec{p}_R| \le \delta$, we have a configuration space spreading of order δ^{-1}. For weak coupling, we expect increased spreading in configuration space, as a bound state grows in size and disappears into the continuum. Thus we expect $\delta \to 0$ as $\lambda \to 0$. The binding forces have characteristic dependences on δ and λ: The dispersive effect is $O(\delta^2)$. In $P(\varphi)_2$ models, we find in perturbation theory that attractive and repulsive effects are $O(\delta)$, times the appropriate dimensionless coupling constants λ_j/m_0^2. We discuss the balance of these forces in Section 3. 4.

3.2 Formal Perturbation Theory

For a $\lambda\varphi^4$ interaction, the first order shift in the two particle energy is given by the Feynman diagram

which is positive for $\lambda > 0$. In second order, we find the shift has two sorts of contributions, a second order mass shift with the disconnected Feynman diagrams

and a second order attractive (negative) contribution of the form

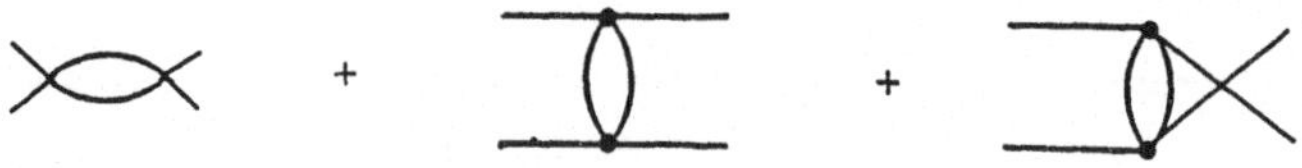

The first order repulsive shift dominates for small λ. Thus we do not expect two particle bound states to occur in weakly coupled φ_2^4 models, and we establish this in Section 3.3.

We remark that the mass shift diagrams above represent the second order mass renormalization of single particle states, i.e., the shift from m_0 to m_2. Of course, to second order, we measure our n-particle forces (energy shifts) with respect to nm_2. We do not include vacuum energy shifts, since they are eliminated by considering perturbations of the exact (coupling λ) ground state.

If we consider three particle interactions, in lowest order, diagrams of the form

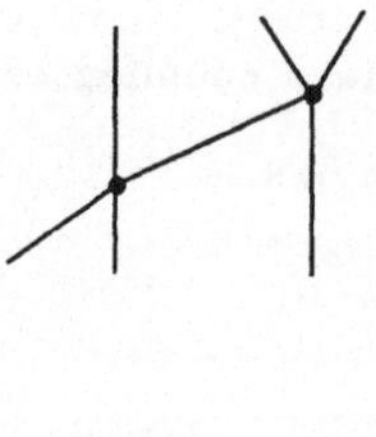

give an attractive three body force. However the diagram

gives a repulsive effect in the two particle subsystems. Since the two body force is first order, and the three body force is second order, we expect the repulsive force to dominate at small coupling. A three particle unstable state (resonance) is possible.

With a φ^3 interaction, the lowest order two body force is attractive

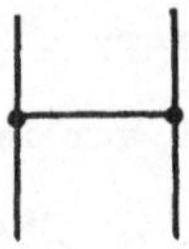

Similarly, n body forces in lowest order are attractive. For instance, in third order we have diagrams of the form

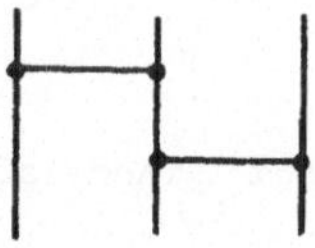

These attractive forces complement the attractive forces in two body subsystems, i.e., in the three body case,

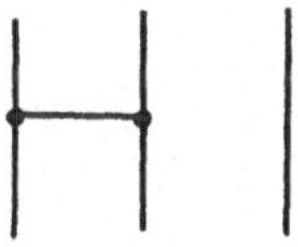

Thus we expect two particle bound states, and bound states of three or more particles if a selection rule prevents their decay. Otherwise, the attractive many body forces should yield many body resonances.

Of course, a pure φ^3 theory does not exist, because the energy is unbounded from below. However, if the φ^3 term in an interaction has a coefficient much larger than the other coupling constants, we expect that the φ^3 effects will dominate. Thus the above qualitative discussion applies to the $\lambda_1\varphi^3 + \lambda_2 P(\varphi)_2$ model, where $\lambda_1 >> \lambda_2$. In this case we expect bound states, and in particular, two particle bound states.

Closely related is the case of a $P(\varphi)_2$ model in an external field, i.e., the $P(\varphi)_2 - \mu\varphi$ model. By the transformation $\varphi \to \varphi + \text{const.}$ (implemented locally by $\exp(i \int \pi))$ we can eliminate the external field. For instance, the $\lambda\varphi^4 - \mu\varphi$ model is transformed into a $\lambda\varphi^4 + a\varphi^3 + b\varphi^2$ model, where $4\lambda a^3 + am^2 = \mu$. The mass term b also grows with μ, but by scaling it can be reduced to unity. Thus we <u>conjecture</u>: Bound states exist in the φ^4 model with a strong external field, $\mu >> \lambda$.

A similar analysis applies to an arbitrary $\lambda P(\varphi)_2 - \mu\varphi$ model. Transforming away the external field, we add to P a lower degree polynomial. For μ large, the dominant coefficients have degree 2 and 3. The degree 2 term gives a mass shift, while the degree 3 term yields an attractive potential in lowest order. Thus we <u>conjecture</u>: Bound states exist in $\lambda P(\varphi)_2$ models in external fields with $\mu \gg \lambda$.

<u>Question</u>: Do bound states occur in Y_2 models? We conjecture that this is the case.

3.3 On the Absence of Bound States

We consider the weakly coupled $\lambda\varphi_2^4$ model, and we prove that two particle bound states do not occur.

Theorem 3.3.1. Let λ/m_0^2 be sufficiently small in the $\lambda\varphi_2^4$ model. Then the mass operator $M = (H^2 - P^2)^{1/2}$ has no spectrum in the two particle bound state interval $(m, 2m)$.

From the uniqueness of the vacuum, we infer that the symmetry $\varphi \to -\varphi$ can be unitarily implemented, and that the Hilbert space $\mathcal{H}$ decomposes into even and odd subspaces $\mathcal{H}_e$, $\mathcal{H}_o$ each invariant under $U(a, \Lambda)$ and φ. Our theorem depends on three facts: Cluster expansions [Gl Ja Sp 1] reduce the problem to the consideration of the two point function for $\mathcal{H}_o$, and the four point function for $\mathcal{H}_e$. Second, an inequality that Lebowitz [Leb 2] proved for Ising models excludes the possibility that mass spectrum in the interval $(0, 2m)$ occurs in the four point function. Finally, cluster bounds exclude mass spectrum in the interval $(m, 2m)$ in the two point function.

The condition of weak coupling in Theorem 3.3.1 concerns the rate γ, of exponential decay $e^{-\gamma d}$, in the error term of the two particle cluster expansion. We show in [Gl Ja Sp 1] that $\gamma \to 3m_0$ and $m \to m_0$ as $\lambda/m_0^2 \to 0$. In Theorem 3.3.1 we require that λ/m_0^2 be sufficiently small to ensure $\gamma \geq 2m$.

More generally, we obtain for even $P(\varphi)_2$ models a larger mass gap on the odd subspace, as suggested in Figure 8 above.

Theorem 3.3.2. Consider an even $\lambda P(\varphi)_2$ model. Given $\epsilon > 0$, let λ/m_0^2 be sufficiently small to ensure $\gamma \geq 3m_0 - \epsilon$, for the rate γ of exponential decay for the error in the two particle cluster expansion. Then $M \restriction \mathcal{H}_o$ has no spectrum in the interval $(m, 3m_0 - \epsilon)$.

Let dq be the Euclidean measure for the $\lambda\varphi_2^4$ model, and for a function A of the Euclidean field Φ let

$$\langle A \rangle \equiv \int A \, dq$$

Proposition 3.3.3. For the $\lambda\varphi_2^4$ model,

$$\langle \Phi(x_1)\Phi(x_2)\Phi(x_3)\Phi(x_4)\rangle - \langle \Phi(x_1)\Phi(x_2)\rangle\langle \Phi(x_3)\Phi(x_4)\rangle$$

(3.3.1)

$$\leq \langle \Phi(x_1)\Phi(x_3)\rangle\langle \Phi(x_2)\Phi(x_4)\rangle + \langle \Phi(x_1)\Phi(x_4)\rangle\langle \Phi(x_2)\Phi(x_3)\rangle .$$

Remark. Since $\langle \Phi(x)\rangle = 0$, this inequality states that the connected (truncated) four point function is negative. This bound is special to φ^4 models. In fact the philosophy of Section 3.2 suggests the presence of two particle bound states in $\varphi^6 - \varphi^4$ models.

The key inequality due to Lebowitz concerns independent spin variables $\sigma_i = \pm 1$ for a ferromagnetic Ising model. The energy of a spin configuration $\underset{\sim}{\sigma} = \{\sigma_i\}$, $1 \leq i \leq n$, is

$$H(\underset{\sim}{\sigma}) = -\sum_{i<j} J_{ij}\,\sigma_i\sigma_j ,$$

where $J_{ij} \geq 0$. For a function $f(\underset{\sim}{\sigma})$, let

$$\langle f \rangle = Z^{-1} \sum_{\underset{\sim}{\sigma}} f(\underset{\sim}{\sigma})\, e^{-H(\underset{\sim}{\sigma})} ,$$

where

$$Z = \sum_{\underset{\sim}{\sigma}} e^{-H(\underset{\sim}{\sigma})} .$$

Lebowitz proves [Leb 2]

$$\langle \sigma_i\sigma_j\sigma_k\sigma_\ell\rangle - \langle \sigma_i\sigma_j\rangle\langle \sigma_k\sigma_\ell\rangle \leq \langle \sigma_i\sigma_k\rangle\langle \sigma_j\sigma_\ell\rangle + \langle \sigma_i\sigma_\ell\rangle\langle \sigma_j\sigma_k\rangle .$$

The inequality (3.3.4) follows immediately, since Griffiths and Simon [Gr Si] have proved that the Euclidean φ_2^4 model is a limit of Ising models of the above form, where $\Phi(x)$ can be expressed as a limit of a sum of spin variable σ_i.

We recall that the relativistic time zero field $\varphi(f)$ equals the time zero Euclidean field $\Phi(f, t=0)$. We let $f_i \in \mathcal{S}(R)$ and define

$$(3.3.2) \quad \theta(f_1, f_2) = \varphi(f_1)\varphi(f_2)\Omega - \langle \Omega, \varphi(f_1)\varphi(f_2)\Omega \rangle \Omega \quad ,$$

where Ω is the vacuum vector.

Corollary 3.3.4. The vectors $\theta(f_1, f_2)$ have energy $\geq 2m$.

Proof. It is no loss of generality to choose f_i real and positive. By the Feynman-Kac formula,

$$\langle \theta, e^{-tH}\theta \rangle = \langle \varphi(f_1)\varphi(f_2)\Omega, e^{-tH}\varphi(f_1)\varphi(f_2)\Omega \rangle - |\langle \Omega, \varphi(f_1)\varphi(f_2)\Omega \rangle|^2$$

$$= \langle \Phi(g_1)\Phi(g_2)\Phi(g_3)\Phi(g_4) \rangle - \langle \Phi(g_1)\Phi(g_2) \rangle \langle \Phi(g_3)\Phi(g_4) \rangle ,$$

where $x = (x_1, x_0)$ and

$$g_1(x) = f_1(x_1)\delta(x_0) \quad , \qquad g_2(x) = f_2(x_1)\delta(x_0) \quad ,$$

$$g_3(x) = f_1(x_1)\delta_t(x_0) \quad , \qquad g_4(x) = f_2(x_1)\delta_t(x_0) \quad .$$

By (3.3.1) and the Feynman-Kac formula,

$$\langle \theta, e^{-tH}\theta \rangle \leq \prod_{i=1}^{2} \langle \varphi(f_i)\Omega, e^{-tH}\varphi(f_i)\Omega \rangle$$

$$+ |\langle \varphi(f_1)\Omega, e^{-tH}\varphi(f_2)\Omega \rangle|^2 \quad .$$

Since $\langle \Omega, \varphi\Omega \rangle = 0$, the spectrum condition yields $|\langle \varphi(f_i)\Omega, e^{-tH}\varphi(f_j)\Omega \rangle| \leq O(1)e^{-tm}$ and therefore

$$\langle \theta, e^{-tH}\theta \rangle \leq O(1)e^{-2tm} \quad ,$$

completing the proof. We remark that only vacuum cluster expansions are necessary to this point.

Next we state a result [Gl Ja Sp 1] which follows from the two particle cluster expansion. We let $\epsilon > 0$, and we let E be the spectral projection for the energy interval $[0, 3m_0 - \epsilon]$ in an even $\lambda P(\varphi)_2$ model. We assume λ/m_0^2 sufficiently small to ensure a decay rate $\gamma = 3m_0 - \epsilon$ in the two particle cluster expansion.

Proposition 3.3.5. With the above assumptions, linear combinations of the vector Ω and $e^{tH} E\theta(f_1, f_2)$ are dense in $E\mathcal{H}_e$. Also the vectors $E\varphi(f)\Omega$ are dense in $E\mathcal{H}_o$.

We remark that in [Gl Ja Sp 1] we prove a weaker result for $E\mathcal{H}_o$, namely that vectors $e^{tH} E\varphi(f)\Omega$ span $E\mathcal{H}_o$. A simple modification of Theorem 4.2, [Gl Ja Sp 1] can be used to bring first degree polynomials in the n-particle cluster expansion to time zero. This yields Proposition 3.3.5, for $n = 2$.

Proof of the Theorems. Suppose that $M \restriction \mathcal{H}_e$ has mass spectrum in the interval $(m, 2m)$. By Lorentz invariance, there is a nonzero vector $\psi \in \mathcal{H}_e$ corresponding to that spectral interval and with energy $< 2m$. By Proposition 3.3.5, ψ is a limit of sums of vectors $e^{tH} E\theta(f_1, f_2)$. By Corollary 3.3.4, $\psi = 0$, proving Theorem 3.3.1 on $\mathcal{H}_e$.

Finally we show $M \restriction E\mathcal{H}_o$ has only one point in its spectrum, namely m. By Proposition 3.3.5, the vectors $\mathcal{K}_0 = \{E\varphi(f)\Omega\}$ span $E\mathcal{H}_o$. We let $\mathcal{K}$ be the closure of the union of Lorentz translates of $\mathcal{K}_0$. Our assertion then follows by Proposition 2.1.2. Theorems 3.3.2 and 3.3.1 then follow by Lorentz invariance.

3.4 On the Presence of Bound States

The ideas of Section 3.2 suggest the presence of bound states in certain $P(\varphi)_2$ models. We give two methods to establish the existence of mass spectrum in the two particle bound state interval $(m, 2m)$. As we mentioned above, there is no physical interpretation of continuous spectrum in this interval, so the existence of spectrum presumably ensures the existence of eigenvalues, i.e., bound states.

Variational Method. The first method is to choose an approximate bound state wave function θ, with the properties: (i) $\|\theta\| \geq 1$; (ii) θ is orthogonal to the vacuum and one particle states; and (iii) $\langle\theta, M\theta\rangle < 2m$. Since $M \leq H$, we may replace the bound on $\langle\theta, M\theta\rangle$ by

$$(3.4.1) \qquad \langle\theta, H\theta\rangle < 2m \quad .$$

In a theory with weak coupling, the cluster expansion shows that the low momentum part of the mass interval $(m, 2m)$ is spanned by vectors

$$(3.4.2) \qquad \theta = \alpha\Omega + a^*(f)\Omega + e^{tH} a^*(f_1)a^*(f_2)\Omega \quad ,$$

see [Gl Ja Sp 1] and Section 3.3. Here a^* is a time zero creation operator. If, in addition, $P(\varphi)$ is even, we may choose $f = 0$ and $\alpha = -\langle\Omega, a^*(f_1)a^*(f_2)\Omega\rangle$. Alternatively, we can replace a^* by the time zero field φ in (3.4.2).

With this variational method, we eliminate H from $\langle\theta, H\theta\rangle$ by using $H\Omega = 0$ and the canonical commutation relations. For instance,

$$[H, a^*(f)] = a^*(\mu f) + [H_I, a^*(f)] \quad , \qquad \text{where} \quad \mu = (\vec{p}^{\,2} + m_0^2)^{1/2} \quad .$$

If $\chi = a^*(f)\Omega$, then

$$(3.4.3) \qquad \begin{aligned} \langle\chi, H\chi\rangle &= \langle f, \mu f\rangle_{L_2} + \langle\Omega, a^*(\mu f)\, a(\overline{f})\Omega\rangle \\ &\quad + \langle a^*(f)\Omega, [H_I, a^*(f)]\Omega\rangle \end{aligned}$$

We estimate vacuum expectation values of Wick ordered monomials

$$W = \int a^*(x_1) \cdots a(x_n) w(x)\, dx \quad ,$$

by the cluster expansion [Gl Ja Sp 1]. In fact, before estimation, we expand $\langle \Omega, W\Omega \rangle$ using integration by parts, to isolate low order dependence in the coupling λ, see Chapter 4. For instance, in second order, we obtain a second order mass-shift correction to $\langle f, \mu f \rangle_{L_2}$.

In this manner, we need not calculate the physical mass m exactly, but we can obtain explicitly the relevant low order corrections to m_0. (Here we assume that m is asymptotic to m_0.) Furthermore, let us assume that f is scaled to give momentum localization $O(\delta)$, namely $f(\vec{p}) = \delta^{-1/2} h(\vec{p}/\delta)$. (In Section 3.1 we explained that $\delta \to 0$ as $\lambda \to 0$.) Then

$$\langle f, \mu f \rangle = m_0 \|f\|^2 + O(\delta^2) \quad ,$$

which exhibits the momentum dispersion about $\vec{p} = 0$ of the single particle state. Similarly, the second order mass correction will equal $m_2\lambda^2 \|f\|^2 + O(\lambda^2\delta^2)$.

We sketch our proof for the $\lambda(\varphi^6 - \varphi^4)$ interaction. We take

$$\theta = a^*(f)^2\Omega - \langle \Omega, a^*(f)^2\Omega \rangle \Omega \quad , \qquad \text{with } \|f\|_{L_2} = 2^{-1/4}$$

which satisfies (i), (ii) above. We study

$$\langle \theta, H\theta \rangle = \langle \theta, \{a^*(\mu f)a^*(f) + a^*(f)a^*(\mu f)\}\Omega \rangle + \left\langle \theta, \left[\lambda \int \varphi^6 - \lambda \int \varphi^4, a^*(f)^2\right]\Omega \right\rangle \quad ,$$

and integrate by parts. We isolate, in closed form, all terms of degree 0, 1 or 2 in λ. The mass terms have the form

$$2\{m_0 + \lambda^2 m_2 + O(\delta^2 + \lambda^3 + \delta^2\lambda^2)\} \quad .$$

The attractive contribution from diagrams of the form

lowers the energy by $-O(\delta\lambda)$. Other contributions are $O(\lambda^2\delta)$ or higher order. We choose $\delta = \lambda^{1+\epsilon}$. Then for small λ, the decrease in energy $-O(\delta\lambda) = -O(\lambda^{2+\epsilon})$ dominates the dispersive effect $O(\delta^2) = O(\lambda^{2+2\epsilon})$ and the repulsive effects $O(\lambda^2\delta + \lambda^3) \le O(\lambda^3)$. The operator parts of these estimates result from a variant of the cluster expansion. This completes our sketch of the proof that bound state spectrum exists in the weakly coupled $\lambda(\varphi^6 - \varphi^4)$ model.

Similar arguments should hold for $\lambda\varphi^6$. In this case, however, the attraction is $O(\lambda^2\delta)$. We must therefore isolate the fourth order mass shift and we set $\delta = \lambda^{2+\epsilon}$. For the interaction $\lambda\varphi^3 + \lambda^6\varphi^4$, we must orthogonalize θ to the one particle states (at least to third order in λ). We would then isolate the fourth order mass renormalization and take $\delta = \lambda^{2+\epsilon}$. We thank B. Simon for observing that an even theory is technically simpler.

Cluster Method. In an even $P(\varphi)_2$ model, for θ of the form (3.3.2),

$$\langle \theta, e^{-tH}\theta\rangle = \langle \Phi(g_1)\Phi(g_2)\Phi(g_3)\Phi(g_4)\rangle_C + \langle \Phi(g_1)\Phi(g_3)\rangle\langle \Phi(g_2)\Phi(g_4)\rangle + \langle \Phi(g_1)\Phi(g_4)\rangle\langle \Phi(g_2)\Phi(g_3)\rangle \quad .$$

where $\langle\cdot\rangle_C$ denotes the connected (truncated) part. Thus $\langle\theta, e^{-tH}\theta\rangle$ exhibits the two particle decay $O(e^{-2mt})$, unless

$$\langle \Phi(g_1)\cdots\Phi(g_4)\rangle_C \ge O(e^{-2(m-\epsilon)t}) \quad . \tag{3.4.4}$$

Using the Bethe-Salpeter equation, we can isolate in $\lambda(\varphi^6 - \varphi^4)$ models a slowly decaying part of $\langle\Phi(g_1)\cdots\Phi(g_4)\rangle_C$, given by (positive) φ^4 contributions. We propose using cluster expansions to estimate the errors. The inequality (3.4.4) would establish the existence of mass spectrum on $\mathcal{H}_e$ in the interval $(0, 2m-\epsilon]$. This proposed calculation appears interesting. However, unlike the variational proof above, we presently have no error estimates using this method. Conversely, we remark that the existence of two-particle bound state spectrum in a weakly coupled even $P(\varphi)_2$ model (as established by the variational method) ensures (3.4.4).

4. PHASE SPACE LOCALIZATION AND RENORMALIZATION

4.1 Results for φ_3^4

In a series of related papers, we have given convergent expansions [Gl Ja Sp 1, 2] and convergent upper bounds [Gl Ja IV, 8] for quantum field models. These expansions and bounds deal with the problem of removing cutoffs $\varkappa$, Λ, namely in taking infinite volume limits in phase space. Most of this conference has dealt with the $\Lambda \rightarrow R^2$ limit. However the $\varkappa \rightarrow \infty$ limit in Y_2 and in higher dimensional models presents the most challenging problems, for both physics and for mathematics; we hope these ultraviolet problems will be the focus of increasing attention in constructive field theory. In this section we describe the results for φ_3^4: Let $d\Phi_C$ be the Gaussian measure with covariance C, and let $d\Phi$ denote the choice $C = (-\Delta + m_0^2)^{-1}$.

Let $V(\Lambda,\varkappa)$ denote the Euclidean action, the sum of the φ^4 interaction V_I and the counterterms V_C. Then

$$V_I = \lambda \int_{\Lambda \subset R^3} :\Phi_\varkappa^4: \, dx$$

and V_C are the Green's function counterterms given in second and third order perturbation theory. The partition function for the action $V = V_I + V_C$, namely

$$Z(\Lambda,\varkappa) = \int e^{-V(\Lambda,\varkappa)} \, d\Phi \quad ,$$

contains the ultraviolet divergent counterterms.

Theorem 4.1.1 [Gl Ja 8]. For $0 \leq \lambda$

$$Z(\Lambda,\varkappa) \leq e^{\alpha|\Lambda|)} \quad , \tag{4.1.1}$$

uniformly in $\varkappa$. For λ bounded, (4.1.1) is uniform in λ also.

We now let $H(\nu)$ denote the renormalized φ_3^4 Hamiltonian, defined formally by

$$H(\mathcal{V}) = H_0 + \lambda \int_{\mathcal{V} \subset R^2} :\varphi^4: \, d\vec{x} - \tfrac{1}{2}\, \delta m_2^2 \int :\varphi^2: \, d\vec{x} - E_2 - E_3 \ .$$

Here δm_2^2, E_2 and E_3 are the Hamiltonian counterterms in second and third order perturbation theory. (These counterterms differ by a constant and a transient from the Green's function counterterms, see [Gl Ja 8].)

Corollary 4. 1. 2. The Hamiltonian $H(\mathcal{V})$ is bounded from below by a constant proportional to the volume $|\mathcal{V}|$,

$$0 \le H(\mathcal{V}) + O(|\mathcal{V}|) \ . \tag{4. 1. 2}$$

The corollary follows from the theorem and the fact that $\langle \Omega(\mathcal{V},\varkappa), \Omega_0 \rangle \neq 0$. In fact

$$\langle \Omega_0, e^{-tH(\mathcal{V},\varkappa)}\, \Omega_0 \rangle = e^{-tE(\mathcal{V},\varkappa) - A(\mathcal{V},\varkappa) + T(\mathcal{V},\varkappa,t)} \ ,$$

where $E(\mathcal{V},\varkappa)$ is the partially renormalized vacuum energy, vanishing in second and third order, and convergent as $\varkappa \to \infty$ for fixed volume. The constant A in $e^{-A(\mathcal{V},\varkappa)} = |\langle \Omega_0, \Omega(\mathcal{V},\varkappa)\rangle|^2$ is the logarithmically divergent wave function renormalization constant, independent of t. Also $T(\mathcal{V},\varkappa,t)$ is a transient that is bounded as $\varkappa \to \infty$ and $o(1)$ as $t \to \infty$.

As $|\mathcal{V}| \to \infty$, the constants $E(\mathcal{V},\varkappa)$, $A(\mathcal{V},\varkappa)$ and $T(\mathcal{V},\varkappa,t)$ diverge. The second order, i. e., the ultraviolet divergent, part of $A(\mathcal{V},\varkappa)$ has been cancelled in $Z(\Lambda,\varkappa)$.

These results have been extended by Joel Feldman [Fel 2], who proved

Theorem 4. 1. 3. The finite volume partition function $Z(\Lambda,\varkappa)$ and

$$Z(\Lambda,\varkappa)\, S(\Lambda,\varkappa;\, f_1,\dots,f_n) = \int \Phi(f_1)\dots\Phi(f_n)\, e^{-V(\Lambda,\varkappa)}\, d\Phi \tag{4. 1. 3}$$

converge as $\varkappa \to \infty$. The limits are continuous in λ and satisfy

$$|Z(\Lambda)\, S(\Lambda;\, f_1,\cdots,f_n)| \le n! \Big(\prod_i \|f_i\|\ e^{O(|\Lambda|)}\Big) \tag{4. 1. 4}$$

for a Schwartz space norm $\|\cdot\|$

From continuity in λ and $Z(\Lambda) = 1$ for $\lambda = 0$, we conclude that for Λ fixed and λ sufficiently small,

$$Z(\Lambda) > 1/2 \quad .$$

Thus for Λ fixed and λ small, the approximate Schwinger functions $S(\Lambda, f_1, \ldots, f_n)$ do not vanish identically and

(4.1.5) $$|S(\Lambda; f_1, \ldots, f_n)| \leq n! \prod_i \|f_i\| \; .$$

Corollary 4.1.4 [Fe 2]. For small λ, volume Λ Schwinger functions are the moments of a unique measure on $\mathcal{S}'(R^3)$, namely

$$dq = \lim_{\varkappa \to \infty} Z(\Lambda, \varkappa)^{-1} e^{-V(\Lambda, \varkappa)} d\Phi$$

$$= \lim_{\varkappa \to \infty} dq_{\Lambda, \varkappa} \quad .$$

The corollary is based on a study of the perturbation of Z in an external Euclidean field, namely on the study of the generating functional

$$Z(h) = \int e^{\Phi(h)} dq_\Lambda$$

for the (disconnected) Schwinger functions. This functional was studied in $P(\varphi)_2$ by Fröhlich [Fr 2]; see also [Gl Ja 13].

Of course, we are interested in the $\Lambda \to R^3$ limit of these Schwinger functions $S(\Lambda; \cdot)$ and of measures dq_Λ, in order to obtain the full relativistic theory. We conjecture that the Kirkwood-Salsburg equations of Part II can be generalized to φ_3^4 and yield the limit. In fact the local estimates of Theorem 4.1.3 and Corollary 4.1.4 are exactly the type of local estimates which the cluster expansion for $P(\varphi)_2$ uses as input. We conjecture that $Z(\Lambda) \geq \exp(-O|\Lambda|)$ for small λ. We expect that such estimates lead to the Wightman axioms for φ_3^4.

4.2. Elementary Expansion Steps

The proof of the estimates for $P(\varphi)_2$, as well as those for φ_3^4, results from the use of four elementary identities and bounds concerning the non-Gaussian measure

$$e^{-V(\Lambda,\varkappa)}\, d\Phi_C \,. \tag{4.2.1}$$

The four steps are

I. Change of covariance C.

II. Change of exponent V.

III. Wick ordering bound.

IV. Integration by parts.

The four steps are combined to yield expansions or bounds. The difficult part of the construction is generally the question of how to combine these steps to isolate the desired property of the model, at the same time to ensure convergence. We use three expansion techniques:

a) Explicit expansions. We prescribe definite elementary steps to yield an expansion, as the expansion for ZS in Part II.

b) Neumann series. The Kirkwood-Salsburg equations of Part II, obtained by explicit expansion of Z, yield a Neumann series $(I - \mathcal{K})^{-1} = \Sigma_n \mathcal{K}^n$ for their solution.

c) Inductively defined expansions. We prescribe for each possible term (integral) in our expansion, rules to expand it into a sum of similar terms.

There is considerable freedom in the definition of our expansions and bounds. The inductively defined expansions leave the widest lattitude of choice, since they are not tied to recovering an expansion expressible in closed form, or to obtaining the inverse of an operator. In addition, the inductive expansions and bounds yield the most detailed information about our models, including the positivity of φ_3^4 [Gl Ja 8] and the φ-bounds for all couplings [Gl Ja IV]. These expansions and bounds are not tied to the use of particular boundary conditions on the covariance

C, but have more general validity.

We now give the elementary steps; the first two steps are merely the fundamental theorem of calculus:

I. Change of Covariance. Let $C_\alpha = \alpha C_1 + (1-\alpha)C_0$ be a family of interpolating covariances, and let

$$\begin{aligned} d\Phi_{C_1} &= d\Phi_{C_0} + \int_0^1 d\alpha \frac{d}{d\alpha} d\Phi_{C_\alpha} \\ &= d\Phi_{C_0} + \frac{1}{2}(C_1 - C_0)\cdot \Delta_\Phi \int_0^1 d\alpha\, d\Phi_{C_\alpha} . \end{aligned}$$

This formula has been used to deal with the infinite volume limit, see Part II. It is established by integration by parts on function space [Di Gl]; see also the proof of IV below. We do not use Step I in this chapter.

II. Change of V. Let $V_\alpha = \alpha \in [0,1]$ be a differentiable family of interpolating Euclidean actions. Then

$$(4.2.2) \qquad \begin{aligned} e^{-V_1} &= e^{-V_0} + \int_0^1 \frac{d}{d\alpha} e^{-V_\alpha} d\alpha \\ &= e^{-V_0} - \int_0^1 \frac{dV_\alpha}{d\alpha} e^{-V_\alpha} d\alpha . \end{aligned}$$

We use this identity to lower an upper momentum cutoff in the action V, in the positivity proofs for $P(\varphi)_2$ [See Gl Ja 7, IV] and φ_3^4 [Gl Ja 8], and we call this formula the perturbation or Duhamel identity.

Iterating (4.2.2) leads to the unrenormalized perturbation series. With an ultraviolet cutoff, this series diverges because of the $O(n!^2)$ diagrams arising in n^{th} order. For example, with one degree of freedom,

$$\int e^{-q^2-\lambda q^4} dq \neq \sum_{n=0}^{\infty} \frac{(-\lambda)^n}{n!} \int e^{-q^2} q^{4n} dq ,$$

since the series on the right side diverges. It is therefore necessary to truncate perturbation theory, for which we use step III below.

III. Wick Bound. For $:\Phi_\varkappa^4:$, we have

$$(4.2.3) \qquad e^{-V(\Lambda,\varkappa)} \le \begin{cases} e^{O(\log^2 \varkappa)|\Lambda|} & d=2 \\ e^{O(\varkappa^2)|\Lambda|} & d=3 \end{cases}$$

This bound follows by integrating

$$:\Phi_\varkappa^4: \;=\; (\Phi_\varkappa - 3c_\varkappa)^2 - 6c_\varkappa^2 \;\ge\; -6c_\varkappa^2$$

over the space time volume Λ. Here

$$c_\varkappa = \int \Phi_\varkappa(x)^2 \, d\Phi = C_\varkappa(x,x) \le \begin{cases} O(\log \varkappa) & d=2 \\ O(\varkappa) & d=3 \end{cases}$$

This Wick bound is used to raise the lower momentum cutoff ρ in the exponent $V(\Lambda,\varkappa,\rho)$. Our expansions terminate if $\varkappa = \rho$.

IV. Integration by Parts.

$$(4.2.4) \qquad \int \Phi(x)\, F(\Phi)\, d\Phi_C = \int dy \int C(x,y) \frac{\partial F}{\partial \Phi(y)} \, d\Phi_C \; .$$

We use this integration by parts formula to exhibit the cancellation of the divergent renormalization counterterms in $V_1 - V_0$ of (4.2.2). In [Gl Ja IV, 8] we use other forms of (4.2.4), called there the pull through and contraction formulas.

It is easy to establish (4.2.4) by studying finite dimensional approximations to the function space integral. We choose a Gaussian measure $d\Phi_N$ converging to $d\Phi_C$,

$$d\Phi_N = N \exp\left(-\frac{1}{2} \sum_{i,j} q_i C_{ij}^{-1} q_j\right)\left(\prod_k dq_k\right)$$

$$\equiv \nu_N dq_N \; ,$$

where C_{ij} is the covariance matrix and N is a normalization constant. Ordinary integration by parts then yields

$$\int \sum_j C_{ij}^{-1} q_j F(q)\, d\Phi_N = -\int F(q) \frac{\partial \nu_N}{\partial q_i}\, dq_N = \int \frac{\partial F(q)}{\partial q_i}\, d\Phi_N \ .$$

Inverting C,

$$\int q_j F(q)\, d\Phi_N = \int \sum_i C_{ji} \frac{\partial F}{\partial q_i}\, d\Phi_N \quad ,$$

which converges to (4.2.4) as $d\Phi_N \to d\Phi_C$.

For Wick ordered monomials, we obtain similarly

$$\int :\Phi(x_1)\ldots\Phi(x_n):F(\Phi)\, d\Phi_C = \int :\Phi(x_2)\ldots\Phi(x_n): \int dx_1'\, C(x_1,x_1') \frac{\partial F}{\partial \Phi(x_1')}\, d\Phi_C$$

As an example, we integrate by parts one $\Phi(x)$ factor in a simple expression,

$$\int :\Phi^4(x): e^{-\int :\Phi^4: dz}\, d\Phi_C = -4\int dxdy \int :\Phi^3(x): C(x-y) :\Phi^3(y): e^{-\int :\Phi^4: dz}\, d\Phi_C \ .$$

Further integration by parts yields

$$\int :\Phi^4(x): e^{-\int :\Phi^4: dz}\, d\Phi_C = 4! \int dx\, dy\, C(x-y)^4 \int e^{-\int :\Phi^4: dz}\, d\Phi_C \tag{4.2.5}$$

$$+ \text{ other terms} \ .$$

4.3. Synthesis of the Elementary Steps

We have two basic aims in combining the elementary expansion steps. First, we desire convergent expansions in a given space-time or phase space volume. Second, we desire polynomial decoupling of different localization regions. In Part II we present the vacuum cluster expansion for $P(\varphi)_2$ models in full detail. In that case, momentum localization is unnecessary (no ultraviolet divergences occur) and our localization regions are unions of unit lattice squares in space time. With no cutoff, distant regions decouple exponentially, and the decay rate determines the physical mass. In this section we present the basic ideas of phase space localization which we used to deal with the ultraviolet divergent φ_3^4 model, and which yield the results summarized in Section 4.1. For smooth cutoffs in momentum space, we obtain polynomial decoupling.

For simplicity we discuss the partition function Z. We fix the volume Λ and investigate how Z depends on the ultraviolet cutoff $\varkappa$. In order to truncate the perturbation expansions, we introduce a lower cutoff ρ into the action V. To ensure that $V(\varkappa, \rho)$ is bounded from below, we introduce the momentum cutoffs in a symmetric fashion: each momentum component in $V(\varkappa, \rho)$ lies in the interval $[\rho, \varkappa]$.

We perform our expansions on integrals of the form

$$\int R(\Phi)\, e^{-V(\varkappa, \rho)}\, d\Phi \quad , \tag{4.3.1}$$

where R is a polynomial function of Φ. Each expansion step replaces (4.3.1) by a sum of similar terms. We use a high momentum (perturbation) expansion step to lower $\varkappa$, and we use a low momentum (truncation) expansion step to raise ρ. At the start of the expansion, $\rho = 0$ and $\varkappa = \varkappa_0$. The expansion terminates when $\rho = \varkappa$, i.e., $V(\varkappa, \rho) = 0$, and (4.3.1) is reduced to a sum of Gaussian integrals $\int R\, d\Phi$. We estimate this sum uniformly in $\varkappa_0$.

The rules for alternating the expansion steps are somewhat complicated. The main idea is to obtain a small contribution from each high energy vertex in R, by performing explicit renormalization cancellations. We avoid the $(n!)^r$ number of terms which would arise from iterating (4.2.2), by truncating the perturbation expansion.

The high momentum expansion. We use Step II to lower κ, taking $V_1 = V(\bar{\kappa}, \rho)$, $V_0 = V(\underline{\kappa}, \rho)$. The first term in (4.2.2) has the desired form. The second term has the same upper cutoff and a new vertex $\delta V = dV_\alpha / d\alpha$ in R. Since δV has a lower momentum cutoff at $\underline{\kappa}$, we desire that δV contributes a convergence factor $\underline{\kappa}^{-\epsilon}$ to our final estimates. We obtain the proof of this fact only after performing the renormalization cancellation of the divergent counterterms δV_C in δV. Using Step IV, we integrate by parts δV_I, namely the ϕ^4 part of δV. We also integrate any new V_I produced in R as a result of differentiating the exponent. We thus obtain in closed form (and we cancel) the ultraviolet divergent part of δV. For instance, in (4.2.5), we displayed the second order vacuum energy contribution. The third order vacuum and the mass counterterms occur among the "other terms" in (4.2.5). The vacuum energy contributions cancel exactly with the corresponding counterterm in δV_C. The mass renormalization diagram, after cancellation, leaves a remainder $O(\underline{\kappa}^{-\epsilon})$. The remaining "other terms" from this procedure are convergent and so contribute $\underline{\kappa}^{-\epsilon}$ to the final estimate.

In this manner, Steps II and IV combine to yield one order (i.e., one δV) in a renormalized perturbation expansion. Because of the large number of terms, it is necessary to truncate this expansion after introducing $\bar{\kappa}^{\delta}$ vertices δV.

The low momentum expansion. We truncate the perturbation series by raising the lower cutoff in the exponent. We use Step III to raise ρ from $\underline{\rho}$ to $\bar{\rho}$. The Wick bound is an L_∞ estimate on path space, so we expect to apply it in space-time cubes Δ on which (4.2.3) remains bounded, i.e., cubes for which

$$\bar{\rho}^2 |\Delta| \leq O(1) \quad .$$

This restriction means that the localization length $L = |\Delta|^{1/3}$ satisfies

$$L \le O(\overline{\rho}^{-2/3}) \quad , \tag{4.3.2}$$

and defines our phase space localization. On the other hand, the uncertainty principle requires that

$$O(\underline{\rho}^{-1}) \le L \tag{4.3.3}$$

for the localization to be proper, i.e., for the spreading $O(\underline{\rho}^{-1})$ of the wave packet (due to momentum localization) to be less than L. We note that (4.3.2) - (4.3.3) are compatible. This compatibility is actually another aspect of super-renormalizability. For the φ_4^4 model, (4.3.2) would be replaced by $L \le O(\overline{\rho}^{-1})$, for which our estimates are borderline.

Our analysis has shown that we must treat separately cubes Δ belonging to a space-time cover $\mathcal{D}$. (Also some Δ's tend to zero as $\kappa_0 \to \infty$.) Thus we actually deal with upper and lower cutoff functions $\kappa(\Delta)$, $\rho(\Delta)$ which are functions of Δ.

Furthermore, we remark that the Wick bound (4.2.3) deals only with the pure low momentum part of $\delta V = V(\kappa, \underline{\rho}) - V(\kappa, \overline{\rho})$. This low momentum part equals $V(\overline{\rho}, \underline{\rho})$, i.e., all momenta are less than $\overline{\rho}$. The cross terms in δV must also be removed in order to raise the lower cutoff in the exponent. We dominate the cross terms by the low momentum terms and by $V(\kappa, \overline{\rho})$, the new exponent. This procedure has some complications, but poses no essential difficulty, see [Gl Ja 8]. In φ_4^4, however, it is such cross terms which yield the charge renormalization divergences, our biggest challenge.

Independence of Phase Cells. Finally we remark that the relevant distance parameter that we must use with phase space localization is the scaled distance d,

$$d = \text{Euclidean distance} \times \text{lower momentum cutoff}.$$

For smooth momentum cutoffs, scaling standard estimates gives $O(d^{-n})$ decay of correlations between different phase cells with proper localization. Any such decay for $n > 3$ is sufficient to control distance factors in sums over phase cells (whose diameter goes to zero as $\kappa_0 \to \infty$). We remark that in the limiting theory without ultraviolet cutoff, we expect to recover exponential decoupling and a mass gap.

REFERENCES

[AHK 1] S. Albeverio and R. Hoegh-Krohn, Uniqueness of the physical vacuum and the Wightman functions in the infinite volume limit for some non-polynomial interactions, Commun. Math. Phys. 30, 171-200 (1973).

[AHK 2] ______, The scattering matrix for some non-polynomial interactions, I and II, Helv. Phys. Acta, to appear.

[AHK 3] ______, The Wightman axioms and the mass gap for strong interactions of exponential type in two dimensional space-time, Oslo reprint.

[AHK 4] ______, Asymptotic series for the scattering operator and asymptotic unitarity of the space cut-off interactions, Oslo preprint.

[Ba] R. Baumel, Private communication, see also [GuRoSi 3].

[BoGr] A. Bortz and R. Griffiths, Phase transitions in anisotropic classical Heisenberg ferromagnets, Commum. Math. Phys. 26, 102-108 (1972).

[Br 1] O. Bratelli, Conservation of estimates in quantum field theory, Commun. Math. Phys., to appear.

[Br 2] ______, Local norm convergence of states on the zero time boson fields, Courant preprint.

[Ca] J. Cannon, Continuous sample paths in quantum field theory, Rockefeller preprint.

[CaJa] J. Cannon and A. Jaffe, Lorentz covariance of the $(\varphi^4)_2$ quantum field theory, Commun. Math. Phys. 17, 261-321 (1970).

[Co 1] S. Coleman, Dilitations, Lectures given at the 1971 International Summer School of Physics "Ettore Majorana," to appear.

[Co 2] ______, There are no Goldstone bosons in two dimensions, Commun. Math. Phys. 31, 259-264 (1973).

[CoWe] S. Coleman and E. Weinberg, Radiative corrections as the origin of spontaneous symmetry breaking, Phys. Rev. D 7, 1888-1910, (1973).

[ConTh] F. Constantinescu and W. Thalheimer, Euclidean Green's functions for Jaffe fields, Frankfurt preprint.

[De Ma] C. De Dominicis and P. Martin, Stationary entropy principle and renormalization in normal and superfluid systems, I, II. Jour. Math. Phys. 5, 14-30, 31-59 (1964).

[Di 1] J. Dimock, Estimates, renormalized currents and field equations for the Yukawa_2 field theory, Ann. Phys. 72, 177-242 (1972).

[Di 2] ______, Spectrum of local Hamiltonians in the Yukawa_2 field theory, J. Math. Phys. 13, 477-481 (1972).

[Di 3] ______, Perturbation series asymptotic to Schwinger functions in $P(\varphi)_2$, Courant preprint.

[DiGl] J. Dimock and J. Glimm, Measures on the Schwartz distribution space and application to $P(\varphi)_2$ field theories, Courant preprint.

[Dob 1] R. L. Dobrushin, Gibbsian random fields for lattice systems with pairwise interactions, Funct. Anal. and Appl., 2, 292-301 (1968).

[Dob 2] ______, The problem of uniqueness of a Gibbsian random field and the problem of phase transitions, Funct. Anal. and Appl., 2, 302-312 (1968).

[Dob 3] ______, The description of the random field by its conditional distributions and its regularity conditions, Theor. Prob. and Appl., 13, 201-229 (1968).

[DobMin] R. Dobrushin and R. Minlos, Construction of a one dimensional quantum field via a continuous Markov field, Moscow preprint.

[DHR] S. Doplicher, R. Haag and J. Roberts, Fields, observables and gauge transformations II, Commun. Math. Phys., 15, 173-200 (1969).

[Ec] J.-P. Eckmann, Representations of the CCR in the $(\varphi^4)_3$ model: independence of space cutoff, Commun. Math. Phys. 25, 1-61 (1972).

[EcOs] J.-P. Eckmann and K. Osterwalder, On the uniqueness of the Hamiltonian and of the representation of the CCR for the quartic boson interaction in three dimensions, Helv. Phys. Acta 44, 884-909 (1971).

[FaPo] L. Faddeev and V. Popov, Feynman diagrams for the Yang-Mills field, Phys. Letters 25B, 29-30 (1967).

[Fed] P. Federbrush, Positivity for some generalized Yukawa models in one space dimension, Michigan reprint.

[Fe 1] J. Feldman, A relativistic Feynman-Kac formula, Nuclear Physics B 52, 608-614 (1973).

[Fe 2] ______, The $\lambda\varphi_3^4$ field theory in a finite volume, Harvard preprint.

[FiWi] M. Fisher and K. Wilson, Critical exponents in 3.99 dimensions, Phys Rev. Lett. 28, 240-243 (1972).

[Fr 1a] J. Fröhlich, On the infrared problem in a model of scalar electrons and massless, scalar bosons, Annales de l'Institut Henri Poincare, to appear.

[Fr 1b] ______, Existence of dressed one electron states in a class of persistent models, Fortschritte der Physik, to appear.

[Fr 2] ______, Schwinger functions and their generating functionals, Harvard preprint.

[FoKaGi] C. Fortuin, P. Kastelyn and J. Ginibre, Correlation inequalities on some partially ordered sets, Commun. Math. Phys. 22, 89-103 (1971).

[Gi] J. Ginibre, On some recent work of Dobrushin, in: Systemes a un nombre infini de degrés de liberté, ed. by L. Michel and D. Ruelle, CNRS, Paris, 1970.

[Gl 1] J. Glimm, Boson fields with nonlinear self-interaction in two dimensions, Commun. Math. Phys. 8, 12-25 (1968).

[Gl 2] ______, Yukawa coupling of quantum fields in two dimensions, I, Commun. Math. Phys. 5, 343-386 (1967).

[Gl 3] ______, Yukawa coupling of quantum fields in two dimensions, II, Commun. Math. Phys. 6, 120-127 (1967).

[Gl 4] ______, Boson fields with the $:\varphi^4:$ interaction in three dimensions, Commun. Math. Phys. 10, 1-47 (1968).

[GlJa I] J. Glimm and A. Jaffe, A $\lambda(\varphi^4)_2$ quantum field theory without cutoffs I, Phys. Rev., 176, 1945-1951 (1968).

[GlJa II] ______, The $\lambda(\varphi^4)_2$ quantum field theory without cutoffs II, The field operators and the approximate vacuum, Ann. Math. 91, 362-401 (1970).

[GlJa III] ______, The $\lambda(\varphi^4)_2$ quantum field theory without cutoffs III, The physical vacuum, Acta Math. 125, 203-261 (1970).

[GlJa IV] ______, The $\lambda(\varphi^4)_2$ quantum field theory without cutoffs IV, Perturbations of the Hamiltonian, J. Math. Phys. 13, 1558-1584 (1972).

[GlJa 5] ______, The energy momentum spectrum and vacuum expectation values in quantum field theory II, Commun. Math. Phys. 22, 1-22 (1971).

[GlJa 6] ______, What is renormalization? in Symposium on Partial Differential Equations, Berkeley 1971, American Mathematical Society, Providence, 1973.

[GlJa 7] ______, Positivity and self-adjointness of the $P(\varphi)_2$ Hamiltonian, Commun. Math. Phys. 22, 253-258 (1971).

[GlJa 8] ______, Positivity of the φ_3^4 Hamiltonian, Fortschritte der Physik 21, 327-376 (1973).

[GlJa 9] ______, Self-adjointness of the Yukawa_2 Hamiltonian, Ann. of Phys. 60, 321-383 (1970).

[GlJa 10] ______, The Yukawa_2 quantum field theory without cutoffs, J. Funct. Analysis 7, 323-357 (1971).

[GlJa 11] ______, Quantum field models, in: Statistical mechanics and quantum field theory, ed. by C. de Witt and R. Stora, Gordon and Breach, New York, 1971.

[GlJa 12] ______, Boson quantum field models, in: Mathematics of contemporary physics, ed. by R. Streater, Academic Press, New York, 1972.

[GlJa 13] ______, Entropy principle for vertex functions in quantum field models.

[GlJaSp 1] J. Glimm, A. Jaffe and T. Spencer, The Wightman axioms and particle structure in the $P(\varphi)_2$ quantum field model, Ann. Math., to appear.

[GlJaSp 2] ______, The particle structure of the weakly coupled $P(\varphi)_2$ models and other applications of high temperature expansions, Part II: The cluster expansion, these Erice lectures.

[Go] J. Goldstone, Field theories with "superconductor" solutions, Nuovo Cimento 19, 154-164 (1961).

[GoSaWe] J. Goldstone, A. Salam and S. Weinberg, Broken symmetries, Phys. Rev. 127, 965-970 (1962).

[Gr 1] R. Griffiths, Phase transitions, in: Statistical mechanics and quantum field theory, Les Houches 1970, ed. by C. De Witt and R. Stora, Gordon and Breach, New York, 1971.

[Gr 2] ______, Rigorous results for Ising ferromagnets for arbitrary spin, J. Math. Phys., 10, 1559-1565 (1969).

[GrHuSh] R. Griffiths, C. Hurst and S. Sherman, Concavity of magnetization of an Ising ferromagnet in a positive external field, J. Math. Phys. 11, 790-795 (1970).

[GrSi] R. Griffiths and B. Simon, The $(\varphi^4)_2$ field theory as a classical Ising model, Commun. Math. Phys. to appear.

[Gro 1] L. Gross, The relativistic polar without cutoffs, Commun. Math. Phys. 31, 25-74 (1973).

[Gro 2] ______, Existence and uniqueness of physical ground states, J. Funct. Anal. 10 52-109 (1972).

[Gro 3] ______, Logarithmic Sobolev inequalities, Cornell preprint.

[Gro 4] ______, Analytic vectors for representations of the canonical commutation relations and non-degeneracy of ground states, Cornell preprint.

[Gu 1] F. Guerra, Uniqueness of the vacuum energy density and Van Hove phenomenon in the infinite volume limit for two dimensional self-coupled Bose fields, Phys. Rev. Letts. 28, 1213 (1972).

[GuRu] F. Guerra and P. Ruggiero, A new interpretation of the Euclidean-Markov field in the framework of physical Minkowski space-time, Salerno preprint.

[GuRoSi 1] F. Guerra, L. Rosen and B. Simon, Nelson's symmetry and the infinite volume behaviour of the vacuum in $P(\varphi)_2$, Commun. Math. Phys. 27, 10-22 (1972).

[GuRoSi 2] ______, The vacuum energy for $P(\varphi)_2$: infinite volume limit and coupling constant dependance, Commun. Math. Phys. 29, 233-247 (1973).

[GuRoSi 3] ______, The $P(\varphi)_2$ Euclidean quantum field theory as classical statistical mechanics, Ann. Math., to appear.

[He 1] K. Hepp, Théorie de la renormalisation, Springer-Verlag, Heidelberg 1969 .

[He 2] ______, Renormalization theory, in Statistical mechanics and quantum field theory, ed. by C. De Witt and R. Stora, Gordon and Breach, New York, 1971.

[He 3] ______, Erice lectures, 1973.

[HKSi] R. Hoegh-Krohn and B. Simon, Hypercontractive semi-groups and two dimensional self-coupled Bose fields, J. Funct. Anal. 9, 121-180 (1972).

[Ja 1] A. Jaffe, The dynamics of a cutoff $\lambda\varphi^4$ field theory, Princeton University Thesis.

[Ja 2] ______, Existence theorems for a cutoff $\lambda\varphi^4$ field theory, in: Mathematical theory of elementary particles, ed. by R. Goodman and I.Segal, M.I.T. Press, 1966.

[JaMcB] A. Jaffe and O. McBryan, What constructive quantum field theory has to say about currents, in: Proceedings of a conference on currents, Princeton,1971.

[JoL] G. Jona-Lasinio, Relativistic field theories with symmetry-breaking solutions, Nuovo Cimento Letters 34, 1790-1795 (1964).

[KaRoSw] D. Kastler, D. Robinson and A. Swieca, Conserved currents and associated symmetries; Goldstone's theorem. Commun. Math. Phys. 2, 108-120 (1966).

[KoWi] J. Kogut and K. Wilson, The renormalization group and the ϵ expansion, Phys. Reports, to appear.

[Leb 1] J. Lebowitz, Bounds on the correlations and analyticity properties of ferromagnetic Ising spin systems, Commun. Math. Phys. 28, 313-321 (1972).

[Leb 2] ______, GHS and other inequalities, Yeshiva preprint.

[Leb Pe] J. Lebowitz and O. Penrose, Decay of correlations, Yeshiva preprint.

[ML] A. Martin-Löf, Mixing properties, differentiability of the free energy and the central limit theorem for a pure phase in the Ising model at low temperature, Commun. Math. Phys. 32, 75-92 (1973).

[McB 1] O. McBryan, The vector currents in the Yukawa_2 quantum field theory with SU_3 symmetry, Harvard thesis.

[McB 2] ______, Generators for the Lorentz group in the $P(\varphi)_2$ theory, Toronto preprint.

[MinSin 1] R.A. Minlos and Y.G. Sinai, The phenomenon of "phase separation" at low temperatures in some lattice models of a gas I, Math Sbornik, Tom 73 (115), 335-395 (1967).

[MinSin 2] ______, The phenomenon of "phase separation" at low temperatures in some lattice models of a gas II, Trans. Moscow Math. Soc., 19, 121-196 (1968).

[Nel 1] E. Nelson, A quartic interaction in two dimensions, in: Mathematical theory of elementary particles, ed. by R. Goodman and I. Segal, M.I.T. Press, 1966.

[Nel 2] ______, Quantum fields and Markoff fields, in: Proceedings of Summer Institute of Partial Differential Equations, Berkeley 1971, Amer. Math. Soc., Providence, 1973.

[Nel 3] ______, Construction of quantum fields from Markoff fields, J. Funct. Anal. 12, 97-112 (1973).

[Nel 4] ______, The free Markoff field, J. Funct. Anal. 12, 211-227 (1973).

[Nel 5] ______, These Erice notes.

[New 1] C. Newman, The construction of stationary two-dimensional Markoff fields with applications to quantum field theory, J. Funct. Anal., to appear.

[New 2] ______, Zeroes of the partition function for generalized Ising systems, Courant preprint.

[OsSch 1] K. Osterwalder and R. Schrader, On the uniqueness of the energy density in the infinite volume limit for quantum field models, Helv. Phys. Acta, 45, 746-754 (1972).

[OsSch 2] ______, Feynman-Kac formula for Euclidean Fermi and Bose fields, Phys. Rev. Letts. 29, 1423-1425 (1972).

[OsSch 3] ______, Axioms for Euclidean Green's functions, Commun. Math. Phys. 31, 83-112 (1973).

[OsSch 4] ______, Euclidean Fermi fields and a Feynman-Kac formula for boson-fermion models, Helv. Phys. Acta, to appear.

[Os] K. Osterwalder, Duality for free Bose fields, Commun. Math. Phys. 29, 1-14 (1973).

[Oz] H. Ozkaynak, Euclidean fields for particles of arbitrary spin, Harvard preprint.

[Pa] Y. Park. Local Lorentz transformations of the $P(\varphi)_2$ model, Indiana preprint.

[Pe] R. Peierls, On Ising's model of ferromagnetism, Proc. Camb. Phil. Soc. 32, 477-481 (1936).

[Ro 1] L. Rosen, A $\lambda\varphi^{2n}$ field theory without cutoffs, Commun. Math. Phys. 16, 157-183 (1970).

[Ro 2] ______, The $(\varphi^{2n})_2$ quantum field theory: higher order estimates, Comm. Pure App. Math. 24, 417-457 (1971).

[Ro 3] ______, The $(\varphi^{2n})_2$ quantum field theory: Lorentz covariance, J. Math. Anal. and Appl. 38, 276-311 (1972).

[Ro 4] ______, Renormalization of the Hilbert space in the mass shift model, J. Math. Phys. 13, 918-927 (1972).

[Ru] D. Ruelle, Statistical mechanics, Benjamin, New York, 1969.

[SaSt] A. Salam and J. Strathdee, A renormalizable gauge model of lepton interactions, Nuovo Cimento, 11A, 397-435 (1972).

[Sch 1] R. Schrader, Yukawa quantum field theory in two space-time dimensions without cutoff, Annals of Physics 70, 412-457 (1972).

[Sch 2] ______, A remark on Yukawa plus boson self-interaction in two space-time dimensions, Commun. Math. Phys. 21, 164-170 (1971).

[Schw] S. Schweber, An introduction to relativistic quantum field theory, Row, Peterson and Co., New York, 1961.

[Se 1] I. Segal, Notes toward the construction of nonlinear relativistic quantum fields I; the Hamiltonian in two space-time dimensions as the generator of a C*-automorphism group, P.N.A.S., 57, 1178-1183 (1967).

[Se 2] ______, Construction of nonlinear local quantum processes I, Ann. Math. 92, 462-481 (1970).

[Si I] B. Simon, Correlation inequalities and the mass gap in $P(\varphi)_2$ I, Domination by the two point function, Commun. Math. Phys. 31, 127-136 (1973).

[Si II] ______, Correlation inequalities and the mass gap in $P(\varphi)_2$ II, Uniqueness of the vacuum for a class of strongly coupled theories, Toulon preprint.

[Sl] A. Sloan, The relativistic polaron without cutoffs in two space dimensions, Cornell thesis.

[Sp 1] T. Spencer, Perturbation of the $P(\varphi)_2$ quantum field Hamiltonian, J. Math. Phys. 14, 823-828 (1973).

[Sp 2] ______, The mass gap for the $P(\varphi)_2$ quantum field model with a strong external field.

[St] R. Streater, Connection between the spectrum condition and the Lorentz invariance of $P(\varphi)_2$, Commun. Math. Phys. 26, 109-120 (1972).

[St Wi] R. Streater and A. Wightman, PCT, spin and statistics and all that, Benjamin, New York, 1964.

[Sw] J. Swieca, Range of forces and broken symmetries in many-body systems, Commun. Math. Phys. 4, 1-7 (1967).

[Sy 1] K. Symanzik, On the many-particle structure of Green's functions in quantum field theory I, J. Math. Phys. 1, 249-273 (1960).

[Sy 2] ______, A modified model of Euclidean quantum field theory, N.Y.U. preprint, 1964.

[Sy 3] ______, Euclidean quantum field theory, in: Local quantum theory, proceedings of the International School of Physics "Enrico Fermi" Course 45, ed. by R. Jost, Academic Press, New York, 1969.

[Sy 4] ______, Renormalizable models with simple symmetry breaking I, Symmetry breaking by a source term, Commun. Math. Phys. 16, 48-80 (1970).

[Sy 5] ______, Small distance behaviour in field theory and power counting, Commun. Math. Phys. 18, 227-246 (1970).

[Sy 6] ______, Small distance behaviour, analysis and Wilson expansions, Commun. Math. Phys. 23, 49-86 (1971).

[Sy 7] ______, Small distance behaviour in field theory, in: Strong interaction physics, Springer tracts in modern physics, Vol. 57, Springer-Verlag, New York, 1971.

[Wi] A. Wightman, Lecture at the 1972 Coral Gables Conference, Gordon and Breach Sci. Pub., to appear.

* Supported in part by the National Science Foundation, Grant NSF-GP-24003.

† Supported in part by the National Science Foundation, Grant NSF-GP-40354X.

‡ Supported in part by the Alfred P. Sloan Foundation.

THE PARTICLE STRUCTURE OF THE WEAKLY COUPLED $P(\phi)_2$ MODEL AND OTHER APPLICATIONS OF HIGH TEMPERATURE EXPANSIONS

Part II: The cluster expansion

James Glimm*
Courant Institute, NYU
New York, N.Y.

Arthur Jaffe†
Harvard University
Cambridge, Mass.

Thomas Spencer‡
Courant Institute, NYU
New York, N.Y.

Contents

*Supported in part by the National Science Foundation, Grant NSF-GP-24003.

†Supported in part by the National Science Foundation, Grant NSF-GP- 40354X

‡Supported in part by the Alfred P. Sloan Foundation.

§1. INTRODUCTION

From the point of view of physics, there are essentially two problems in the study of quantum field models. The first problem is to find the phenomena of interest to physics (e.g. particles, bound states, resonances, broken symmetries, and asymptotic power series expansions) in the two and three dimensional quantum field models.[1] This problem includes obtaining the mathematical structure underlying these phenomena, and determining qualitatively which interactions and/or parameter values give rise to these phenomena and which do not. We believe this problem is feasible at the present time; indeed the initial progress is the subject of the present lectures. The second problem is the construction of nontrivial fields in four space time dimensions. Here it seems that some new ideas are needed to supplement those which have already been used in the construction of quantum fields.

These two problems are also interesting from the point of view of mathematics. In addition we mention a third problem, which is to place the solutions already constructed in a more general conceptual framework. In particular, elliptic equations with an infinite number of independent variables, Markov fields, non Gaussian integrals over functions of several variables, and C*-algebras arise naturally in the construction of quantum fields. The quantum field results may suggest directions for further development of these theories.

In Part II of these lectures, we give in detail a high temperature (cluster) expansion, in a form which yields an exponential cluster property for the Schwinger functions, as well as analyticity in the coupling constant λ, for λ in a bounded sector

$$0 < |\lambda| < \lambda_0 , \qquad -\pi/2 < \arg \lambda < \pi/2 ,$$

for the $P(\phi)_2$ quantum field theory. Other applications of this type of expansion to the study of quantum field models were developed in part I of these lectures; see also [4,5,6,14].

The expansion itself is related to the virial and cluster expansions [13, Chapter 4] of statistical mechanics. The analogy between statistical mechanics and $P(\Phi)$ quantum field theories has been known for some time. For a discussion of the relation of statistical mechanics to quantum field theory, see the lectures of Guerra, Rose

[1] For three space time dimensions, it is still necessary to complete the construction of the quantum fields. However enough progress has been made to eliminate any serious doubt as to their eventual existence.

and Simon, as well as Part I of these lectures.

These lectures develop a cluster expansion in the case of small coupling. We believe expansion techniques will apply for any parameter values which are sufficiently far from the critical point. Our belief is based partly on low temperature expansions developed by Peierls and by Minlos and Sinai [10] in statistical mechanics as well as work of Dobrushin and Minlos mentioned in part I. Furthermore Spencer [14] has extended the expansion presented here to the case of large external field.

Within their region of convergence expansion methods tend to give the most detailed information available. In fact these methods yield the main information known about the energy spectrum, above the energy level $E = 0$ of the vacuum. Correlation inequalities, the Lee-Yang theorem and other analyticity methods complement these techniques and often yield results outside the region of convergence. See [7,10,14] and other lectures of this series.

For the cluster expansion, the relevant formula from statistical mechanics is the following expansion for the density of the Gibbs ensemble:

$$(1.1) \quad \prod_{i<j} e^{-\beta V(x_i - x_j)} = \prod_{i<j} \left[1 + \left(e^{-\beta V(x_i - x_j)} - 1\right)\right]$$

$$= \sum_{\Gamma} \prod_{(i,j)\in\Gamma} \left(e^{-\beta V(x_i - x_j)} - 1\right) .$$

Here Γ is a set of distinct unordered pairs (i,j), i.e. Mayer graphs, and the sum extends over all such graphs. This formula expresses the interaction between the distinct i and j $(i \neq j)$ particles, $e^{-\beta V(x_i - x_j)}$, as a sum of a zero interaction term, 1, for which there is no coupling between the particles and a perturbation $e^{-\beta V(x_i - x_j)} - 1$ which is small at high temperatures $kT = \beta^{-1}$. Heuristically, the role of the Gibbs density in the $P(\phi)_2$ field theory is replaced by the measure

$$(1.2) \quad e^{-\int [\mathcal{H}_0(x) + \lambda P(\Phi(x))] dx} \prod_{x \in R^2} d\Phi(x) .$$

Here

(1.3) $$\mathcal{H}_o(x) = \frac{1}{2}\left(\nabla\Phi(x)^2 + m_o^2\Phi(x)^2\right)$$

and the formal expression

(1.4) $$e^{-\int\mathcal{H}_o(x)dx} \prod_{x\in R^2} d\Phi(x)$$

denotes the Gaussian measure on $\mathcal{S}'(R^2)$ with mean zero and covariance

(1.5) $$C_\emptyset = (-\Delta + m_o^2)^{-1} .$$

We also use the notation $d\Phi_C$ for the Gaussian measure with covariance C, so that (1.4) $= d\Phi_{C_\emptyset}$. We recognize (1.2) as the canonical Gibbs density for transverse vibrations of an elastic membrane subject to the nonlinear restoring force

$$F = -m_o^2\Phi(x) - \lambda P'\left(\Phi(x)\right) ,$$

after integration over the momentum variables $\Pi(x)$.

In (1.2), the coupling between distinct points comes entirely from the $\nabla\Phi$ term in $\mathcal{H}_o(x)$ and so $e^{-\int\nabla\Phi^2}$ in (1.2) plays the role of $e^{-\beta V}$ in (1.1). Our cluster expansion is constructed in the spirit of (1.1). In the completely decoupled theory, the Laplacian in

$$\int\mathcal{H}_o(x)dx = \frac{1}{2}\int\Phi(x)\left(-\Delta + m_o^2\right)\Phi(x)dx$$

is replaced by zero. The resulting ultralocal theory [9,12] seems to be very singular relative to the theory defined by (1.2). We reduce and control the singularity of the difference between the coupled and decoupled theories in two separate steps. The first step introduces a lattice structure into R^2 and into the expansion generalizing (1.1). We do not introduce the lattice into the measure (1.2) and so the resulting expansion is an expansion for the continuum infinite volume $P(\phi)_2$ theory, rather than for a lattice approximation to this theory. Let Γ denote a set of lattice lines, joining nearest neighbor lattice points in Z^2, let Δ_Γ be the Laplace operator with Dirichlet[2] boundary conditions on Γ, and let

(1.6) $$C_\Gamma = (-\Delta_\Gamma + m_o^2)^{-1} .$$

[2]From the point of view of this discussion, Neumann data might seem more appropriate, but we use Dirichlet data because it is technically easier.

Then $d\Phi_{C_\Gamma}$ plays the role of the decoupled measure, with decoupling along the curve Γ.[3] In summary, the lattice structure gives discrete variables in the sum and product

$$\sum_{\Gamma} \prod_{(i,j)\in\Gamma}$$

in (1.1), even when this formula is applied to the continuum $P(\Phi)_2$ model.

The second step regularizes the differences corresponding to

$$e^{-\beta V(x_i - x_j)} - 1$$

in (1.1). The difference between two Gaussian measures can be expressed as

$$d\Phi_{C_1} - d\Phi_{C_2} = \int_0^1 \frac{d}{ds} d\Phi_{C(s)}$$

where

$$C(s) = sC_1 + (1-s)C_2 .$$

This formula applies equally well to the non Gaussian measures

$$e^{-\lambda \int P(\Phi(x))dx} d\Phi_C$$

of interest to us. There is a simple formula for the evaluation of $\frac{d}{ds} d\Phi_{C(s)}$, namely [1]

$$\text{(1.7)} \qquad \frac{d}{ds} \int F d\Phi_{C(s)} = \frac{1}{2} \int \left(C'(s)\cdot\Delta_\Phi\right) F \, d\Phi_{C(s)}$$

where

$$\text{(1.8)} \qquad C'(s)\cdot\Delta_\Phi F = \frac{d}{ds} \int C(s,x,y) \frac{\delta^2}{\delta\Phi(x)\delta\Phi(y)} F \, dxdy$$

$$= \int [C_1(x,y) - C_2(x,y)] \frac{\delta^2}{\delta\Phi(x)\delta\Phi(y)} F \, dxdy$$

and $C(s,x,y)$ is the kernel of $C(s)$.

The proof of this formula contains an integration by parts in

---

[3]In comparison with (1.1), Γ should be replaced by $(Z^2)^* \sim \Gamma$, where $(Z^2)^*$ is the set of all nearest neighbor bonds in Z^2.

the function space integral. In the case of interest to us, $C_1 = C_{\Gamma_1}$ and $C_2 = C_{\Gamma_2}$, and the integration by parts regularizes[4] $\frac{d}{ds} d\Phi_{C(s)}$.

---

[4]Formally, $\frac{d}{ds} d\Phi_{C(s)} = J(\Phi) d\Phi_{C(s)}$, where the Jacobean J is

$$J(\Phi) = \frac{1}{2} \int :\Phi(x)[C(s)^{-1}C'(s)C(s)^{-1}](x,y)\Phi(y): dx\, dy$$

$$= -\frac{1}{2} \int :\Phi(x)[\frac{d}{ds} C^{-1}(s)](x,y)\Phi(y): dx\, dy \; .$$

Since C_{Γ_1} and C_{Γ_2} are mutually singular, for $\Gamma_1 \neq \Gamma_2$, some singularity in $J(\Phi)$ should be anticipated.

§2. THE MAIN RESULTS

For small values of λ/m_o^2, we prove an exponential cluster property on the Schwinger functions. We prove the cluster property in a finite volume with bounds independent of the volume. From these results, it follows easily that the Schwinger functions converge in the infinite volume limit, and that the infinite volume Schwinger functions satisfy an exponential cluster property. (They are also independent of the boundary conditions.) See [6]. Using the Osterwalder-Schrader reconstruction theorem, the infinite volume $P(\phi)_2$ field theory is constructed from the Schwinger functions, and in this theory the Wightman axioms are satisfied and the physical mass is strictly positive. We also show that the Schwinger functions are analytic in λ, for λ in the bounded sector

(2.1) $$0 < |\lambda| < \varepsilon \;, \qquad -\pi/2 < \arg \lambda < \pi/2 \;.$$

Let $\mathcal{C}$ be the set of convex combinations of Dirichlet covariance operators (1.6). The important properties of the measures $d\Phi_C$, $C \in \mathcal{C}$, are summarized in §9 (as preparation for the estimate of §10). In order to make elementary definitions, we mention merely that for Λ a bounded measurable set and for $C \in \mathcal{C}$,

(2.2) $$V(\Lambda) = \int_\Lambda :P(\Phi(x)):dx \in L_p(\mathcal{S}', d\Phi_C)$$

and

(2.3) $$e^{-\lambda V(\Lambda)} \in L_p(\mathcal{S}', d\Phi_C)$$

for all $p \in [1,\infty)$ and $\mathrm{Re}\,\lambda \geq 0$. (See Prop. 9.3, Th. 9.5.) In (2.2) the polynomial P is arbitrary, while in (2.3), P is semibounded. The Wick ordering in (2.2) is taken relative to the fixed measure $d\Phi_{C_\emptyset}$ of (1.4), even when $C \neq C_\emptyset$. To simplify the discussion, we take Λ to be a union of lattice squares, and in the limits $\Lambda \to \infty$, we further restrict Λ to be a square centered at the origin.[5] Let

(2.4) $$Z(\Lambda) = Z(\Lambda, C) = \int e^{-\lambda V(\Lambda)} d\Phi_C .$$

In Theorem 6.1 below we show that $Z(\Lambda) \neq 0$. Let

(2.5) $$dq_\Lambda = dq_{\Lambda,C} = Z(\Lambda,C)^{-1} e^{-\lambda V(\Lambda)} d\Phi_C .$$

[5] More general cutoffs, as required for the proof of Euclidean covariance of the limit $\Lambda \to \infty$, can be introduced by inserting a nonnegative L_∞ function h in the integral (2.2). Such an h does not affect the subsequent discussion.

The finite volume Schwinger functions are by definition the moments of the measure dq_Λ:

$$(2.6) \qquad S_\Lambda(x_1,\ldots,x_n) = \int \Phi(x_1)\ldots\Phi(x_n)\, dq_\Lambda \ .$$

S_Λ is defined as a tempered distribution in $\mathcal{S}'(R^{2n})$. To see this, take a test function of the form $w(x) = w_1(x_1)\ldots w_n(x_n)$ and integrate against (2.6). The n+1 factors

$$\Phi(w\ ) = \int \Phi(x_i)w_i(x_i)\, dx_i \qquad \text{and} \qquad e^{-\lambda V(\Lambda)}$$

in the integrand are bounded by Hölder's inequality, (2.3), and the following consequence of Prop. 9.3:

$$\|\Phi(w_i)\|_{L_p(\mathcal{S}')} \leq M_p\|w_i\|_{L_2(R^2)} \ .$$

In addition to the monomials in (2.4), it is useful to integrate products of Wick ordered polynomials, namely

$$(2.7) \qquad A = \int :\Phi(x_1)^{n_1}:\ldots:\Phi(x_j)^{n_j}:w(x_1,\ldots,x_j)dx \ .$$

We assume $w \in \mathcal{S}(R^{2j})$, although weaker bounds would suffice. We define suppt A to be the intersection of all closed subsets $C \subset R^2$ with

$$(2.8) \qquad \text{suppt } w \subset C \times \ldots \times C \qquad (j \text{ factors})$$

<u>Theorem 2.1</u>. Let λ belong to the closure of the half circle (2.1) and let ε/m_0^2 be sufficiently small. Let A and B be functions on $\mathcal{S}'$ of the form (2.7). Let d be the width of a strip in R^2 separating suppt A and suppt B. There is a constant $M = M_{A,B}$ and a positive constant m independent of A and B such that

$$\left| \int AB\, dq_{\Lambda,C_\emptyset} - \int A\, dq_{\Lambda,C_\emptyset} \int B\, dq_{\Lambda,C_\emptyset} \right| \leq M_{A,B}\, e^{-md} \ ,$$

uniformly in Λ, as $\Lambda \to \infty$. Furthermore, M is independent of translations in either A or B.

<u>Theorem 2.2</u>. Let λ belong to the closure of the half circle (2.1), and let ε/m_0^2 be sufficiently small. For A of the form (2.7), $\left|\int A\, dq_{\Lambda,C}\right|$ is bounded uniformly in λ and in Λ, as $\Lambda \to \infty$.

From analyticity of the finite volume Schwinger functions, from

Vitali's theorem and from the convergence, as $\Lambda \to \infty$, for λ real and small, we have

Corollary 2.3. The Schwinger functions are analytic in λ in (2.1) for ε/m_o^2 small.

§3. THE CLUSTER EXPANSION

The proof of the main theorems are based on a cluster expansion, which we now derive. Let $\mathcal{B}$ be a set of line segments in R^2. We are interested in two examples: Either $\mathcal{B} = (Z^2)^*$, the set of all lattice lines (bonds) joining nearest neighbor lattice sites in Z^2 or $\mathcal{B} = (Z^2)^* \sim \Gamma$ with Γ a finite subset of $(Z^2)^*$. We identify a subset $\Gamma \subset \mathcal{B}$ with the subset

$$\Gamma = \bigsqcup_{b \in \Gamma} b \subset R^2 .$$

The subsets $\Gamma \subset \mathcal{B}$ label terms in our expansion. The term labeled by Γ is decoupled across

$$\Gamma^c = \mathcal{B} \sim \Gamma \tag{3.1}$$

and formed by choosing a Gaussian measure with Dirichlet covariance on Γ^c. Thus the lines b in Γ^c are called Dirichlet lines. For the lines $b \in \Gamma$ there are differences between coupled and decoupled measures as in (1.1). These differences are expressed in terms of derivatives by the fundamental theorem of calculus. Thus the $b \in \Gamma$ are called derivative lines.

The covariance operators we consider are convex combinations of the operators C_Γ. For each $b \in \mathcal{B}$, we introduce a parameter $s_b \in [0,1]$ to measure the strength of the coupling across b. The parameter value $s_b = 0$ corresponds to zero Dirichlet data on b, hence to zero coupling across b, while $s_b = 1$ corresponds to full coupling across b. With

$$s = (s_b)_{b \in \mathcal{B}} , \tag{3.2}$$

we define the multiparameter family of covariance operators

$$C(s) = \sum_{\Gamma \subset \mathcal{B}} \prod_{b \in \Gamma} s_b \prod_{b \in \Gamma^c} (1 - s_b) C_{\Gamma^c} . \tag{3.3}$$

Since the coefficients in (3.3) are the terms in the expansion of

$$1 = \prod_{b \in \mathcal{B}} 1 = \prod_{b \in \mathcal{B}} [s_b + (1-s_b)] ,$$

(3.3) is a convex sum as asserted. The free covariance is now denoted

$$C_\emptyset = C(1,1,\ldots) = (-\Delta + m_o^2)^{-1}$$

and the completely decoupled covariance is

$$C_{\mathcal{B}} = C(0,0,\ldots) = (-\Delta_{\mathcal{B}} + m_o^2)^{-1} .$$

The Schwinger functions and the partition function are, in a natural way, functions of s, and we use the notation

$$(3.4)\qquad \begin{cases} Z(s)S_s(x) = Z(\Lambda,s)S_{\Lambda,s}(x) = \int \Pi\Phi(x_i)e^{-\lambda V(\Lambda)}d\Phi_s \\ Z(s) = Z(\Lambda,s) = \int e^{-\lambda V(\Lambda)}d\Phi_s \end{cases}$$

where

$$d\Phi_s \;=\; d\Phi_{C(s)}$$

The goal of the cluster expansion is to express coupled quantities ($s_b \equiv 1$) in terms of decoupled (and hence finite volume) quantities. The decoupled quantities have $s_b = 0$ for many $b \in \mathcal{B}$, and in order to formulate them, it is convenient to define

$$s(\Gamma) \;=\; \left(s(\Gamma)_b\right)_{b\in\mathcal{B}}$$

by the formula

$$s(\Gamma)_b \;=\; \begin{cases} s_b, & b \in \Gamma \\ 0, & b \notin \Gamma . \end{cases}$$

For Γ finite, $s(\Gamma)$ specifies Dirichlet boundary conditions at large distances (on Γ^c), while s can be thought of as giving general boundary conditions on Γ^c, and agreeing with $s(\Gamma)$ on Γ. Thus the following definition is a way of saying that F is independent of the boundary conditions at ∞.

<u>Definition 3.1.</u> A function F(s) is called regular at infinity if for each s,

$$(3.6)\qquad F(s) \;=\; \lim_{\{\Gamma\nearrow\mathcal{B}:\ \Gamma\ \text{finite}\}} F\left(s(\Gamma)\right) .$$

<u>Proposition 3.1.</u> The functions (3.4) are regular at infinity. Here S converges in $\mathcal{S}'$, and $\mathcal{B} \subseteq (Z^2)^*$.

Because Λ is fixed and bounded, the limit (3.6) is elementary. We omit the proof, since it is a consequence of techniques to be developed in Chapters 7,9.

The first step in the cluster expansion is to apply the fundamental theorem of calculus to the finite number of nonzero parameters in

$F(s(\Gamma))$. Define

(3.7) $$\partial^{\Gamma} = \prod_{b\in\Gamma} \frac{d}{ds_b}$$

and for two coordinate values s and σ define order coordinatewise, so that

$$\sigma \le s \Leftrightarrow \sigma_b \le s_b , \qquad \forall b \in \mathcal{B}.$$

Proposition 3.2. Let $F(s)$ be smooth and regular at infinity. Then

(3.8) $$F(s) = \sum_{\{\Gamma\subset \mathcal{B}:\Gamma \text{ is finite}\}} \int_{0\le\sigma\le s(\Gamma)} \partial^{\Gamma} F(\sigma(\Gamma))\, d\sigma .$$

Proof. Let $G(s)$ denote the right side of (3.8). We assert that $F(s(B)) = G(s(B))$, for B any finite subset of $\mathcal{B}$. Now $G(s(B))$ is just the sum in (3.8), restricted to sets $\Gamma \subset B$. Since $F(s)$ is regular at infinity, convergence of the sum in (3.8) follows from the assertion. Also (3.8) follows by limits in the assertion, as $B \nearrow \mathcal{B}$.

For a function $f(s_b)$ of a single variable, let

$$(\delta^b f)(s_b) = f(s_b) - f(0) = \int_0^{s_b} \partial^b f(\sigma_b)\, d\sigma_b$$

$$(E_0^b f)(s_b) = f(0) .$$

Then $I = E_0^b + \delta^b$ (fundamental theorem of calculus) and so

(3.9) $$I = \prod_{b\in B} (E_0^b + \delta^b) = \sum_{\Gamma\subset B} E_0^{B\sim\Gamma}\, \delta^{\Gamma}$$

where

$$\delta^{\Gamma} = \prod_{b\in\Gamma} \delta^b , \qquad E_0^{B\sim\Gamma} = \prod_{b\in B\sim\Gamma} E_0^b .$$

It is easy to see that (3.9) yields the desired identity $F(s(B)) = G(s(B))$.

The next step in the cluster expansion is a factorization and partial resummation of (3.8). Write

(3.10) $$R^2 \sim \Gamma^c = X_1 \cup X_2 \cup \ldots \cup X_r$$

so that each X_i is a union of connected components such that $X_i \cap X_j = \emptyset$ for $i \neq j$.

Definition 3.2. A function $F(\Lambda,s)$ decouples at $s=0$ provided

$$(3.11)\qquad F(\Lambda,s(\Gamma)) = \prod_{i=1}^{r} F(\Lambda\cap X_i,\ s(\Gamma\cap X_i))$$

whenever (3.10) holds.

Proposition 3.3. Given (3.10), the measures $d\Phi_{s(\Gamma)}$ and $e^{-\lambda V(\Lambda)}d\Phi_{s(\Gamma)}$ each factor into a product of r measures, and the ith factor measure is defined on $\mathcal{S}'(X_i)$. To avoid regularity questions, we assume $\mathcal{B}\subseteq (Z^2)^*$.

Sketch of Proof. Since $C(s(\Gamma))$ has zero Dirichlet data on Γ^c, $C(s(\Gamma))$ is the direct sum of the operators

$$C(s(\Gamma)) \restriction L_2(X_i) \quad .$$

The factorization of $d\Phi_{s(\Gamma)}$ follows from this fact, and can be seen from the standard formula for evaluation of a Gaussian integral of a polynomial (Theorem 9.1). Because $:P(\Phi(x)):$ is a local function, $\exp(-\lambda V(\Lambda))$ also factors, and so does $\exp(-\lambda V)d\Phi_{s(\Gamma)}$.

Corollary 3.4. The functions ZS and Z of (3.4) decouple at $s=0$.

The cluster expansion represents F as a sum of products of contributions from connected graphs. We start with some set X_o of interest (e.g. $X_o = \{x_1,\dots,x_n\}$ in (3.4)). The graphs which do not meet X_o are resummed, giving a single factor of F in an exterior region.

We now carry out this resummation in more detail. Substitute (3.11) in the expansion (3.8). This yields

$$(3.12)\qquad F(\Lambda,s) = \sum_{\Gamma}\prod_{i=1}^{r}\int_0^{s(\Gamma_i)} \partial^{\Gamma_i} F(\Lambda\cap X_i,\sigma(\Gamma_i))\,d\sigma$$

where $\Gamma_i = \Gamma\cap X_i$. If the X_i are connected, this is the sum over products of connected graphs. Now we choose X_1 in (3.10) to be the union of all components meeting X_o and let X_2 be the union of the remaining components. The resummation consists of holding X_1 and Γ_1 fixed, while summing over all choices of Γ_2. Explicitly,

$$F(\Lambda,s) = \sum_{X_1,\Gamma_1} \int_0^{s(\Gamma_1)} \partial^{\Gamma_1} F(\Lambda \cap X_1, \sigma(\Gamma_1))\, d\sigma$$

$$\times \sum_{\Gamma_2} \int_0^{s(\Gamma_2)} \partial^{\Gamma_2} F(\Lambda \cap X_2, \sigma(\Gamma_2))\, d\sigma \ .$$

The Γ_2 sum runs over all finite sets Γ_2 of bonds in $\mathcal{B} \sim X_1^-$. For this reason, the Γ_2 sum can be evaluated by (3.8) as $F(\Lambda \cap X_2, s(\mathcal{B} \sim X_1^-))$. Setting $X = X_1^-$ and writing Γ for Γ_1, the expansion has the form

$$(3.13) \qquad \begin{cases} F(\Lambda,s) = \sum_X K(X_o,X) F(\Lambda \sim X, s(\mathcal{B} \sim X)) \\ K(X_o,X) = \sum_\Gamma \int_0^{s(\Gamma)} \partial^\Gamma F(\Lambda \cap X, \sigma(\Gamma))\, d\sigma \ . \end{cases}$$

In these sums, X ranges over finite unions of closed lattice squares while Γ ranges over finite subsets of $\mathcal{B}$ such that

(i) Each component of $X \sim \Gamma^c$ meets X_o

(ii) $\Gamma \subset \text{Int } X$.

If no such Γ exists, for a given X, then $K(X_o,X) \equiv 0$.

Theorem 3.5. Let X_o be bounded and let F be smooth, regular at infinity and decouple at $s=0$. Then the cluster expansion (3.13) holds.

Example 1. Let $\mathcal{B} = (Z^2)^*$ be the set of all lattice lines and let $X_o = \{x_1, \ldots, x_n\}$. We substitute $ZS = F$. See (3.4). Observe that $s=1$

$$(3.14) \qquad F(\Lambda \sim X, s(\mathcal{B} \sim X)) = \int e^{-\lambda V(\Lambda \sim X)} d\Phi_{s(\mathcal{B} \sim X)}$$

$$= Z(\Lambda \sim X, s(\mathcal{B} \sim X))$$

$$= Z(\Lambda \sim X, C_{\partial X}) \equiv Z_{\partial X}(\Lambda \sim X)$$

since a change in the data within X does not affect the integral. Thus if we divide by Z, (3.13) yields

$$(3.15)\quad S_\Lambda(x) = \sum_{X,\Gamma} \int \partial^\Gamma \int \Pi_i\ \Phi(x_i)\ e^{-\lambda V(\Lambda\cap X)}\ d\Phi_{s(\Gamma)}\ ds(\Gamma)$$

$$\times\ Z_{\partial X}(\Lambda\sim X)/Z(\Lambda)\ .$$

If we write $X \sim \Gamma^c$ as a union of connected components, the integral in (3.15) can be factored, as in (3.12).

Example 2. Let $\Gamma_1 \subset (Z^2)^*$ and $\Gamma_2 = \Gamma_1 \sim b_1$ be finite sets of lattice lattice bonds, where b_1 is the first element of Γ_1 in some lexicographic order of $(Z^2)^*$. We use the cluster expansion (in all bonds $b \neq b_1$) to study the difference $Z_{\Gamma_1} - Z_{\Gamma_2}$ where

$$Z_\Gamma = \int e^{-\lambda V(\Lambda)}\ d\Phi_{C_\Gamma}$$

and so we take $\mathcal{B} = (Z^2)^* \sim b_1$, $X_0 = b_1$. Z is now a function of the pair (s, s_{b_1}). We define

$$F(\Lambda,s) = \begin{cases} Z(\Lambda,s(\Gamma^c),0) - Z(\Lambda,s(\Gamma^c),1)\ , & b_1 \subset \Lambda \\ Z(\Lambda,s(\Gamma^c),0) & ,\ b_1 \cap \Lambda = \emptyset \end{cases}$$

Ths definition is arranged so that

$$F(s=1) = Z_{\Gamma_1} - Z_{\Gamma_2}\ ,\quad \text{for}\ \ b \subset \Lambda\ .$$

Also F is smooth, regular at infinity and decouples at $s = 0$. (The last property depends on the exclusion of b_1 from $\mathcal{B}$.) We note that F is independent of the variables s_b for $b \in \Gamma_2$, and so $\partial^{b_j} F = 0$, for $b \in \Gamma_2$. Thus in all nonzero terms, Γ_2 consists of Dirichlet lines in the cluster expansion for F, and in (3.13), we may impose the further restriction on $\sum_\Gamma$:

(iii) $\Gamma \cap \Gamma_2 = \emptyset$.

From (3.13), we have

$$F(\Lambda,s=1) = Z_{\Gamma_1}(\Lambda) - Z_{\Gamma_2}(\Lambda)$$

$$= \sum_X K(b_1,\Gamma_1,X)F\big(\Lambda\sim X,\ s(\mathcal{B}\sim X)\big)$$

$$= \sum_X K(b_1,\Gamma_1,X)Z_{\Gamma_1 \cup X^*}(\Lambda\sim X)$$

Here X^* is the set of lattice lines in X and the last line is justified by the fact that the second factor above is independent of s_b, $b \in \text{Int } X$. We multiply and divide by

$$Z_{\partial\Delta}(\Delta)^{|\Lambda \cap X|}$$

where Δ is a single lattice square. Since

$$Z_{\partial\Delta}(\Delta)^{|\Lambda\cap X|}\, Z_{\Gamma_1 \cup X^*}(\Lambda\sim X) = Z_{\Gamma_1 \cup X^*}(\Lambda) ,$$

we obtain (with a new K)

$$(3.16)\quad Z_{\Gamma_1} = Z_{\Gamma_1\sim b_1} + \sum_X K(b_1,\Gamma_1,X)Z_{\Gamma_1 \cup X^*}(\Lambda)$$

and now

$$(3.17)\quad K(b_1,\Gamma_1,X) = Z_{\partial\Delta}(\Delta)^{-|\Lambda\cap X|}\int \sum_\Gamma \partial^{\Gamma \cup b_1} Z\big(\Lambda\cap X, s(\Gamma \cup b_1)\big)\, ds(\Gamma \cup b_1).$$

These equations have a structure intermediate between the Kirkwood-Salsburg and the Mayer-Montroll equations. They will be studied in §6 in order to obtain bounds on Z_Γ/Z, i.e. the second factor in (3.15).

§4. CLUSTERING AND ANALYTICITY: PROOF OF THE MAIN RESULTS

We prove the cluster property and analyticity -- the main results of these lectures -- in this section, using as hypothesis convergence of the cluster expansion, (3.15). Let $T(x,\Lambda,X,\Gamma)$ denote the X-Γ term in (3.15), so that

$$S_\Lambda(x) = \sum_{X,\Gamma} T(x,\Lambda,X,\Gamma) \ . \tag{4.1}$$

For each test function $w \in \mathcal{S}(R^{2n})$, the series

$$\int S_\Lambda(x)\ w(x)\ x = \sum_{X,\Gamma} \langle w,T\rangle \tag{4.2}$$

converges absolutely. The rate of convergence is governed by $|X|$, the area of X. With $K > 0$, we prove that

$$\sum_{\{X,\Gamma:|X|\geq D\}} |\langle w,T\rangle| \leq |w|\ e^{-K(D-n)} \ . \tag{4.3}$$

where $|w|$ is some (n-dependent) $\mathcal{S}$-norm on w.

<u>Theorem 4.1</u>. Let $K > 0$ be given. Let m_0 be large and ε be small (depending on K) and let λ belong to the closure of (2.1). There is an $\mathcal{S}$-norm w such that (4.3) holds uniformly in λ, m_0 , Λ and D, and $|w|$ is invariant under translations in any of its variables.

The proof allows integrands A of the form (2.7) to be substituted in (3.15). Thus Theorem 2.2 and Corollary 2.3 follow from the case $K = 1$, $D = 1$ of (4.3).

Theorem 2.1 also follows from the convergence of the cluster expansion. The proof for the two point function with an even interaction is conceptually easier, but still illustrates the main idea, so we consider that case first.

<u>Proof of Theorem 2.1</u>, for $n = 2$, P even, assuming Theorem 4.1. Because P, the interaction polynomial in (1.2), is even, $\Phi(x) \to -\Phi(x)$ is a symmetry of the theory. For Gaussian integrals defined by the factorizing measure $\exp(-\lambda V)\ d\Phi_{C(s(\Gamma))}$ (cf. Proposition 3.3), more is true:

$$\Phi(x) \to \sigma(x)\ \Phi(x) \tag{4.4}$$

is a symmetry, where

$$\sigma(x) = \pm 1; \quad \sigma(x) = \text{const. on each } X_i \ .$$

Because of the symmetry $\Phi \to -\Phi$, $S_\Lambda(x_1,\ldots,x_n) = 0$ for n odd. Similarly,

$$\int \Pi\ \Phi(x_i)\ e^{-\lambda V(\Lambda\cap X)}\ d\Phi_{C(s(\Gamma))} = 0$$

unless each of the connected components $X_1,\ldots,X_r$ contains an even number of x_i's. Since $\partial^\Gamma 0 = 0$, the same restriction $T(x,\Lambda,X,\Gamma) = 0$ applies unless each X_j has an even number of x_i's. Since each X_j has at least one x_i by (i) of §3, each X_j must have at least two x_i's. Thus for $n = 2$, there is only one X_j. In other words $X\sim\Gamma^c$ is connected. Let d be defined by Theorem 2.1 and let $w = w_1 \otimes w_2$. Because X is connected and suppt $w_1 \cap X \neq \emptyset$, we have $d \le |X| + 1$. Thus by (4.2-3),

$$\left|\int S_\Lambda(x_1,x_2)w_1(x_1)w_2(x_2)dx\right| \le |w|e^{-K(D-2)} \le M_w e^{-Kd}$$

Since the one-point functions $S_\Lambda(x_1)$ vanish for P even, this bound completes the proof of Theorem 2.1.

Proof of Theorem 2.1 (general case) assuming Th. 4.1. The idea, as before, is to reduce the cluster expansion to terms involving only one connected component $X_1 = X$. The terms with two or more connected components are to vanish, because of some symmetry. Since the general case (P not even) does not have such a symmetry, we follow Ginibre [2] and introduce a new theory containing an artificial symmetry.

To construct the new theory, let $d\Phi^*_{C^*}$ denote an isomorphic copy of the measure $d\Phi_C$, $(C^*\cong C)$ defined on an isomorphic copy $\mathcal{S}'^*$ of $\mathcal{S}'$. The new theory has free measure $d\Phi_C \times d\Phi^*_{C^*}$, covariance

$$C \otimes I + I \otimes C^* = C^\sim$$

and normalized physical measure

$$Z^{\sim -1} e^{-V(\Lambda)} e^{-V(\Lambda)^*} d\Phi_C \times d\Phi^*{}_{C^*} = dq^\sim$$

and a field

$$\Phi^\sim = \Phi \otimes I + I \otimes \Phi^* .$$

This theory is invariant (even) under the symmetry $\Phi \leftrightarrow \Phi^*$, which interchanges the two factors.

We now apply the cluster expansion to the expression

$$Z^\sim \int (A-A^*)(B-B^*)dq^\sim$$

The covariance operators which arise in this expansion have the form

$$C(s)^\sim = C(s) \otimes I + I \otimes C(s)^*$$

so that the symmetry $\Phi \leftrightarrow \Phi^*$ is preserved in the Gaussian measure of each term of the expansion. (However the expansion is modified by choosing $\mathcal{B} = (Z\ )^* \sim \Gamma$, where Γ is the set of lattice lines in two connected sets, one containing suppt A = suppt A-A*, and the other containing suppt B. Thus for each component X_i, $X_i \supset$ suppt A or $X_i \cap$ suppt A = $\emptyset$, and the same for suppt B. With this restriction, there are at most two components. Since the n in (4.3) is a bound on the number of components, as one can check, n = 2 in (4.3).) Now consider a term in (3.15) with components $X_1, X_2, \ldots$ satisfying

$$\text{suppt } A \subset X_i \; , \quad \text{suppt } B \subset X_j \; , \quad i \neq j \; . \qquad (4.5)$$

For this term, the symmetry $\Phi \leftrightarrow \Phi^*$ can be applied separately in each X_i. However A-A* is odd for the X_i symmetry, so terms satisfying (4.5) must vanish. The nonvanishing terms, which violate (4.5) contain a component X_1 = X with $d \leq 0|X|$, and so for $d \geq 4$,

$$\left|\int (A-A^*)(B-B^*)dq^{\sim}\right| \leq M_{AB} e^{-md} \; .$$

Expanding the integrand on the left yields four terms, and in each term the dq integral factors. We evaluate the result as

$$2\left|\int AB dq_{\Lambda,C} - \int A dq_{\Lambda,C} \int B dq_{\Lambda,C}\right| \; ,$$

and Theorem 2.1 follows.

§5. CONVERGENCE: THE MAIN IDEAS

There are two main ideas in part II of these lectures. The first is the formula (3.17) giving the cluster expansion for the Schwinger functions. The second is the estimates which lead to the convergence of this expansion, uniformly as $\Lambda \to \infty$. In this section we state these estimates as a series of three propositions, and using them, give the proof of Theorem 4.1. The simpler estimates are then proved, while the harder ones are postponed for later sections. The most difficult of these estimates is contained in Prop. 5.3. Since it is in some sense the core of the paper, we discuss the ideas involved in its proof at the end of this section.

The proof of convergence depends on estimates of the following types:

(a) Combinatoric estimates to count the number of terms in the expansion, and especially to count or bound the number of terms of some special type.

(b) Single particle estimates, to bound kernels of covariance operators, and their derivatives, $\partial^{\Gamma} C(s)$.

(c) Estimates on function space integrals. Typically estimates of type (a) and (b) will be components of an estimate of type (c).

Returning to specifics, (4.3) or (3.15) is a sum in which each term is a product of two factors: a ratio of partition functions times a function space integral. The first proposition is purely combinatoric - type (a)- and it counts the number of terms with $|X|$ = area X held fixed. The second and third propositions bound the two factors making up a term in (3.15). These latter two propositions are hybrids, since their proof uses estimates of types (a), (b) and (c).

Proposition 5.1. The number of terms in (3.15-3.16) with a fixed value of $|X|$ is bounded by $e^{K_1|X|} \equiv O(1)e^{19|X|}$.

Proposition 5.2. There is a constant K_2, independent of λ in the closure of (2.1), and of Λ and m_0, such that for ε small and m_0 large,

$$\left|Z_{\partial X}(\Lambda \sim X)/Z(\Lambda)\right| \leq e^{K_2|X|} .$$

Proposition 5.3. There is a constant K_3 and a norm $|w|$ on test functions such that for any $K > 0$, for any Λ and for λ in the closure of (2.1) with m_0 large (the m_0 bound depends on K)

$$| < \int \partial^{\Gamma} \int \prod_{i=1}^{n} \Phi(x_i)\, e^{-\lambda V(\Lambda)} d\Phi_{C(s(\Gamma))}\, ds(\Gamma), w> | \leq e^{-K|\Gamma|+K_3|\Lambda|} |w| .$$

(The m_0 bound depends on K, and $|w|$ is invariant under translation in any of the variables of w.)

Remark. Wick polynomials, as in (2.7), are also allowed in the integrand above.

Proof of Theorem 4.1. We replace Λ by $\Lambda \cap X$ in Prop. 5.3. For the X in (3.15), we have $X = \bigcup_{i=1}^{r} X_i^-$ with $r \leq n$ and X_i connected. Moreover

$$\Gamma \subset \bigcup_{i=1}^{r} \text{Int } X_i^-$$

and so "many" of the lattice lines in X_i^- belong to Γ. In fact since $X_i^- \sim \Gamma^c = X_i$ is connected,

(5.1) $\quad |X_i| - 1 \leq 2|\Gamma \cap \text{Int } X_i^-|$ and $|X| - n \leq 2|\Gamma|$.

Thus we replace the upper bound in Prop. 5.3 by

$$e^{-K(|X|-n)} \; |w| ,$$

with a new choice of K and $|w|$. Theorem 4.1 follows directly from this bound combined with Prop. 5.1 and 5.2.

Proof of Proposition 5.1. We consider (3.15). First we bound the number of ways of choosing the component X_i containing a particular x_j, $1 \leq j \leq n$. We identify each lattice square Δ with a lattice point located at its center, and we identify each lattice line $b \in \mathcal{B}$ lying between two squares Δ and Δ' with the line joining the center of Δ to the center of Δ'. Hence we must count the number of ways of drawing a connected graph with nearest neighbor bonds. We assert that each such graph may be constructed by starting at x_j and drawing an oriented path formed of unit line segments which traverses each segment at most twice. In fact if we regard the lattice sites as islands and the line segments as bridges, this simply follows from the solution to the Konigsberg bridge problem[6]. The number of connected paths of length ℓ formed from lattice line segments and starting at x_j is at most 4^{ℓ}. Since $\ell \leq 8|X_i|$, the number of choices for X_i is at most $O(1)2^{16|X_i|}$. Since the number of choices for Γ is at most $4^{|X|}$, the number of choices for pairs X, Γ is at most

$$O(1)2^{|X|}4^{|X|}\prod_i 2^{16|X_i|} = O(1)2^{19|X|} ,$$

since $2^{|X|}$ bounds the number of choices of $|X_i|$ with $|X|$ given. This completes the proof. The case (3:16) is similar.

Proposition 5.2 is proved in §6 using equations related to the

[6] We thank O. Bratteli for this observation. For a solution of the Konigsberg bridge problem see, D.W. Blackett, Elementary Topology, Academic Press, New York, 1967, page 159.

Kirkwood-Salsburg equations. We only remark here that the change in partition function involves both a change in the interaction region and a change in covariance ($\Lambda\sim X \to \Lambda$ and $C_{\partial X} \to C$).

Discussion of Prop. 5.3. Each derivative d/ds_b in ∂^{Γ} differentiates either the measure or the integrand. Derivatives of the measure are evaluated by (1.7), and produce s-dependent kernels $C'(s)$ in the integrand. Repeated differentiation yields

(a) A sum of terms coming from the iterated functional derivatives $\partial^2/\partial\Phi^2$ in (1.7).

(b) A sum over terms arising from possible repeated derivatives $\partial^{\Gamma_1}C(s)$ of each $C'(s)$ introduced by (1.7).

Each derivative d/ds_b also produces convergence in one of two ways. Differentiation of the measure introduces a kernel C', and

(c) C' and C are small in the sense that

$$(c_1)\quad \|C'(x,y)\|_{L_p} \leq O(m_0^{-\varepsilon})$$

$$(c_2)\quad 0 \leq C'(x,y) = \partial^b C(x,y) \leq e^{-m_0(\text{dist}(x,b)+\text{dist}(y,b))}$$

$$(c_3)\quad 0 \leq C(x,y) \leq e^{-m_0|x-y|}, \quad |x-y| \geq 1 .$$

Repeated differentiation of $C(s)$ produces the further convergence

(d) $\partial^{\Gamma}C$ is small in the sense that

$$(d_1)\quad \|\partial^{\Gamma}C(x,y)\|_{L_p} \leq O(m_0^{-\varepsilon|\Gamma|})$$

$$(d_2)\quad 0 \leq \partial^{\Gamma}C(x,y) \leq e^{-m_0 d}$$

where d is the length of the shortest path connecting x and y and passing through each $b\in\Gamma$. In fact from the Weiner integral representation for $(-\Delta+m_0^2)^{-1}$, $\partial^{\Gamma}C(s)$ is the integral over Wiener paths from x to y passing through each $b\in\Gamma$, see §7-8.

We use (c_2) to control (a). In fact by (c_2), only x and y near b contribute to the Laplacian $C'(s)\cdot\Delta_{\Phi}$ of (1.7). By the nature of the expansion, there is at most one Laplacian Δ_{Φ} per bond $b\in\mathcal{B}$, hence a bounded power, Δ_{Φ}^{n}, locally. Evaluation of Δ_{Φ}^{n} leads to $K(n) < \infty$ terms, and this evaluation, repeated over disjoint local regions leads to $e^{O(\text{Vol})}$ terms. We use (d_2) to control (b). The basic convergence, $m_0^{-\varepsilon|\Gamma|}$, comes from (d_1).

After performing these steps, there remains a function space integral of the form $\int Re^{-\lambda V}d\Phi$. Then

$$|\int Re^{-\lambda V(\Lambda)}\, d\Phi| \le \left(\int R^2\, d\Phi\right)^{1/2} \left(\int e^{-2\mathrm{Re}\lambda V(\Lambda)}\, d\Phi\right)^{1/2}$$

and each factor on the right is $e^{O(\mathrm{Vol})}$. Now (c_3) produces the decoupling of distant local regions in R^2.

With some practice, estimates such as Proposition 5.3 can be done by inspection. In Sections 7-9 we develop some general machinery, the goal of which is to make these estimates more routine.

§6. AN EQUATION OF KIRKWOOD-SALSBURG TYPE

Equations (3.16-3.17) can be rewritten as a Banach space equation

$$\rho = Z(\Lambda)1 + \mathcal{K}\rho \tag{6.1}$$

with a unique solution

$$\rho = (1-\mathcal{K})^{-1} Z(\Lambda)1$$

satisfying the bound

$$\|\rho\| \leq \|(I-\mathcal{K})^{-1}\| \, |Z(\Lambda)| \leq 4|Z(\Lambda)| \tag{6.2}$$

This bound is essentially Proposition 5.2, as we shall see.

Let $\mathcal{X}$ be the Banach space of functions f defined on finite subsets $\Gamma \subset (Z^2)^*$. For $f \in \mathcal{X}$ let $(f_n)_{n\geq 0}$ denote the restriction of f to n element subsets. The norm on $\mathcal{X}$ is defined to be

$$\|f\| = \sup_{\substack{n,\ \Gamma \\ |\Gamma|=n}} 2^{-n} |f_n(\Gamma)| .$$

We define $\rho_\Lambda \equiv (\rho_{\Lambda,n})_{n\geq 0}$ to be the function

$$\Gamma \to Z_\Gamma(\Lambda) .$$

Theorem 6.1. Let $|\lambda| \leq \varepsilon$, $\operatorname{Re}\lambda \geq 0$ where $\varepsilon > 0$ is small and let m_0 be sufficiently large. Then $Z(\Lambda) \neq 0$, and ρ_Λ defined above is the unique solution in $\mathcal{X}$ of equation (6.1). Moreover ρ_Λ satisfies the bounds (6.2).

By the dominated convergence theorem, $Z_{\partial\Delta}(\Delta) \to 1$ as $\varepsilon \to 0$. Thus for small ε,

$$\tfrac{1}{2} \leq |Z_{\partial\Delta}(\Delta)| \leq 2 .$$

This restriction fixes the ε in Theorem 6.1, and in fact is the only restriction on ε in these lectures.

Proof of Proposition 5.2, assuming Theorem 6.1.

$$\begin{aligned} |Z_{\partial X}(\Lambda \sim X)/Z(\Lambda)| &= |Z_{X^*}(\Lambda)/Z(\Lambda)| \; |Z_{\partial\Delta}(\Delta)|^{-|\Lambda\cap X|} \\ &\leq 2^{|X^*|} \|(1-\mathcal{K})^{-1}\| \; |Z_{\partial\Delta}(\Delta)|^{-|\Lambda\cap X|} \\ &\leq e^{K_2|X|} \end{aligned}$$

Proof of Theorem 6.1. Let $1 = (1,0,0,\ldots) \in \mathcal{X}$, and define $\mathcal{K}$ by the equations

$$(\mathcal{K} f)_n(\Gamma) = f_{n-1}(\Gamma \sim b_1) + \sum_{m=1}^{\infty} \sum_{X,|X|=n} K(b_1,\Gamma,X)\, f_{|X^* \cup \Gamma|}(X^* \cup \Gamma) \tag{6.4}$$

for $n \geq 2$ where the sum over X (as in (3.13) and (3.17)) ranges over connected unions of lattice squares such that $b_1 \in X$. For $n = 1$, we omit the first term on the right of (6.4) and for $n = 0$, $(\mathcal{K} f)_0 \equiv 0$.

We now show that $\mathcal{K}$ is a contraction, $\|\mathcal{K}\| \leq 3/4$, in order to obtain (6.2). It suffices to show that

$$\frac{1}{2} + \sup_{\Gamma \subset \mathcal{B}} 2^{-|\Gamma|} \sum_{m=1}^{\infty} \sum_{X,|X|=m} |K(b_1,\Gamma,x)|\, 2^{|\Gamma \cup X|} \leq \frac{3}{4} \tag{6.5}$$

The proof of (6.5) is similar to the proof of Theorem 4.1 given in §5. In particular it depends on Propositions 5.1 and 5.3. By Proposition 5.1 there are at most $e^{K_1|X|}$ terms for each fixed value of $|X|$. From (3.17) each term has at least one derivative ∂^{b_1}. Using this fact and (5.1) the bound from Propositions 5.1 and 5.3 is

$$|K(b_1,\Gamma,X)| \leq e^{-K(|X|+1)}\, e^{K_3|X|} .$$

For K sufficiently large (6.5) follows. We complete the proof by showing $Z(\Lambda) \neq 0$. By our choice of ε, $Z_{\Lambda^*} = |Z_{\partial\Lambda}(\Delta)|^{|\Lambda|} \neq 0$. Thus $\rho_\Lambda \neq 0$ and by (6.2), $Z(\Lambda) \neq 0$.

§7. COVARIANCE OPERATORS

The basic facts for the kernel $C_\emptyset(x,y)$ of $C_\emptyset = (-\Delta+m_0^2)^{-1}$ are

(a) $C_\emptyset(x,y)$ is a function of the difference $|x-y|$.

(b) $0 < C_\emptyset(x,y)$

(c) for $|x-y| \geq 1$, $C_\emptyset \leq e^{-m_0(1-\delta)|x-y|}$

(d) as $|x-y| \to 0$, $C_\emptyset \sim \log |x-y|$

(e) for $r > 2/(2-\delta)$, $\delta > 0$, $C_\emptyset \in \mathcal{B}^{loc}_{r,\delta}(R^4)$.

By definition, the space $\mathcal{B}^{loc}_{r,\delta}(R^4)$ is the space of distributions $\psi \in \mathcal{S}'$ which satisfy

$$\|\zeta\psi\|_{\mathcal{B}_{r,\delta}} = \|(1+k^2)^{\delta/2} (\zeta\psi)^{\sim}\|_{L_r} < \infty$$

whenever $\zeta \in C_0^\infty$. For r and r' dual indices and for $1 \leq r' \leq 2$, $\mathcal{B}^{loc}_{r,\delta}$ is a space of functions with fractional $L_{r'}$ derivatives, by the Hausdorff-Young inequality. See [8, Chapter 2] for a general theory of such spaces. (e) follows from the fact that as a function of x-y, $C_\emptyset$ has the Fourier transform $(k^2+m^2)^{-1}$.

Property (a), which is based on the translation invariance of the Laplacian, does not extend to the operators $C \in \mathcal{C}$. Recall that $\mathcal{C}$ is the set of all convex combinations of the Dirichlet covariance operators (1.6), or in other words, operators of the form C(s). Properties (b) and (c) are valid for all $C \in \mathcal{C}$, while (d) is valid as an upper bound for all $C \in \mathcal{C}$.

Proposition 7.1. The kernel $C(x,y)$ of $C \in \mathcal{C}$ satisfies

$$0 \leq C(x,y) \leq C_\emptyset(x,y) .$$

Proof. Let $dz^T_{x,y}$ be the conditional Wiener density on paths $z(\tau)$ starting at x at $\tau = 0$ and ending at y at $\tau = T$. Let J^T_b be the function

$$J^T_b(z) = \begin{cases} 0 & \text{if } z(\tau) \in b, \ 0 \leq \tau \leq T \\ 1 & \text{otherwise} \end{cases} \tag{7.1}$$

defined on Wiener paths. Then C_Γ and $C(s)$ have the representations

$$C_\Gamma(x,y) = \int_0^\infty e^{-m_0^2 T} \int \Pi_{b\in\Gamma} J^T_b(z)\, dz^T_{x,y} \,.dT \tag{7.2}$$

$$C(s)(x,y) = \int_0^\infty e^{-m_0^2 T} \int \Pi_{b\in\mathcal{B}}(s_b + (1-s_b)J^T_b)\, dz^T_{x,y}\, dT . \tag{7.3}$$

See [0, §5.3, 6.1 and 6.3].

The inequality of Proposition 7.1 comes from substituting

$$0 \le (s_b + (1-s_b)J_b^T) \le 1$$

in the formula above for $C(s)$.

The next estimate combines the exponential decay (c) with a local bound coming from (d). We label each lattice square $\Delta = \Delta_j \subset R^2$ by the lattice point $j \in Z^2$ in the lower right corner of Δ. Any lattice square (or any set $X \subset R^2$), will be identified with the multiplication on $L_2(R^2)$ by its characteristic function. Now let $j = (j_1, j_2) \in Z^4$ be a pair of lattice points. We define the localized covariance operator

$$C(j) = \Delta_{j_1} C \Delta_{j_2} \tag{7.4}$$

and the distance

$$d(j) = \text{Dist}\,(\Delta_{j_1}, \Delta_{j_2})\ , \tag{7.5}$$

which measures the nonlocality of $C(j)$.

Proposition 7.2. For $1 \le q < \infty$ and $\delta > 0$, there is a constant $K_4(q,\delta)$ independent of $m_0 \ge 1$ and $C \in \mathcal{C}$ such that

$$\| C(j) \|_{L_q(\Delta_{j_1} \times \Delta_{j_2})} \le K_4 m_0^{-2/q} \exp(-m_0(1-\delta) d(j))\ .$$

Proof. Because of Proposition 7.1, we may assume $C = C_\emptyset$. For $d(j) > 0$, necessarily $d(j) \ge 1$, and the required bound follows from property (c). The factor $m_0^{-2/q}$ in the proposition comes from choosing δ in (c) smaller than the δ in the proposition. Next consider the case $d(j) = 0$ (i.e. equal or adjacent squares). We have

$$\begin{aligned} \| C(x,y) \|_{L_q(\Delta_{j_1} \times \Delta_{j_2})} &\le K_3 \| C(\cdot,0) \|_{L_q(R^2)} \\ &\le K_4(q) m_0^{-2/q}\ . \end{aligned}$$

The last inequality follows from the fact that $(-\Delta+1)^{-1}(x,0) \in L_q$ for $q \in [1,\infty)$ and from the identity

$$(-\Delta+m_0^2)^{-1}(x,0) = (-\Delta+1)^{-1}(m_0 x, 0)\ .$$

Remark. The differentiated covariance $\partial^\Gamma C$ also satisfies the bound of Proposition 7.1, as we can see from (7.3) and the inequality

$$0 \le \frac{d}{ds_b}(s_b + (1-s_b)J_b) = 1-J_b \le 1 .$$

Hence $\partial^\Gamma C$ also satisfies the bound of Proposition 7.2.

Local regularity of the covariance $C \in \mathcal{C}$ is used to justify the definition and basic formal properties (e.g. integration by parts) of Gaussian integrals and it is used in the bound $|Z| \le e^{O|\Lambda|}$. We obtain fractional derivatives of $C \in \mathcal{C}$, with bounds slightly weaker than (e).

Proposition 7.3. $C \in \mathcal{B}^{loc}_{2,\delta}$ for $0 \le \delta < 1/2$, with bounds independent of C.

Proof. We start with the observation that

$$0 \le -\Delta \le -\Delta_\Gamma .$$

Consequently $C_\emptyset^{-1/2} C_\Gamma C_\emptyset^{-1/2}$ is a bounded operator, with norm at most one. By convex combinations, $\|B\| \le 1$ also, where

$$B = C_\emptyset^{-1/2}\, C\, C_\emptyset^{-1/2} .$$

Let

$$A = C_\emptyset^{-(1/4)+\varepsilon} \zeta C_\emptyset^{1/2},$$

for $\zeta \in C_0^\infty(R^2)$. We use the fact that $A\,A^*$ is Hilbert Schmidt to calculate the Hilbert Schmidt norm

$$(7.6)\quad \|(-\Delta+m_0^2)^{-\varepsilon+1/4}\, \zeta\, C\, \zeta(-\Delta+m_0^2)^{-\varepsilon+1/4}\|^2_{HS}$$

$$= \mathrm{Tr}\, A\, B\, A^*A\, B\, A^* \;=\; \mathrm{Tr}\, B\, A^*A\, B\, A^*A$$

$$\le \mathrm{Tr}\, (B\, A^*A)^* B\, A^*A = \mathrm{Tr}\, A^*A\, B^2 A^*A$$

$$\le \|B\|^2\, \mathrm{Tr}\, (A^*A)^2 \;\le\; \|A^*A\|^2_{HS} .$$

This inequality proves the proposition (and slightly more).

The reader may skip the next result, since it is not needed in what follows.

Proposition 7.4. For any $q \in (4/3,\infty)$, there is a $\delta = \delta(q) > 0$ with $C \in \mathcal{B}^{loc}_{q,\delta}$ and the bounds are independent of $C \in \mathcal{C}$.

Proof. With $k = (k_1,k_2) \in R^4$, define

$$f(k) = 1+k^2 , \quad g(k) = (1+k_1^2)(1+k_2^2) ,$$

$$h(k) = (\zeta\, C\, \zeta)^\sim(k) .$$

Then

$$g^{-\varepsilon-1} \in L_1 , \quad f/g \in L_\infty .$$

By Proposition 7.1 and the fact that $\zeta C_\emptyset \zeta \in L_1$,

$$h \in L_\infty$$

and by (7.6),

$$g^{-\varepsilon+1/4} h \in L_2 \quad \text{and} \quad h \in L_2 .$$

We apply these facts to the integral

$$\int |f^{\delta/2} h|^q = \int |g^{-\varepsilon+1/4} h| \, |\tfrac{f}{g}|^{\delta q/2} |g|^{(\delta q/2)+\varepsilon-1/4} |h|^{q-1} . \tag{7.7}$$

First take $\delta > 0$ and suppose that $h \in L_{q_1}$, $q_1 > q > 4/3$. The factors in (7.7) belong to L_2, $L_{4+\varepsilon}$, and $L_{q_1/q-1} \cap L_\infty$. For ε small and q near q_1, this implies $h \in L_q$ also. Since $h \in L_q$ for q in an interval, $h \in L_q$ for $q \in (4/3,\infty]$. Now with $h \in L_q$, $q > 4/3$, (7.7) shows that $f^{\delta/2} h \in L_q$ also, for small δ. This completes the proof.

<u>Proposition</u> 7.5. For $1 \le q < \infty$,

$$c(x) = \lim_{y\to x} C(x,y) - C_\emptyset(x,y) \in L_q^{loc} .$$

There is a constant $K_5(q)$, independent of $m_0 \ge 1$ and $C \in \mathcal{C}$ such that for any lattice square Δ,

$$\|c\|_{L_q(\Delta)} \le K_5 \, m_0^{-1/q} .$$

<u>Proof</u>. By scaling, we may take $m_0 = 1$, as in Proposition 7.2. Let $\Gamma = \mathcal{B}$ be the set of all lattice lines. For $x \notin \Gamma$,

$$0 \le - c(x) \le C_\emptyset(x,x) - C_\Gamma(x,x)$$
$$\le O(1 + |\log \operatorname{dist}(x,\Gamma)|) .$$

This inequality completes the proof. It is proved in [7] as follows. For $x \notin \Gamma$, $y \notin \Gamma$,

$$(-\Delta_y + m_0^2)[C_\emptyset(x,y) - C_\Gamma(x,y)] = 0 .$$

Hence by the maximum principle, and the fact that $C_\Gamma(x,y) = 0$ for $y \in \Gamma$,

$$0 \le C_\emptyset(x,y) - C_\Gamma(x,y) \le \sup_{y\in\Gamma} C_\emptyset(x,y) \le O(1+|\log \operatorname{dist}(x,\Gamma)|).$$

§8. DERIVATIVES OF COVARIANCE OPERATORS

For the differentiated covariance operator $\partial^\gamma C$, there should be a strong decay, $\sim e^{-m_0 d}$, where d is the length of the shortest path in R^2 joining x to y and passing through each lattice line segment $b \in \gamma$. This can be seen by inspection from the Wiener integral representation

$$(8.1)\quad (\partial^\gamma C(s))(x,y) = \int_0^\infty e^{-m_0^2 T} \int \Pi_{b\in\gamma}(1-J_b^T) \times \Pi_{b\in \mathcal{B}\sim\gamma}[s_b + (1-s_b)J_b^T]\, dz_{x,y}^T\, dT .$$

We need the improved bounds on $\partial^\gamma C$ for two reasons. The first is to localize x and y, with γ given. For this purpose

$$(8.2)\qquad d(j,\gamma) = \sup_{b\in\gamma} \{\mathrm{Dist}(\Delta_{j_1},b) + \mathrm{Dist}(\Delta_{j_2},b)\}$$

is sufficient, as a crude lower bound on d.

We now explain the second use of bounds on $\partial^\gamma C$. Let $\mathcal{P}(\Gamma)$ be the set of all partitions π of the set of lattice line segments Γ. In Proposition 5.3, we are called on to bound $\partial^\Gamma \int F\, d\Phi_s$, which by Leibnitz' rule and by (1.7) is just

$$(8.3)\qquad \partial^\Gamma \int F\, d\Phi_s = \sum_{\pi\in\mathcal{P}(\Gamma)} \int \left(\Pi_{\gamma\in\pi} \tfrac{1}{2}\, \partial^\gamma C\cdot\Delta_\Phi\right) F\, d\Phi_s .$$

The second use of the bounds on $\partial^\gamma C$ is to control $\sum_{\pi\in\mathcal{P}(\Gamma)}$. As in §7, we also find a factor $m_0^{-0|\gamma|}$, which yields the overall convergence of the expansion.

<u>Proposition</u> 8.1. Let $1 \le q < \infty$ and let m_0 be sufficiently large. There are constants $K_6(q,\gamma)$ and $K_7(q)$, independent of m_0, such that

$$(8.4)\quad \|\partial^\gamma C\|_{L_q(\Delta_{j_1}\times\Delta_{j_2})} \le K_6(q,\gamma)\, m_0^{-|\gamma|/2q} \exp(-m_0\, d(j,\gamma)/2)$$

$$(8.5)\qquad \sum_{\pi\in\mathcal{P}(\Gamma)} \Pi_{\gamma\in\pi} K_6(q,\gamma) \le e^{K_7(q)|\Gamma|} .$$

<u>Proof</u>. We use the Wiener integral representation (8.1) for $\partial^\gamma C$. The proof consists of estimates on the Wiener measure of paths $z(\tau)$ which cross the lattice lines $b \in \gamma$ in some definite order together with combinatoric arguments to count the number of ways the lines

$b \in \gamma$ can be so ordered.

Let $L(\gamma)$ be the set of all possible linear orderings of the lines $b \in \gamma$, and for $\ell \in L(\gamma)$, let $\mathcal{W}(\ell)$ be the set of Wiener paths which cross all lines $b \in \gamma$, and whose order of first crossing is ℓ. Then

$$(8.6) \qquad 0 \leq \partial^{\gamma} C(s) \leq \int_0^{\infty} e^{-m_0^2 T} \int \prod_{b\in\gamma} (1-J_b^T(z)) \, dz_{x,y}^T \, dT \; = \; \partial^{\gamma} C_{\emptyset} \, ,$$

and

$$(8.7) \qquad \partial^{\gamma} C_{\emptyset}(x,y) = \sum_{\ell\in L(\gamma)} \int_0^{\infty} e^{-m_0^2 T} \int_{\mathcal{W}(\ell)} dz_{x,y}^T \, dT \, .$$

Let $b_1, b_2, \ldots$ be the elements of γ, as ordered by ℓ. Let b_2' be the first of the b's not touching $b_1 = b_1'$, let b_3' be the first of the b's after b_2' and not touching b_2', etc. Set

$$a_j = \mathrm{Dist}(b_{j+1}', b_j') \, , \qquad 1 \leq j \leq m$$

and define

$$|\ell| = \textstyle\sum_{i=1}^m a_i \, .$$

If there is no such b_2', we set $|\ell| = 0$, by convention.

With these definitions, we bound the $\ell \in L(\gamma)$ term in (8.7), for $|\ell| \geq 1$, by

$$\int e^{-m_0^2 T} \int_{\sum t_i = T} \Pi_i (2\pi t_i)^{-1} \exp\left(-\tfrac{1}{2} \textstyle\sum_{i=1}^m \dfrac{a_i^2}{t_i}\right) dt \, dT$$

$$\leq K_8^m \sup_{T, \sum t_i = T} e^{-(m_0^2-2)T} \, e^{-\frac{1}{3} \sum_{i=1}^m a_i^2/t_i}$$

since

$$\int_{\sum t_i = T} dt \leq e^T \, , \qquad \int e^{-T} \, dT \leq 1 \, ,$$

and for all $a_i \geq 1$,

$$(2\pi t_i)^{-1} e^{-a_i^2/6t_i} \leq K_8 < \infty \, ,$$

thereby defining K_8.

Using the method of Lagrange multipliers to evaluate the maxima, we bound the $\ell \in L(\gamma)$ term by

$$K_8^m \exp\left((m_0^2-2)^{1/2} \, |\ell|\right) \, ,$$

for $|\ell| \geq 1$. For $|\ell| = 0$, we use the remark following Prop. 7.2.

There is an entirely similar estimate, based on the distance $d(j,\gamma)$ of (8.2), and taking geometric means of these two bounds yields

$$\|\partial^\gamma C\|_{L_q(\Delta_1\times\Delta_2)} \le \sum_{\ell\in L(\gamma)} K_8^{|\gamma|} e^{-m_0|\ell|/(2+\delta)} e^{-m_0 d(j,\gamma)/(2+\delta)} \tag{8.8}$$

for m_0 large. If $|\ell| \ge 1$ for all $\ell \in L(\gamma)$, then we can include a factor $m_0^{-|\gamma|}$ on the right side of (8.8), by increasing δ. If $|\ell| < 1$ for some ℓ, then $|\ell| = 0$, and in this case $|\gamma| \le 4$. With $|\gamma| \le 4$ and $d(j,\gamma) \ge 1$, we can still include the factor $m_0^{-|\gamma|}$ in (8.8) by increasing δ. Finally for $|\gamma| \le 4$ and $d(j,\gamma) = 0$, the factor $m_0^{-|\gamma|/2q} \le m_0^{-2/q}$ in (8.4) comes from scaling, as in Proposition 7.2.

We define

$$K_6(q,\gamma) = K_4 \sum_{\ell\in L(\gamma)} K_8^{|\gamma|} e^{-m_0|\ell|/(2+\delta)} \tag{8.9}$$

With this definition, (8.4) follows; in the case $d(j,\gamma) = 0$ and $|\ell| = 0$ for some ℓ, $K_6 \ge K_4$, and we use the bound of Proposition 7.2 to establish (8.4).

We complete the proof by establishing (8.5) as a separate proposition.

Proposition 8.2. For m_0 sufficiently large,

$$\sum_{\pi\in\mathcal{P}(\Gamma)} \prod_{\gamma\in\pi} \sum_{\ell\in L(\gamma)} e^{-m_0|\ell|/3} \le e^{K_9|\Gamma|} . \tag{8.10}$$

Proof. Let $\mathcal{L}(\Gamma)$ be the set of linear orderings defined on subsets of Γ. Thus $L(\Gamma) \subsetneq \mathcal{L}(\Gamma)$. As before, we define $|\ell|$ for $\ell \in \mathcal{L}(\Gamma)$. We assert that the number of $\ell \in \mathcal{L}(\Gamma)$ with $|\ell| \le r$ is bounded by

$$|\Gamma|\, e^{K_{10}(r+1)} . \tag{8.11}$$

Using (8.11), we complete the proof. Let $A_\ell = \exp(-m_0|\ell|/3)$. Expanding $\sum\prod\sum A_\ell$ in (8.10), we get a sum of terms of the form

$$A_{\ell_1} A_{\ell_2} \cdots A_{\ell_j} .$$

where the ℓ_j are distinct elements of $\mathcal{L}(\Gamma)$. Adding all terms of this form, we bound (8.10) by

$$\sum\prod\sum A_\ell \le \sum A_{\ell_1}\cdots A_{\ell_j} = \prod_{\ell\in\mathcal{L}(\Gamma)} (1+A_\ell)$$

$$\le \prod_{\ell\in\mathcal{L}(\Gamma)} \exp A_\ell = \exp \sum_{\ell\in\mathcal{L}(\Gamma)} A_\ell \le \exp(O(1)|\Gamma|) .$$

Here in the last expression, we used the bound (8.11) to estimate $\sum_{\ell\in\mathcal{L}(\Gamma)} A_\ell$ and we choose m_0 sufficiently large.

Next we establish (8.11). Suppose the integer part $[a_i]$ of the distances a_i are given. We choose $b_1 = b_1'$ in $|\Gamma|$ ways, and we choose the b's between b_1' and b_2' in O(1) ways, since they all must overlap b_1. Next b_2' is chosen in O(1) $[a_1]$ ways, namely from the lattice line segments b with

$$[a_1] \le \mathrm{Dist}(b,b_1') < [a_1]+1 .$$

Continuing in this fashion, we choose all the b's in

$$|\Gamma| \ \Pi_i \ O(1) \ [a_i] \le |\Gamma| \ e^{O(1) \sum [a_i]} \le |\Gamma| e^{O(1)r}$$

ways. Finally we count the number of choices of the $[a_i]$. This is the number of ways of choosing integers $r_i \ge 1$ with $\sum r_i \le r$, namely 2^r. In fact suppose $\sum r_i = r$, and we distribute the r units in $\sum r_i$ as follows: The first 1 goes into a_1 (no choice). The second 1 goes to a_1 or a_2 (one binary choice). If the jth 1 goes to a_i, the j+1st goes to a_i or a_{i+1} (one binary choice). Thus there are r-1 binary choices, or 2^{r-1} ways to choose r_i with $\sum r_i = r \ge 1$. Summing $j = \sum r_i$ gives $\sum_{j=1}^{r} 2^{j-1} = 2^j - 1$. Finally, we get one more choice from $|\ell| = 0$ (no a_i's).

§9. GAUSSIAN INTEGRALS

The integral of a polynomial with respect to a Gaussian measure can be evaluated in closed form. The closed form expression is a sum, and each term in the sum is labelled by a graph. We will encounter complicated polynomials of high degree, and the resulting graphs will also be complicated. However we present some very simple estimates for such polynomials; the structure of these estimates can be seen easily from the associated graphs.

We define a localized monomial to be a polynomial of the form

$$(9.1) \qquad R = \int \prod_{i=1}^{r} :\Phi(x_i)^{n_i}: \, w(x)\, dx \, ,$$

where $w(x)$ is supported in a product $\Delta_{j_1} \times \ldots \times \Delta_{j_r}$ of lattice squares. We also require $w \in L_{1+\varepsilon}$ and it is convenient but not essential to assume a bound

$$(9.2) \qquad n_i \le \bar{n} \, , \quad 1 \le i \le r \, .$$

the bound (9.2) does not restrict r, nor the total degree of R. Polynomials which arise naturally are not usually of this form because the kernels w are not localized. However any polynomial can be written as a sum of localized monomials.

Associated with R of (9.1) is a graph G(R) consisting of r vertices and at the ith vertex, we draw n_i legs. See Figure 1.

Fig. 1. $G\left(\int :\Phi(x_1)^4: \, :\Phi(x_2)^4: \, w(x)\, dx \right)$.

In order to evaluate $\int R \, d\Phi_C$, we integrate by parts:

$$\int \Phi(x) R \, d\Phi_C = \int C(x,y) \frac{\delta R}{\delta \Phi(y)} \, d\Phi_C \, dy \, .$$

This formula can be proved by passing to the Fock space[7] of the measure $d\Phi_C$, expanding Φ as a sum of a creation and an annihilation operator and using the canonical commutation relations. See also Theorem 9.1 below. We integrate by parts to reduce the degree of the monomial R we want to integrate. After $(\sum_{i=1}^{r} n_i)/2$ partial integrations, the monomial is replaced by a sum of constants, and since

[7] See for example Theorem 3.5 of J. Glimm and A. Jaffe, Boson quantum field models, in Mathematics of Contemporary Physics. Ed. by R. Streater, Academic Press, New York, 1972.

$\int d\Phi_C = 1$, the integral is evaluated explicitly.

In applying this procedure, we encounter Φ's in a Wick ordered factor $:\Phi(x_i)^{n_i}:$ in R. For such Φ's, we use the formula

$$(9.3)\quad \int :\Phi(x)^n: R\, d\Phi_C = (n-1)\, c(x) \int :\Phi(x)^{n-2}: R\, d\Phi_C + \int :\Phi(x)^{n-1}: C(x,y) \frac{\delta R}{\delta \Phi(y)}\, d\Phi_C\, dy\ ,$$

with

$$c(x) = C(x,x) - C_\emptyset(x,x)$$

defined by Proposition 7.5. The first term arises from the difference between the covariance $C_\emptyset$ in : : and the covariance in $\int \ldots d\Phi_C$. The second term is exactly as before.

The integration by parts formula (9.3) has a simple expression in terms of graphs. In case the $\Phi(x)$ is a factor in $:\Phi(x_i)^{n_i}:$, we label the terms on the right side of (9.3) by drawing a line connecting one leg of the x_i-vertex to a distinct leg at the same or a distinct vertex. The graph with a line from the x_i to the x_j vertex labels each of the n_j terms

$$(9.4)\quad \int C(x_i,x_j) :\Phi(x_i)^{n_i-1}: :\Phi(x_j)^{n_j-1}: \prod_{\ell\neq i,j} :\Phi(x_\ell)^{n_\ell}: w(x)\, dx$$

coming from a single integration by parts in (9.1). See Figure 2.

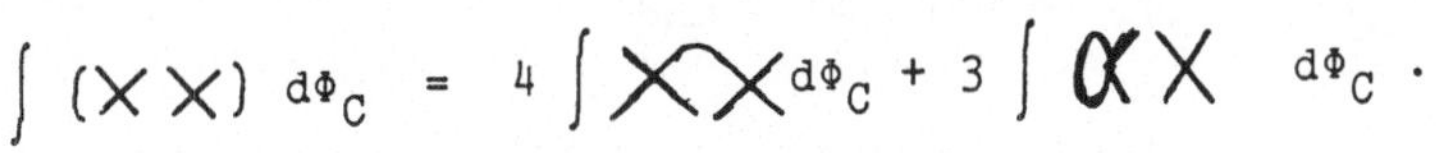

Fig. 2. Integration by Parts.

As an example, we evaluate the integral of Figure 2. After four integrations by parts, we have

$$\iint :\Phi(x_1)^4: :\Phi(x_2)^4: w\, dx\, d\Phi_C = 4! \int C(x_1,x_2)^4 w\, dx + 2!\binom{4}{2}^2 \int c(x_1)\, c(x_2)\, C(x_1,x_2)^2 w\, dx + \binom{4}{2}^2 \int c(x_1)^2 c(x_2)^2 w\, dx\ .$$

The absolute value of the first term is bounded by

$$4!\ \left| \int C(x_1,x_2)^4\, w(x_1,x_2)\, dx \right| \le 4!\ \| C \|^4_{L_8(\Delta_{j_1}\times\Delta_{j_2})} \| w \|_{L_2} \le 4!\ K_4 m_0^{-1}\, e^{-m_0(1-\delta)4d(j)}\ \| w \|_{L_2}$$

if suppt $w \subset \Delta_{j_1} \times \Delta_{j_2}$. In the last line, we used Proposition 7.2 to bound $\|C\|_{L_q(\Delta_{j_1}\times\Delta_{j_2})}$. See Figure 3.

Fig. 3. Evaluation of a Gaussian Integral.

To evaluate $\int R\, d\Phi_C$ in the general case, we form the set $\mathcal{V}(R)$ of all vacuum graphs. A vacuum graph is a graph obtained from $G(R)$ by joining pairwise distinct legs until all legs are so joined. If the total number of legs is odd, $\mathcal{V}(R) = \emptyset$ (and $\int R\, d\Phi_C = 0$). For each $G \in \mathcal{V}(R)$, define

$$(9.5)\quad I(G,w,C) = \int w(x)\ \Pi_{\ell'}\ c(x_{i(\ell')})\Pi_\ell\ C(x_{i_1(\ell)},x_{i_2(\ell)})\ dx\ .$$

The first product, $\Pi_{\ell'}$, runs over all lines ℓ' of G which connect two legs of a single vertex, labeled $i(\ell')$. The second product runs over all lines ℓ joining pairs $(i_1(\ell),i_2(\ell))$ of distinct vertices.

We now state the formula for evaluation of Gaussian integrals. The above discussion of integration by parts and its graphical interpretation gives the formal derivation.

<u>Theorem</u> 9.1. Let w be localized and let $w \in L_q$ for some $q > 1$. Then $R \in L_r(\mathcal{S}', d\Phi_C)$, $r \in [1,\infty)$, and

$$(9.6)\qquad \int R\, d\Phi_C = \sum_{G\in\mathcal{V}(R)} I(G,w,C)\ .$$

The proof has a combinatoric aspect (sketched above) and an analytic aspect. The latter is an approximation argument, which uses a modified (momentum cutoff) form of (9.6) to prove that the approximants converge. For this reason, we bound the integrals $I(G,w,C)$ before sketching the proof of Theorem 9.1.

We consider a function of the form

$$F(x_1,\dots,x_n) = \Pi_\ell\ F_\ell(x_{i_1(\ell)},x_{i_2(\ell)})\ \Pi_i\ \Delta_i(x_i)$$

where $\Delta_i(x)$ is the characteristic function of a set of area 1, and suppose that for each index ℓ,

$$1 \le i_1(\ell) < i_2(\ell) \le n\ .$$

<u>Lemma</u> 9.2. With the above notation,

$$\|F\|_{L_q} \le \Pi_\ell\ \|F_\ell\|_{L_{q_\ell}(\Delta_{i_1(\ell)},\Delta_{i_2(\ell)})}$$

provided that for each i,

$$q^{-1} \geq \sum \left\{ q_\ell^{-1};\ i_1(\ell) = i \text{ or } i_2(\ell) = i \right\} .$$

Proof: With the substitution $F \to F^q$, we are reduced to the case $q = 1$. We apply Hölder's inequality successively in each of the n variables (vertices) of F, and we use the language of graphs, in order to visualize this process. Each factor F_ℓ corresponds to a line joining the $i_1(\ell)$ vertex to the $i_2(\ell)$ vertex. Let

$$\mathcal{L}_i = \{\ell :\ i_1(\ell) < i = i_2(\ell)\}$$

$$\mathcal{L}_i' = \{\ell :\ i_1(\ell) = i < i_2(\ell)\} .$$

The hypothesis concerning q_ℓ is

$$\sum_{\ell \in \mathcal{L}_i \cup \mathcal{L}_i'} q_\ell^{-1} \geq 1$$

($\mathcal{L}_i = \emptyset = \mathcal{L}_i'$ is also allowed, trivially.) Thus

$$\int_{\Delta_i} \prod_{\ell \in \mathcal{L}_i} |F_\ell(x_{i_1(\ell)}, x_i)| \ \prod_{\ell \in \mathcal{L}_i'} \|F_\ell(x_i, \cdot)\|_{L_{q_\ell}(\Delta_{i_2(\ell)})} \, dx_i$$

$$\leq \prod_{\ell \in \mathcal{L}_i} \|F_\ell(x_{i_1(\ell)}, \cdot)\|_{L_{q_\ell}(\Delta_i)} \prod_{\ell \in \mathcal{L}_i'} \|F_\ell\|_{L_{q_\ell}(\Delta_{i_1(\ell)} \times \Delta_{i_2(\ell)})}$$

by Hölder's inequality. A finite induction on i (decreasing from $i = n$ to $i = 1$) now completes the proof.

Remark. An obvious modification allows some of the factors F_ℓ to depend on a single variable, i.e. $F_\ell = F_\ell(x_{i(\ell)})$.

Sketch of proof of Theorem 9.1. We introduce momentum cutoffs, $\Phi(x_i) \to \int \Phi(y)\, j_k(y-x_i)\, dy$, and omit the x-integration. This replaces R by a polynomial cylinder function $R_k(x)$, based on a finite dimensional subspace of $\mathcal{S}$. $R_k(x)$ is integrable by the definition of $d\Phi_C$, and for $R_k(x)$ the formulas (9.3) and (9.6) are valid by the above combinatorial arguments. Thus (9.6) holds for $R_k = \int R_k(x)\, w(x)\, dx$, and can be used to evaluate the right side of the inequality

$$\int |R_k - R_{k'}|\, d\Phi_C \leq \left(\int (R_k - R_{k'})^2\, d\Phi_C \right)^{1/2} .$$

Removing the momentum cutoff in one linear factor Φ at a time, the estimates of Propositions 7.2 and 7.3 are sufficient. For details see [1].

We remark that the stronger hypothesis of [1], $C \in \mathcal{B}^{\text{loc}}_{q,\delta(q)}$, $\forall q > 1$ (used in [1] to control C-Wick ordering) is not required for the present lectures, since only $C_\emptyset = (-\Delta+m_0^2)^{-1}$ Wick ordering is used. See also Proposition 7.5.

Proposition 9.3. Let w be localized. Then

$$(9.7)\quad \left|\int R\, d\Phi_C\right| \le \|w\|_{L_p} \sum_{G\in\mathcal{V}(R)} \Pi_{\ell'} \|c\|_{L_q(\Delta_{j(\ell')})} \times \Pi_\ell \|C\|_{L_q(\Delta_{j_1(\ell)},\Delta_{j_2(\ell)})}$$

where $q \ge p'\bar{n}$, $\bar{n}$ is defined by (9.2) and $j(\ell) = j_{i(\ell)}$ is the localization of the ith vertex.

Proof. By Theorem 9.1, it is sufficient to bound each integral $I(G,w,C)$. We use the Hölder inequality to separate w from the c and C factors. The integral of the c and C factors is bounded by Lemma 9.2 and the remark following it. This completes the proof.

There are $O(\sum n_i/2)!$ graphs in $\mathcal{V}(R)$. Because of the exponential decay in C as $|x-y| \to \infty$, in general most of these graphs are very small. Efficient estimates must take advantage of this fact. For each lattice square Δ, we define

$$N(\Delta) = \sum \{n_i : \Delta_{j_i} = \Delta\}$$

as the number of legs (linear factors $\Phi(x_i)$) of R localized in Δ.

Theorem 9.4. Let w be localized. There is a constant K_{11} such that

$$\left|\int R\, d\Phi_C\right| \le \|w\|_{L_p} [\Pi_\Delta N(\Delta)!\ (K_{11}\, m_0^{-1/2}\, q)^{N(\Delta)}]$$

for $q = p'\bar{n}$, $\bar{n}$ defined by (9.2).

Proof. We bound the L_q norm in (9.7) by Propositions 7.2 and 7.5. Having done this, it is sufficient to show that

$$\sum_{G\in\mathcal{V}(R)} \Pi_\ell\, e^{-m_0(1-\delta)d(j(\ell))} \le \Pi_\Delta(\text{const.}^{N(\Delta)}\, N(\Delta)!)\ .$$

Let ν denote a leg of R. Let Δ_ν be the square in which ν is localized and given $G \in \mathcal{V}(R)$, and let Δ'_ν be the square of the leg joined to ν by G. Also let $d(\nu) = \text{dist}\,(\Delta_\nu,\Delta'_\nu)$. Then the sum over $G \in \mathcal{V}(R)$ can be written as a sum over the choices Δ'_ν for each ν and a sum over the $N(\Delta'_\nu)$ possible contractions in Δ'_ν, for each ν.

The summand is independent of the choice of possible contractions within each square Δ ($\nu \leftrightarrow \Delta'_\nu$ held fixed). To estimate this sum we have only to count the number of terms. Since an arbitrary term can be obtained (nonuniquely, in general) from a single given term by permutation of the legs in each square Δ, there are at most $\Pi_\Delta N(\Delta)!$ terms in this sum. Hence it remains to show that

$$\sum_{\{\Delta'_\nu\}} \Pi_\nu \, e^{-m_0(1-\delta)d(\nu)/2} \leq \Pi_\nu (\text{const.}) ,$$

since each $d(j(\ell))$ occurs as a $d(\nu)$ for exactly two ν's. The summation index $\{\Delta'_\nu\}$ is simply a set of functions from legs to lattice squares. We increase the left side by enlarging the set of summation indices to include all such functions. Then we can interchange the sum and product, obtaining

$$\sum_{\{\Delta'_\nu\}} \Pi_\nu \, e^{-cd(\nu)} \leq \Pi_\nu \sum_{\{\Delta'_\nu :\ \nu \text{ fixed}\}} e^{-cd(\nu)} = \Pi_\nu \ \text{const.}$$

We note that K_{11} is independent of s and m_0, for m_0 bounded away from zero.

Theorem 9.5. Let Λ be a union of lattice squares and let $\text{Re}\,\lambda \geq 0$. Then $e^{-V(\Lambda)} \in L_p(\mathcal{S}', d\Phi_C)$ for all $p \in [1,\infty)$. There is a constant K_{12} independent of C such that

$$|Z| \equiv \left| \int e^{-\lambda V} \, d\Phi_C \right| \leq e^{K_{12}|\Lambda|} . \tag{9.8}$$

With $\text{Re}\,\lambda$ bounded and m_0 bounded away from zero, K_{10} can also be chosen independent of λ and m_0.

Remark. The simple proof of [1, §II.3, p. 22-27] is self contained. In fact, using a slight generalization Theorem 9.4 (incorporating Proposition 7.4), the proof of (9.8) is nearly identical to the (standard) proof that $e^{-V(\Lambda)} \in L_1(\mathcal{S}', d\Phi_C)$. This proof is close in structure to that of [3].

Corollary 9.6. For $p > 1$ and $q \geq p'\bar{n}$,

$$\left| \int R e^{-V(\Lambda)} \, d\Phi_C \right| \leq e^{K_{12}|\Lambda|} \, \|w\|_{L_p} \, [\Pi_\Delta \, N(\Delta)! \, (2K_{11} \, m_0^{-1/2q})^{N(\Delta)}].$$

Proof. By the Schwartz inequality,

$$\left| \int R \, e^{-V(\Lambda)} \, d\Phi_C \right| \leq \left(\int R^2 \, d\Phi_C \right)^{1/2} \left(\int e^{-2V(\Lambda)} \, d\Phi_C \right)^{1/2} .$$

The factors on the right are estimate by Theoresm 9.4 and 9.5.

Theorem 9.7. Let w be a localized kernel in L_p, $p > 1$, let Λ be a union of lattice squares and let $F = Re^{-V(\Lambda)}$ in (1.7). Then (1.7) is valid.

Sketch of Proof: In order to present the formal ideas, we suppose first that F is a polynomial. Then $\int F \, d\Phi_{C(s)}$ is given explicitly in terms of graphs by (9.5) - (9.6). Differentiating with respect to s_b in these formulas yields

$$(9.9) \quad \frac{d}{ds_b} \int F \, d\Phi_{C(s)} = \sum_{G \in \mathcal{V}(F)} \int \sum_{\ell} \left(\frac{d}{ds_b} C_{\ell}\right) \Pi_{\ell' \neq \ell} C_{\ell'} \; w \, dx$$

where C_{ℓ} denotes $C(x_{i_1(\ell)}, x_{i_2(\ell)})$.

The product $\Pi_{\ell' \neq \ell}$ in effect removes one line from the vacuum graph $G \in \mathcal{V}(F)$. Equivalently, one could remove from F the two legs joined by the line ℓ. However removing legs from F is the same as removing linear factors from F, or the same as differentiating F with respect to these linear factors. Thus we see that $\sum_{\ell}$, the sum over lines removed from $G \in \mathcal{V}(F)$, is equivalent to a sum over mixed second derivatives of F with respect to pairs of linear factors. Such a sum is just $\frac{1}{2} \Delta_{\Phi} F$, so we identify the right side of (9.9) as

$$\frac{1}{2} \int \left[\left(\frac{d}{ds_b} C\right) \cdot \Delta_{\Phi}\right] F \, d\Phi_{C(s)} \, .$$

The proof in the general case, $F = R\,e^{-V}$, is based on approximations, starting with F a polynomial. The control over these approximations is given by Corollary 9.6. For details see [1].

§10. CONVERGENCE: THE PROOF COMPLETED

Proof of Proposition 5.3. Without loss of generality, the kernel w is localized, and in this case we take $\|w\| = \|w\|_2$. The expression we want to estimate is

$$(10.1) \qquad < \int \partial^{\Gamma} \int \Pi_{i=1}^{n} \Phi(x_i)\, e^{-\lambda V(\Lambda)}\, d\Phi_{s(\Gamma)}\, ds(\Gamma),\, w> \quad .$$

Let $\mathcal{P}(\Gamma)$ be the set of all partitions π of Γ. By Leibnitz' rule and (1.7), (10.1) equals

$$(10.2) \quad <\int \Sigma_{\pi\in \mathcal{P}(\Gamma)} \int \left(\Pi_{\gamma\in\pi} \tfrac{1}{2}\, \partial^{\gamma} C\cdot\Delta_{\Phi}\right) \Pi\, \Phi(x_i) e^{-\lambda V(\Lambda)}\, d\Phi_{s(\Gamma)}\, ds(\Gamma), w>$$

where $C = C(s(\Gamma))$.

As in (7.3), we define

$$\partial^{\gamma} C(j_{\gamma}) = \Delta_{j_{1,\gamma}}\, \partial^{\gamma} C\, \Delta_{j_{2},\gamma}$$

where $j = (j_{1,\gamma}, j_{2,\gamma}) \in Z^4$, so that the two derivatives in $\partial^{\gamma} C(j_{\gamma})\cdot\Delta_{\Phi}$ are localized in Δ_{j_1} and Δ_{j_2} respectively, and

$$\partial^{\gamma} C = \Sigma_{j_{\gamma}}\, \partial^{\gamma} C(j_{\gamma}) \ .$$

We substitute this identity into (10.2) and expand.

The resulting sum is now indexed by localizations $\{j_{\gamma}\}$ and partitions $\pi \in \mathcal{P}(\Gamma)$. For a given term let $M = M(\pi, \{j_{\gamma}\})$ be the number of terms resulting from the differentiations Δ_{Φ} in (10.2). By Corollary 9.6 each of the resulting terms can be estimated by

$$\|w'\|_{L_p}\, e^{K_{12}|\Lambda|}\, \Pi_{\Delta}\, N(\Delta)!\, (2K_{11})^{N(\Delta)}$$

for $m_0 \geq 1$. Here w' is the w of (10.1) multiplied by the kernels $\partial^{\gamma} C$ arising in (10.2). From Proposition 8.1 and Lemma 9.2 we have (for $p < 2$, and q large),

$$\begin{aligned} \|w'\|_{L_p} &< \|w\|_{L_2}\, \|\Pi_{\gamma\in\pi}\, \partial^{\gamma} C(j_{\gamma})\|_{L_q} \\ &\leq \|w\|_{L_2}\, m_0^{-|\Gamma|/2q}\, \Pi_{\gamma\in\pi}\, K_6(q,\gamma)\, e^{-m_0 d(j_{\gamma},\gamma)/2} \ . \end{aligned}$$

Now using (8.5) to control the sum over $\pi \in \mathcal{P}(\Gamma)$ we can bound

(10.2) by

$$\|w\|_{L_2} \; e^{K_7|\Gamma|} \; m_0^{-|\Gamma|/2q} \sum_{\{j_\gamma\}} \max_{\pi\in\mathcal{P}(\Gamma)} \; M \prod_{\gamma\in\pi} e^{-m_0 d(j_\gamma,\gamma)/2} \; \prod_{\Delta} N(\Delta)!$$

The proof of Proposition 5.3 follows from two lemmas which control M and the sum over $\{j_\gamma\}$ respectively. Let $M(\Delta)$ be the number of elements in the set $\{j_{i,\gamma}: \Delta_{j_{i,\gamma}} = \Delta,\ i = 1 \text{ or } 2,\ \gamma \in \pi\}$.

<u>Lemma</u> 10.1. There exists a constant K_{13}, independent of m_0, such that

$$M \le e^{K_{13}|\Gamma|} \prod_{\Delta} (M(\Delta)!)^p$$

and

$$\prod_{\Delta} N(\Delta)! \le e^{K_{13}|\Gamma|} \prod_{\Delta} (M(\Delta)!)^p .$$

where p is the degree of the interaction polynomial P.

<u>Lemma</u> 10.2. Given $\pi \in \mathcal{P}(\Gamma)$ and $r > 0$, there exists a constant K_{14}, independent of m_0, such that

$$\sum_{\{j_\gamma\}} \prod_{\gamma\in\pi} e^{-m_0 d(j_\gamma,\gamma)/2} \prod_{\Delta} M(\Delta)!^r \le e^{K_{14}|\Gamma|} .$$

<u>Proof of Lemma</u> 10.1: Let $N_0(\Delta)$ be the number of x_i, $1\le i\le n$, which are localized in Δ. The number of terms resulting from $M(\Delta)$ differentiations in Δ is bounded by

$$(N_0(\Delta)+1)(N_0(\Delta)+p+1) \ldots (N_0(\Delta)+p(M(\Delta)-1)+1) .$$

Since $N_0(\Delta) \le \sum_\Delta N_0(\Delta) = n$, we have M, the total number of terms resulting from all $\partial/\partial\Phi(y)$ differentiations, bounded by

$$M \le \Pi_\Delta p^{pM(\Delta)} (N_0(\Delta)+1+pM(\Delta))^{N_0(\Delta)+1} \; M(\Delta)!^p \;;$$

using the inequalities

$$(a+b)! \le (a+b)^a (b!) \quad \text{and} \quad (ab)! \le a^{ab}(b!)^a .$$

Furthermore with $N(\Delta)$, as defined in §9, the number of legs in Δ, after differentiation, we have

$$N(\Delta) \le N_0(\Delta) + (p-1)M(\Delta)$$

and so $\Pi N(\Delta)!$ is bounded as above.

Proof of Lemma 10.2: The sum $\sum_{\{j_\gamma\}}$ is controlled by the exponentially decreasing distance factor, so it is sufficient to show

$$\Pi_\Delta M(\Delta)!^r \le \Pi_\gamma e^{\text{const.}|\gamma|} e^{\text{const.}\sum_\gamma d(j,\gamma)}$$

with constants independent of m_0 , λ, $\{j_\gamma\}$ and π.

Recall that $d(j,\gamma)$, defined by (8.2), contains the distance from $j_{1,\gamma}$ and $j_{2,\gamma}$ to some $b \in \gamma$. Thus for fixed Δ, there are at most $O(1)r^2$ values of γ within a fixed partition π such that

$$\Delta_{j_\nu,\gamma} = \Delta \,, \qquad \nu = 1 \text{ or } 2, \tag{10.4}$$

and $d(j,\gamma) \le r$. By definition there are $M(\Delta)$ γ's which satisfy (10.4). The most distant half $(= M(\Delta)/2)$ of these γ's must also satisfy

$$M(\Delta)^{1/2} \le \text{const. } d(j,\gamma) + \text{const.}$$

because the γ's are nonoverlapping. Hence

$$M(\Delta)^{3/2} \le \text{const. } \sum_\gamma \{d(j,\gamma) : \Delta_{j_\nu,\gamma} = \Delta\} + \text{const.}$$

and so the proof is completed by the inequality

$$\begin{aligned}\Pi_\Delta M(\Delta)!^r &\le \exp\left(r \sum_\Delta M(\Delta) \ln M(\Delta)\right) \\ &\le \exp\left(O \sum_\Delta \{M(\Delta)^{1+\delta} : M(\Delta) > 0\}\right) \\ &\le \exp\left(O \sum_\gamma d(j,\gamma)\right) \exp\left(O|\Gamma|\right) .\end{aligned}$$

REFERENCES

0. Z. Ciesielski, Lectures on Brownian motion, heat conduction and potential theory, Aarhus Universitet, 1965.

1. J. Dimock and J. Glimm. Measures on the Schwartz distribution Space and Applications to $P(\phi)_2$ field theories.

2. J. Ginibre. General formulation of Griffiths inequalities. Comm. Math. Phys. 16 (1970) 310-328.

3. J. Glimm and A. Jaffe. The $\lambda(\phi)_2^4$ quantum field theory without cutoffs. III The physical vacuum. Acta Math. 125 (1970) 203-261.

4. J. Glimm and A. Jaffe. The $\lambda(\phi)_2^4$ quantum field theory without cutoffs, IV. Perturbations of the Hamiltonian. J. Math. Phys. 13 (1972) 1568-1584.

5. J. Glimm and A. Jaffe. Positivity of the ϕ_3^4 Hamiltonian. Fort. der Physik. To appear.

6. J. Glimm, A. Jaffe and T. Spencer. The Wightman axioms and particle structure in the $P(\phi)_2$ quantum field model. To appear.

7. F. Guerra. L. Rosen and B. Simon. The $P(\phi)_2$ quantum field theory as classical statistical mechanics.

8. L. Hörmander. Linear Partial Differential Operators. Springer-Verlag, Berlin, 1964.

9. J. Klauder. Ultralocal scalar field models. Comm. Math. Phys. 18 (1970) 307-318.

10. J. Lebowitz and O. Penrose. Decay of correlations. Preprint.

11. R. Minlos and Ja. Sinai. The phenomenon of phase separation at low temperatures in some lattice models of a gas II. Trans. Moscow Math. Soc. Vol. 19 (1968), 121-196.

12. C. Newman. Ultralocal quantum field theory in terms of currents. Comm. Math. Phys. 26 (1972) 169-204.

13. D. Ruelle. Statistical Mechanics. Benjamin, New York, 1969.

14. T. Spencer. The mass gap for the $P(\phi)_2$ quantum field model with a strong external field. Preprint.

15. K. Wilson and J. Kogut. The renormalization group and the ε-expansion. Phys. Reports, to appear.

BOSE FIELD THEORY AS CLASSICAL STATISTICAL MECHANICS. I. THE VARIATIONAL PRINCIPLE AND THE EQUILIBRIUM EQUATIONS

FRANÇESCO GUERRA
Institute of Physics[x], University of Salerno, Salerno, Italy

1. INTRODUCTION

In the last two years new powerful methods have been exploited for the advancement of the constructive quantum field theory program, [10,11,43] of Glimm and Jaffe and their followers. These new methods rely on ideas from Euclidean field theory [23,35] and use probabilistic techniques and concepts. They have been mostly advocated by Nelson [24,25,26], following earlier proposals by Symanzik [42]. In particular Nelson isolated the crucial Markov property of the Euclidean fields, which plays a very important role for the construction of the Euclidean theory and its physical interpretation.

One of the most attractive features of the Euclidean-Markov field theory for Bosons is that all physical quantities are expressed by means of commutative fields. Moreover the vacuum expectation values for interacting fields have a remarkable similarity with the expectation values in Gibbsian ensembles of classical statistical mechanics. This similarity, very well known since many years, suggests the possibility to exploit the modern techniques of rigorous statistical mechanics [33,14] for the study of constructive field theory. Such a program has been advocated by Guerra, Rosen and Simon [17,18,19], with further developments by Nelson [28], Simon [36,37] and Griffiths and Simon [39,40].

In these lectures we present part of this program, dealing mainly with the variational principle for the entropy and the equilibrium equations for infinite volume systems of the type proposed by Dobrushin, Lanford and Ruelle in statistical mechanics. Our main concern will be to provide a characterization of the infinite volume states associated to a given interaction, independently of limiting procedures on volume cut off theories. In the following lectures Lon Rosen [31] and Barry Simon [38] will present the other parts of the program and its developments, mainly the lattice approximation, the correlation inequalities with applications and the Lee-Yang theorem with its important consequences.

For other applications of statistical mechanics ideas to constructive field theory we refer to the talks by Nelson [27]. The powerful techniques of

[x]Postal Address : Istituto di Fisica dell'Università
Via Vernieri 42, 84100, Salerno, Italy.

contour expansion [13], also related to statistical mechanics ideas, with their important applications to the infinite volume limit and the particle spectrum are presented in the lectures by Glimm [8] and Jaffe [20].

The content of these lectures is the following. In Sections 2 and 3 we review the basic properties of the free and interacting Euclidean-Markov field, in order to fix notations and introduce the motivations of the statistical mechanics analogy which will be exploited in the following Sections. In Section 4 we introduce the infinite volume limit of the pressure associated to a given interaction. The structure of the infinite volume states and their entropy density are discussed in Section 5. Finally in Section 6 we introduce the variational principle for the entropy density and in Section 7 the equilibrium equations of the Dobrushin-Lanford-Ruelle type.

For the basic concepts of probability theory and stochastic processes we refer to Reed's talks [30] and to [7].

2. THE FREE EUCLIDEAN-MARKOV FIELD.

We consider fields on the Euclidean space R^d. The physical case is $d=4$ (three space-one time dimensions). The harmonic and anharmonic oscillators correspond to $d=1$, the $P(\varphi)_2$ theory to $d=2$ and the φ_3^4 theory [12] to $d=3$.

First of all we introduce the Sobolev Hilbert space N of real temperate distributions f on R^d, with symmetric scalar product

$$<f,g>_N = <f,(-\Delta+m^2)^{-1}g>$$

where $<\ ,\ >$ is the usual Lebesgue scalar product on Fourier transforms, Δ is the Laplacian in d dimensions and m^2 is a positive constant.

<u>Definition 1</u>. The free Euclidean-Markov field is the real Gaussian random field $\varphi(f)$, indexed by N and defined by the expectations

$$E(\varphi(f))=0\ , \qquad E(\varphi(f)\varphi(g)) = <f,g>_N .$$

We call (Q,Σ,μ) the underlying probability space, then the fields $\varphi(f)$ are represented as $L^p(Q,\Sigma,\mu)$ functions on Q, $1\le p<\infty$, which we still call $\varphi(f)$, in such a way that the expectations can be expressed as integrals

$$E(\varphi(f_1)\dots\varphi(f_n)) = \int_Q \varphi(f_1)\cdots\varphi(f_n)\,d\mu .$$

We assume that Σ is the smallest σ-algebra with respect to which all fields $\varphi(f)$, $f\in N$, are measurable. Due to the Euclidean invariance of the scalar product in N, the full Euclidean group $E(d)$ (including reflections) can be represented in the natural way as a group of measure preserving automorphisms of the σ-algebra Σ.

To each closed region A of R^d we associate the sub-σ-algebra Σ_A of Σ

generated by fields $\varphi(f)$ with $\mathrm{supp} f \subseteq A$. We call E_A the conditional expectation with respect to Σ_A.

Proposition 2. Let π be a smooth (d-1)-dimensional closed manifold dividing R^d in two closed regions A and B, such that $A \cup B = R^d$ and $A \cap B = \pi$,then

$$E_\pi = E_A E_B .$$

In order to describe the connection with the Hamiltonian theory, let us introduce the Sobolev Hilbert space F of real temperate distributions in R^{d-1} with scalar product

$$< f, g >_F = \frac{1}{2} < f, (-\Delta + m^2)^{-\frac{1}{2}} g > ,$$

where Δ is the (d-1)-dimensional Laplancian. Then the time zero physical free field of mass m in d-dimensional space-time is the real Gaussian random process $\bar{\varphi}(f)$, indexed by F and defined by the expectations

$$E(\bar{\varphi}(f)) = 0 , \qquad E(\bar{\varphi}(f)\bar{\varphi}(g)) = < f, g >_F .$$

If $(\bar{Q}, \bar{\Sigma}, \bar{\mu})$ is the underlying minimal probability space, then the фок space is represented as $L^2(\bar{Q}, \bar{\Sigma}, \bar{\mu})$ and the фок vacuum corresponds to the function $\Omega_0 \equiv 1$ on $\bar{Q}$.

Let us now define, for $t \in R$, the operators J_t

$$J_t : L^p(\bar{Q}, \bar{\Sigma}, \bar{\mu}) \longrightarrow L^p(Q, \Sigma, \mu) , \qquad 1 \leq p \leq \infty ,$$

such that

$$J_t \Omega_0 = \omega_0 , \qquad J_t \bar{\varphi}(f_1) \ldots \bar{\varphi}(f_n) = \varphi(f_1 \delta_t) \ldots \varphi(f_n \delta_t) ,$$

where ω_0 is the function $\omega_0 \equiv 1$ on Q and $f\delta_t$ is, for $f \in F$, the distribution in R^d belonging to N and defined by

$$(f\delta_t)(x) = f(\vec{x})\,\delta(x_d - t) , \qquad x \in R^d , \quad \vec{x} \in R^{d-1} , \quad x = \{\vec{x}, x_d\} .$$

The operators J_t provide the connection between the Hamiltonian theory and the Euclidean-Markov structure. By means of J_t $L^p(\bar{Q}, \bar{\Sigma}, \bar{\mu})$ is embedded isometrically into $L^p(Q, \Sigma, \mu)$ with image $L^p(Q, \Sigma_t, \mu)$, where Σ_t is the sub-σ-algebra of Σ generated by the fields with test functions having support on the hyperplane $x_d = t$.

The physical free Hamiltonian H_0 can be expressed in the form

$$e^{-tH_0} = J_0^* J_t \qquad , \quad t \geq 0 ,$$

or equivalently

$$< u, e^{-tH_0} v > = \int_Q (J_0 u)^* (J_t v)\, d\mu$$

where u and v are $\bar{Q}$ space functions.

The well known hypercontractivity [41] of the free Hamiltonian can be expressed in the best Nelson form [26] as

$$\| e^{-tH_0} \|_{p,q} \leq 1 \qquad \text{for} \qquad e^{-2tm} \leq (p-1)/(q-1) ,$$

with $t \geq 0$, $1 \leq p \leq q \leq \infty$, where $\| \ \|_{p,q}$ denotes the norm of a map from $L^p(\bar{Q},\bar{\Sigma},\bar{\mu})$ to $L^q(\bar{Q},\bar{\Sigma},\bar{\mu})$.

At the level of the Euclidean-Markov theory the following general version of the hypercontractive property can be proved [18].

Theorem 3. [18] (Basic hypercontractive estimate). Let Λ_1 and Λ_2 be two regions in R^d separated by a distance τ . If u_1 and u_2 are two Q space functions measurable with respect to Λ_1 and Λ_2 respectively, then the following estimate holds

$$\| u_1 u_2 \|_1 \leq \| u_1 \|_{p_1} \| u_2 \|_{p_2}$$

provided $(p_1 - 1)(p_2 - 1) \geq \alpha(\tau)$, where $\alpha(\tau) = O(\tau^{d-1} e^{-2m\tau})$.

This estimate is an improvement with respect to plain Hölder's inequality, corresponding to $\alpha(\tau) = 1$, because we can take p_1 and p_2 as near to 1 as we like provided τ is large enough. On the other hand, should distant regions be stochastically independent, we would have

$$\| u_1 u_2 \|_1 \leq E(|u_1||u_2|) = E(|u_1|)\, E(|u_2|) = \| u_1 \|_1 \, \| u_2 \|_1 .$$

But the free measure couples different regions of R^d , in fact $E_{\Lambda_1} E_{\Lambda_2} \neq E$. Therefore the hypercontractive estimate tells us that stochastic independence is almost realized as $\tau \to \infty$. Thus we speak of exponential decoupling of distant regions and we expect a kind of thermodynamic behaviour for the local interacting theories.

We conclude this section with the following checkerboard estimate, stated for the two-dimensional case.

Theorem 4. [18] (Checkerboard estimate). Consider two orthogonal systems of parallel lines at distances $\ell + 2a$, partitioning R^2 in squares of side $\ell + 2a$. Let Λ_i , $i = 1, .., n$, be n distinct squares concentric to squares of the partition and with sides of length ℓ parallel to the lines of the partition. If f_i are Σ_{Λ_i} measurable then the following estimate holds

$$\| f_1 .. f_n \|_1 \leq \| f_1 \|_{\beta^2} .. \| f_n \|_{\beta^2} ,$$

with $\beta = (1 + e^{2ma}) / (e^{2ma} - e^{-m(\ell + 2a)})$.

Remark. If both ℓ and a become very large then $\beta \to 1$, thus the regions Λ_i become stochastically independent.

3. THE INTERACTING THEORY

When the interaction is turned on it is expected that the interacting fields $\hat{\varphi}$ are equal to the free fields φ as functions on Q space, but there is a change in the measure, so that the expectations of the interacting fields $\hat{\varphi}$ are given by

$$E(\hat{\varphi}(f_1)\ldots\hat{\varphi}(f_n)) = \int_Q \varphi(f_1)\ldots\varphi(f_n)\,d\hat{\mu},$$

where $\hat{\mu}$ is a new measure depending on the interaction.

For the two-dimensional case the interaction is specified by a polynomial $P(X)$ with real coefficients, bounded below and (without loss of generality) normalized to $P(0)=0$.

To each compact region Λ of R^2 we associate the Euclidean action defined by

$$U_\Lambda = \int_\Lambda :P(\varphi(x)):\,dx,$$

where the local limit is obtained through the removal of an ultraviolet cutoff (see [18]) and the normal Wick product can be introduced in a purely stochastic fashion as explained in Nelson talks [27]. The main properties of the Euclidean action are summarized in the following theorem.

Theorem 5. (Properties of the Euclidean action U_Λ).

a) $U_\Lambda \in L^p(Q,\Sigma_\Lambda,\mu)$, $1 \le p < \infty$, (smoothness and locality).

b) $U_\Lambda = U_{\Lambda_1} + U_{\Lambda_2}$, if $\Lambda = \Lambda_1 \cup \Lambda_2$ and Λ_1 and Λ_2 have intersection with zero Lebesgue measure (additivity).

c) $E(U_\Lambda) = \int_Q U_\Lambda\,d\mu = 0$.

d) $e^{-U_\Lambda} \in L^p(Q,\Sigma_\Lambda,\mu)$, $1 \le p < \infty$.

The Gibbsian factor plays a very important role in two respects. First of all it allows to write the cut off interacting Hamiltonian H_ℓ in Фок space [10,41], defined by

$$H_\ell = H_0 + V_\ell, \quad V_\ell = \int_{-\frac{\ell}{2}}^{\frac{\ell}{2}} :P(\bar{\varphi}(x)):\,dx,$$

in the following particularly simple form

$$\langle u, e^{-tH_\ell} v\rangle = \int_Q (J_0 u)^*(J_t v)\, e^{-U_{\Lambda(\ell,t)}}\,d\mu$$

where u and v are $\bar{Q}$ space functions and $\Lambda(\ell,t)$ is the rectangle $-\ell/2 \le x_1 \le \ell/2$, $0 \le x_2 \le t$. We call this expression the Feynman-Kac-Nelson formula [24,18].

In the particular case $u = v = \Omega_0$, taking into account the Euclidean covariance of the free Markov field, we obtain the following celebrated Nelson's symmetry

$$\langle \Omega_0, e^{-tH_\ell}\Omega_0\rangle = \langle \Omega_0, e^{-\ell H_t}\Omega_0\rangle,$$

which is at the basis of [15,16].

On the other hand, formal reasoning, based on the analytic continuation of the well known Gell-Mann-Low formula, suggests that for the volume cutoff interacting theory we must take as new measure

$$d\mu_\Lambda = e^{-U_\Lambda} d\mu \Big/ \int_Q e^{-U_\Lambda} d\mu$$

which is still a Markov measure [27,18] even though it is not covariant. The fully covariant interacting measure $\hat{\mu}$ must be obtained taking the limit as $\Lambda \to R^2$ of μ_Λ in a suitable sense.

The volume cut off vacuum expectation values of products of interacting fields are given by

$$S_\Lambda(f_1, \ldots f_n) = \int_Q \varphi(f_1) \ldots \varphi(f_n)\, e^{-U_\Lambda} d\mu \Big/ \int_Q e^{-U_\Lambda} d\mu$$

This expression is the starting point of the statistical mechanics analogy, exploited in these lectures and summarized in the following table.

Configuration space	Q space
Basic observables	fields $\varphi(f)$
Free expectations	$E(A) = \int_Q A\, d\mu$
Gibbsian factor	e^{-U_Λ}
Partition function	$Z_\Lambda = \int_Q e^{-U_\Lambda} d\mu$
Gibbs expectations in Λ	$Z_\Lambda^{-1} \int_Q A\, e^{-U_\Lambda} d\mu$
Pressure	$P_\Lambda \doteq \lvert\Lambda\rvert^{-1} \log Z_\Lambda$
Correlation functions	$S_\Lambda(x_1, \ldots, x_n) = Z_\Lambda^{-1} \int_Q \varphi(x_1) \ldots \varphi(x_n) e^{-U_\Lambda} d\mu$
State	family $\{f_\Lambda\}$ of positive, normalized consistent densities on Q .
Entropy	$S_\Lambda(f) = - \int_Q f_\Lambda \log f_\Lambda\, d\mu$.

This analogy can be further deepened if we go to the lattice approximation [18] as explained in the talks by Nelson [27] , Rosen [31] and Simon [38]. In this case a kind of nearest neighbor interaction can be extracted from the free measure, so that distinct regions become stochastically independent but coupled through this interaction. In this approximation the free theory is represented as an array of Gaussian spins with nearest neighbor ferromagnetic interaction, when the interaction P is turned on then only the distributions of the single spins are affected but not the nearest neighbor coupling. From this point of view, in the lattice approximation the local interaction U_Λ acts like a kind of chemical potential.

In the rest of these lectures we consider the problem of characterizing the states of the infinite volume system in terms of the interaction, using statistical mechanics ideas. We follow two main lines of development connected with the variational principle for the entropy density, as introduced by Ruelle [32,33] , and the equilibrium equations of the type considered by Dobrushin [3,4,5] and Lanford and Ruelle [21] . We show also the connections between the two lines of development.

4. THE PRESSURE

Since in the following we need fields in a region with zero boundary conditions, we begin this section presenting some basic facts related to the general conditioning theory for which we refer to [18,19].

Consider a closed region A of R^d, let N_A be the subspace of N made of distributions with support on A, e_A the orthogonal projection on N_A and E_A the conditional expectation with respect to the sub-σ-algebra Σ_A of Σ generated by fields $\varphi(f)$ with $f \in N_A$, as in Section 2. We call (Q_A, Σ_A, μ_A) the probability space of the random field $\varphi(f)$, with $f \in N_A$.

For the free Markov field $\varphi(f)$, $f \in N$, we can write

$$\varphi(f) = \varphi((1-e_A)f) + \varphi(e_A f)$$

and define the two independent fields φ_A^o and φ_A through

$$\varphi_A^o(f) = \varphi((1-e_A)f) \ , \quad \varphi_A(f) = \varphi(e_A f) \ , \quad f \in N .$$

Obviously $\varphi_A^o(f)$ is zero if $f \in N_A$, therefore we call φ_A^o the field obtained from φ by conditioning it to be zero on the region A. Let us call $(Q_A^o, \Sigma_A^o, \mu_A^o)$ the probability space of φ_A^o, then we have

$$\varphi(f) = \varphi_A^o(f) + \varphi_A(f) \ , \quad Q = Q_A^o \times Q_A \ , \quad \Sigma = \Sigma_A^o \times \Sigma_A \ , \quad \mu = \mu_A^o \times \mu_A \ ,$$

and $\quad E_A u = \int_{Q_A^o} u \, d\mu_A^o .$

Let Λ' be the closure of the complement in R^d of the compact region Λ, then for $f \in N_\Lambda$ we call $\varphi_{\Lambda'}^o(f)$ the Dirichlet field in Λ with zero boundary conditions on $\partial\Lambda$, the boundary of Λ. As a consequence of Markov property if $f \in N_\Lambda$ then $\varphi_{\partial\Lambda}^o(f) = \varphi_{\Lambda'}^o(f)$.

For a two-dimensional field theory we introduce the Euclidean action in Λ with zero boundary conditions on $\partial\Lambda$, defined by

$$U_\Lambda^o = \int_\Lambda : P(\varphi_{\partial\Lambda}^o(x)) : dx = \int : P(\varphi_{\Lambda'}^o(x)) : dx$$

Using the properties of Wick ordering it can be shown [18] that

$$U_\Lambda^o = \int_{Q_{\Lambda'}} U_\Lambda \, d\mu_{\Lambda'} .$$

Given a normalized interaction P, let us introduce the following definitions.

Definitions 6

Partition function $\qquad Z_\Lambda = \int_Q e^{-U_\Lambda} \, d\mu ,$

Dirichlet partition function $\quad Z^{\circ}_{\Lambda} = \int_{Q} e^{-U^{\circ}_{\Lambda}} d\mu = \int_{Q^{\circ}_{\Lambda'}} e^{-U^{\circ}_{\Lambda}} d\mu^{\circ}_{\Lambda'}$.

Pressure $\quad P_{\Lambda} = |\Lambda|^{-1} \log Z_{\Lambda}$.

Dirichlet pressure $\quad P^{\circ}_{\Lambda} = |\Lambda|^{-1} \log Z^{\circ}_{\Lambda}$.

Our main objective of this section is to investigate the behaviour of P_{Λ} and P°_{Λ} as $\Lambda \to \infty$.

Lemma 7. For the partition functions we have

a) $1 \leq Z^{\circ}_{\Lambda} \leq Z_{\Lambda}$

b) If $\Lambda = \Lambda_1 \cup \Lambda_2$ and $\Lambda_1 \cap \Lambda_2$ has zero measure, then

$$Z^{\circ}_{\Lambda} \geq Z^{\circ}_{\Lambda_1} Z^{\circ}_{\Lambda_2} ,$$

if Λ_1 and Λ_2 are disjoint then the equality holds.

c) $\log Z_{\Lambda}$ and $\log Z^{\circ}_{\Lambda}$ are convex functions of P for fixed Λ , i.e. for two interaction polynomials P_1 and P_2 we have

$$\log Z_{\Lambda}(\lambda_1 P_1 + \lambda_2 P_2) \leq \lambda_1 \log Z_{\Lambda}(P_1) + \lambda_2 \log Z_{\Lambda}(P_2),$$

for $0 \leq \lambda_i \leq 1$, $\lambda_1 + \lambda_2 = 1$.

Proof. a) We use Jensen's inequality

$$\int \exp A \, d\mu \geq \exp \int A \, d\mu \qquad , \text{ for } \int d\mu = 1 \quad ,$$

and the relation

$$\int_{Q^{\circ}_{\Lambda'}} U^{\circ}_{\Lambda} \, d\mu^{\circ}_{\Lambda'} = 0 ,$$

following from the properties of Wick ordering and the normalization of P , to write

$$Z_{\Lambda} = \int_{Q} e^{-U_{\Lambda}} d\mu = \int_{Q^{\circ}_{\Lambda'}} \int_{Q_{\Lambda'}} e^{-U_{\Lambda}} d\mu_{\Lambda'} \, d\mu^{\circ}_{\Lambda'} \geq \int_{Q^{\circ}_{\Lambda'}} e^{-\int_{Q_{\Lambda'}} U_{\Lambda} d\mu_{\Lambda'}} d\mu^{\circ}_{\Lambda'} =$$

$$= \int_{Q^{\circ}_{\Lambda'}} e^{-U^{\circ}_{\Lambda}} d\mu^{\circ}_{\Lambda'} = Z^{\circ}_{\Lambda} \geq e^{-\int_{Q^{\circ}_{\Lambda'}} U^{\circ}_{\Lambda} d\mu^{\circ}_{\Lambda'}} = 1 .$$

b) We can write

$$\varphi^{\circ}_{\Lambda'} = \varphi^{\circ}_{\Lambda'_1} + \varphi^{\circ}_{\Lambda'_2} + \varphi_{\sigma} ,$$

where the three fields on the right hand side are independent and φ_{σ} describes the degrees of freedom associated to the intersection of Λ_1 and Λ_2 (thus $\varphi_{\sigma} = 0$ if Λ_1 and Λ_2 are disjoint). Therefore

$$Z^{\circ}_{\Lambda} = \int_{Q^{\circ}_{\Lambda'}} e^{-U^{\circ}_{\Lambda}} d\mu^{\circ}_{\Lambda'} = \int_{Q^{\circ}_{\Lambda'_1} \times Q^{\circ}_{\Lambda'_2}} \int_{Q_{\sigma}} e^{-U^{\circ}_{\Lambda}} d\mu_{\sigma} \, d\mu^{\circ}_{\Lambda'_1} \, d\mu^{\circ}_{\Lambda'_2} \geq$$

$$\geq \int_{Q^{\circ}_{\Lambda'_1} \times Q^{\circ}_{\Lambda'_2}} e^{-U^{\circ}_{\Lambda_1}} e^{-U^{\circ}_{\Lambda_2}} d\mu^{\circ}_{\Lambda'_1} \, d\mu^{\circ}_{\Lambda'_2} = Z^{\circ}_{\Lambda_1} Z^{\circ}_{\Lambda_2} ,$$

where we have used Jensen's inequality and

$$\int_{Q_\sigma} U^\circ_\Lambda \, d\mu_\sigma = U^\circ_{\Lambda_1} + U^\circ_{\Lambda_2} \, .$$

If Λ_1 and Λ_2 are disjoint then we have directly

$$U^\circ_\Lambda = U^\circ_{\Lambda_1} + U^\circ_{\Lambda_2} \, ,$$

and therefore the equality in b) .

c) Follows trivially from Hölder's inequality.

Let us now introduce the infinite volume energy density α_∞ [15,16], defined by

$$\alpha_\infty = \lim_{\ell \to \infty} - E_\ell / \ell = \sup_\ell - E_\ell / \ell \, ,$$

where E_ℓ is the ground state energy of the volume cut off Hamiltonian H_ℓ introduced in Section 3.

Lemma 8. $p^\circ_\Lambda \leq p_\Lambda \leq \alpha_\infty$.

Proof. The first bound follows from a) of Lemma 7. The second bound is obvious for rectangles $\Lambda(\ell,t)$ of sides ℓ and t, since by the Feynman-Kac-Nelson formula we have

$$Z_{\Lambda(\ell,t)} = \langle \Omega_0, e^{-tH_\ell} \Omega_0 \rangle \leq e^{-tE_\ell} \leq e^{t\ell\alpha_\infty} \, .$$

More general regions can be handled through a limiting procedure as explained in [18].

Now we can take the thermodynamic limit for rectangles $\Lambda(\ell,t)$.

Theorem 9. a) $p_{\Lambda(\ell,t)}$ is monotone increasing in ℓ and t and

$$\lim_{\ell,t \to \infty} p_{\Lambda(\ell,t)} = \alpha_\infty \, .$$

b) For fixed ℓ $\log Z^\circ_{\Lambda(\ell,t)}$ is superadditive in t, the following limit exists

$$\lim_{\ell,t \to \infty} p^\circ_{\Lambda(\ell,t)}$$

and is equal to

$$\alpha^\circ_\infty \equiv \sup_{\ell,t} p^\circ_{\Lambda(\ell,t)} \, .$$

c) $0 \leq \alpha^\circ_\infty \leq \alpha_\infty$.

d) α_∞ and α°_∞ are convex functions of P .

Proof. a) For any $k \geq 1$ we have

$$\langle \Omega_0, e^{-tH_\ell} \Omega_0 \rangle^k \leq \langle \Omega_0, e^{-ktH_\ell} \Omega_0 \rangle \, ,$$

therefore

$$Z^k_{\Lambda(\ell,t)} \leq Z_{\Lambda(\ell,kt)} \qquad \text{and} \qquad p_{\Lambda(\ell,t)} \leq p_{\Lambda(\ell,kt)} \, .$$

On the other hand by the spectral theorem we have $\lim_{t \to \infty} p_{\Lambda(\ell,t)} = - E_\ell / \ell$ and therefore

$$\lim_{\ell,t\to\infty} p_{\Lambda(\ell,t)} = \alpha_\infty .$$

b) The superadditivity of $\log Z^0_{\Lambda(\ell,t)}$ follows from part b) of Lemma 7 , in fact

$$Z^0_{\Lambda(\ell,t)} \geq Z^0_{\Lambda(\ell,t_1)} Z^0_{\Lambda(\ell,t_2)} \qquad \text{for} \quad t = t_1 + t_2 ,$$

then, by a standard argument using superadditivity, we have

$$\lim_{\ell,t\to\infty} p^0_{\Lambda(\ell,t)} = \sup_{\ell,t} p^0_{\Lambda(\ell,t)} \equiv \alpha^0_\infty .$$

c) and d) follow from Lemma 8 and Lemma 7 (part c).

The following theorem, whose proof can be found in [19], establishes the convergence of the pressures for more general regions and the equality of the limits.

Theorem 10. As $\Lambda \to \infty$ (Van Hove)

$$\lim p_\Lambda = \lim p^0_\Lambda = \alpha_\infty = \alpha^0_\infty .$$

In this way we have introduced the first basic thermodynamical quantity, the pressure $\alpha_\infty(\mathcal{P})$ associated to a given interaction.

5. STATES AND ENTROPY.

The basic objectives of constructive Euclidean field theory are to prove the existence of states associated to a given interaction and to study their physical properties. In general a state will be given by some probability measure $\hat{\mu}$ on Q space and the problem arises to see how $\hat{\mu}$ is related to the free measure μ and the interaction $\mathcal{P}$.

First of all let us remark that if $\hat{\mu}$ is translation invariant then it cannot be absolutely continuous with respect to the free measure unless it is trivial. In fact, by the ergodicity of the translation group on $L^1(Q,\Sigma,\mu)$, the relation $d\hat{\mu} = f d\mu$, with $f \in L^1(Q,\Sigma,\mu)$, would imply $f = 1$. This is the Euclidean version of Haag's theorem. We can also look at it from the point of view of Van Hove phenomenon [15].

Consider a volume cut off two-dimensional interacting theory. Then the expectation value of a general observable A can be expressed as

$$\int_Q A \, d\mu_\Lambda = Z_\Lambda^{-1} \int_Q A \, e^{-U_\Lambda} d\mu = \langle \psi_\Lambda , A \psi_\Lambda \rangle ,$$

where we have introduced the Q space normalized wave function of the approximate vacuum

$$\psi_\Lambda = Z_\Lambda^{-\frac{1}{2}} e^{-\frac{1}{2} U_\Lambda} , \quad \psi_\Lambda \in L^p(Q,\Sigma,\mu) , \quad 1 \leq p < \infty .$$

Then using the same techniques of [15] and [16] we can prove the following Euclidean version of the Van Hove phenomenon.

Theorem 11. In the limit $\Lambda \to R^2$ the Van Hove phenomenon holds, in the sense that

$$\psi_\Lambda \to 0 \qquad \text{, weakly in } L^2(Q, \Sigma, \mu) .$$

In general we have that $\|\psi_\Lambda\|_p$ tends to zero if $1 \le p < 2$ and tends to ∞ if $2 < p < \infty$.

Even though $\hat{\mu}$ cannot be absolutely continuous with respect to μ , there is no general principle to prevent the occurrence of such absolute continuity locally, i.e. when the measures $\hat{\mu}$ and μ are restricted to every sub-σ-algebra Σ_Λ , associated to a compact region Λ , at least for two dimensional theories. In fact there are the following indications that such absolute continuity does occur

a) it holds for spatially cut off theories,
b) it can be explicitely verified in the exactly solvable linear and quadratic models,
c) it is true in perturbation theory
d) it is suggested by the locally $\Phi o \kappa$ property of the Hamiltonian theory proved by Glimm and Jaffe [9].
e) it follows form recent results of Newman [29] for the $P(\varphi)_2$ in the small coupling region.

Thus we are led to the following definition.

Definition 12. A smooth state f on (Q, Σ) is a family $\{f_\Lambda\}$ of Q space functions f_Λ , labelled by compact regions Λ of R^2 , so that
a) each f_Λ is Σ_Λ measurable, almost everywhere non negative, in $L^1(Q, \Sigma_\Lambda, \mu)$ and normalized

$$\|f_\Lambda\|_1 = \int_Q f_\Lambda \, d\mu = 1 ,$$

b) the family $\{f_\Lambda\}$ is compatible in the sense that $E_\Lambda f_{\Lambda'} = f_\Lambda$ for regions Λ, Λ' such that $\Lambda \subseteq \Lambda'$, where E_Λ is the conditional expectation with respect to Σ_Λ .

We say that the state is p-smooth for some $p > 1$ if in addition we have

$$f_\Lambda \in L^p(Q, \Sigma_\Lambda, \mu) \qquad \text{for each } \Lambda .$$

By definition the physical expectation $E^{(f)}(A)$ of an observable A , which is Σ_Λ measurable, in the state f is given by

$$E^{(f)}(A) = E(A f_\Lambda) = \int_Q A f_\Lambda \, d\mu .$$

It is clear that the family $\{f_\Lambda\}$ defines a unique measure $\hat{\mu}$ on the smallest σ-algebra containing all Σ_Λ for Λ compact.

The similarity with the corresponding definition in statistical mechanics [33] is evident.

In analogy with statistical mechanics [33] and information theory, we now introduce the concept of entropy for these states.

Definition 13. Let f be a p-smooth state. For each compact region Λ the entropy of f in Λ is given by

$$\mathcal{S}_\Lambda(f) = -\int_Q f_\Lambda \log f_\Lambda \, d\mu = -E^{(f)}(\log f_\Lambda),$$

when f is fixed we write simply $\mathcal{S}(\Lambda)$ in place of $\mathcal{S}_\Lambda(f)$.

Theorem 14. The following inequalities hold

a) (boundedness) $-\infty < \mathcal{S}(\Lambda) \leq 0$,

b) (decrease) $\mathcal{S}(\Lambda') \leq \mathcal{S}(\Lambda)$ if $\Lambda \subseteq \Lambda'$,

c) (weak subadditivity) if $\Lambda = \bigcup_{i=1}^{n} \Lambda_i$, then

$$\mathcal{S}(\Lambda) \leq \sum_{i=1}^{n} \mathcal{S}(\Lambda_i) + \log \| f_{\Lambda_1} \cdots f_{\Lambda_n} \|_1 .$$

Proof. We use the elementary inequality $-\log x \leq x^{-1} - 1$ and Jensen's inequality

$$\int \log f \, d\mu \leq \log \int f \, d\mu$$

where μ is any probability measure and f is a non-negative L^1 function.

a) By Jensen's inequality

$$\int f_\Lambda \log f_\Lambda \, d\mu = (p-1)^{-1} \int f_\Lambda \log f_\Lambda^{p-1} \, d\mu \leq (p-1)^{-1} \log \int f_\Lambda^p \, d\mu < \infty .$$

On the other hand

$$-\int f_\Lambda \log f_\Lambda \, d\mu \leq \int f_\Lambda (f_\Lambda^{-1} - 1) \, d\mu = 0 .$$

b) Using the compatibility condition on f we have

$$\mathcal{S}(\Lambda') - \mathcal{S}(\Lambda) = \int f_\Lambda \log f_\Lambda \, d\mu - \int f_{\Lambda'} \log f_{\Lambda'} \, d\mu =$$

$$= \int f_{\Lambda'} \log (f_\Lambda f_{\Lambda'}^{-1}) \, d\mu \leq \log \int f_\Lambda \, d\mu = 0 .$$

c) As in the proof of b) we have

$$\mathcal{S}(\Lambda) - \sum_{i=1}^{n} \mathcal{S}(\Lambda_i) = \int f_\Lambda \log (f_{\Lambda_1} \cdots f_{\Lambda_n} f_\Lambda^{-1}) \, d\mu \leq \log \int f_{\Lambda_1} \cdots f_{\Lambda_n} \, d\mu .$$

This theorem is very similar to analogous results in statistical mechanics [33] . The main difference is in the correction term given by the logarithm in the weak subadditivity condition. This term is clearly related to the lack of stochastic independence of disjoint regions.

As in statistical mechanics we are interested in the limit of the entropy density $|\Lambda|^{-1} \mathcal{S}(\Lambda)$ as $\Lambda \to \infty$, where $|\Lambda|$ is the area of Λ .

Definition 15. A p-smooth state f is called weakly tempered if there are constants $\eta < 1$ and $a > 0$ such that for all regions Λ with sufficiently large diameter $d(\Lambda)$ we have

$$\log \| f_\Lambda \|_p \leq \exp [a \, d(\Lambda)^\eta] .$$

Weak temperedness is all we need for the control of the infinite volume limit of the entropy density, but we expect that the infinite volume $P(\varphi)_2$ states satisfy a stronger condition:

Definition 16. If for some constant A we have

$$\log \| f_\Lambda \|_p \leq A\, d(\Lambda)^2$$

for all large $d(\Lambda)$ then f is called tempered.

Now we can state the main result of this section.

Theorem 17. Let f be a translation invariant, weakly tempered, p-smooth state. Then the following limit exists

$$s(f) = \lim |\Lambda|^{-1} S_\Lambda(f)$$

as $\Lambda \to \infty$ (Fisher). The functional $s(f)$ is affine in f, i.e. for $0 \leq \lambda_i \leq 1$, $\lambda_1 + \lambda_2 = 1$, we have

$$s(\lambda_1 f_1 + \lambda_2 f_2) = \lambda_1 s(f_1) + \lambda_2 s(f_2).$$

We refer to [18] for a complete proof of this theorem. Here we give all details of the proof in the case of Λ square, in order to show how hypercontractivity, in particular the checkerboard estimate (Theorem 4), allows us to decouple distant regions and get a thermodynamic behaviour.

Proof. Fix η' so that $\eta < \eta' < 1$. For each fixed ℓ, let $d = \ell^{\eta'}$ and $\bar{\ell} = \ell + 2d$. Given ℓ', write $\ell' = n\bar{\ell} + \xi$, with n integer and $0 \leq \xi < \bar{\ell}$. The square Λ' of side ℓ' can be partially filled by n^2 squares $\bar{\Lambda}_{ij}$, $i,j = 1, \ldots, n$, of side $\bar{\ell}$. Each square $\bar{\Lambda}_{ij}$ contains a concentric square Λ_{ij} of side ℓ surrounded by a border of width d. Now we use the decrease and weak subadditivity of the entropy to write

$$S(\Lambda') \leq S(\bigcup_{i,j} \Lambda_{ij}) \leq \sum_{i,j} S(\Lambda_{ij}) + \log \int \prod_{i,j} f_{\Lambda_{ij}}\, d\mu .$$

Then by translation invariance of f and the checkerboard estimate we have

$$S(\Lambda') \leq n^2 S(\Lambda) + n^2 \log \| f_\Lambda \|_{\beta^2},$$

where Λ is a square of side ℓ and

$$\beta = (1 + e^{2md}) / (e^{2md} - e^{-m\bar{\ell}}), \text{ so that}$$

$$\beta^2 - 1 \leq \exp(-c\,\ell^{\eta'}) \quad \text{for large } \ell .$$

Now we let $\ell' \to \infty$ with ℓ fixed. Since $|\Lambda'| = (n\bar{\ell} + \xi)^2$, $|\bar{\Lambda}| = \bar{\ell}^2$, $|\Lambda| = \ell^2$, we have

$$\lim_{\ell' \to \infty} |\Lambda'|^{-1} n^2 = \bar{\ell}^{-2} = |\bar{\Lambda}|^{-1} .$$

Therefore, with the definition

$$\bar{s}(f) = \overline{\lim_{\ell' \to \infty}} |\Lambda'|^{-1} S(\Lambda'), \text{ we have}$$

$$\bar{s}(f) \leq |\bar{\Lambda}|^{-1} S(\Lambda) + \tau(\ell)$$, where

$$\tau(\ell) = |\bar{\Lambda}|^{-1} \log\|f_\Lambda\|_{\beta^2} .$$

Since $|\Lambda||\bar{\Lambda}|^{-1} \to 1$ as $\ell \to \infty$, if we can show that in the same limit $\tau(\ell) \to 0$, then we have

$$\bar{s}(f) \leq \lim_{\ell \to \infty} |\Lambda|^{-1} S(\Lambda)$$

and therefore the existence of the following limit is proved

$$s(f) = \lim_{\ell \to \infty} |\Lambda|^{-1} S(\Lambda) .$$

Now for large ℓ we have $\beta^2 \to 1$, therefore by interpolation

$$\log \|f_\Lambda\|_{\beta^2} \leq \beta^{-2}(\beta^2 - 1)\, p\, (p-1)^{-1} \log \|f_\Lambda\|_p .$$

But by weak temperedness $\log\|f_\Lambda\|_p$ increases at most as $\exp(a\ell^\eta)$, on the other hand we have $\beta^2 - 1 \leq \exp(-c\ell^{\eta'})$, with $\eta < \eta'$, therefore $\tau(\ell) \to 0$ as $\ell \to \infty$.

To complete the proof of the theorem we must show the affinity of $s(f)$. If we put

$$\Delta_\Lambda = \lambda_1 S_\Lambda(f_1) + \lambda_2 S_\Lambda(f_2) - S_\Lambda(\lambda_1 f_1 + \lambda_2 f_2) ,$$

then we find as in the statistical mechanics case [33]

$$\lambda_1 \log \lambda_1 + \lambda_2 \log \lambda_2 \leq \Delta_\Lambda \leq 0 ,$$

where the first inequality follows from the monotonicity of $\log t$ and the second from Jensen's inequality. Dividing by $|\Lambda|$ and taking $\Lambda \to \infty$ we obtain the affinity relation.

6. THE VARIATIONAL PRINCIPLE

In statistical mechanics, a variational principle for the entropy density [32,33] provides a very elegant characterization of the infinite volume equilibrium states associated to a given interaction. In this section we introduce similar ideas in Euclidean field theory.

The variational principle involves three quantities : the pressure, $\alpha_\infty(P)$, which depends only on the interaction P , introduced in Section 4 ; the entropy density, $s(f)$, which depends only on the state f , introduced in Section 5; and finally the mean interaction, $\rho(f, P)$, which depends on both the interaction P and the state f .

A complete discussion of the variational principle should involve the following steps.

a) Isolate a class $\mathcal{F}$ of states f , which is expected to contain all states of physical interest, in particular those associated to a given interation P .

b) Prove that for all states in $\mathcal{F}$ the following inequality holds

$$s(f) - \varrho(f,P) \leq \alpha_\infty(P) \qquad \text{(Gibbs variational inequality).}$$

c) Prove that

$$\sup\big(s(f) - \varrho(f,P)\big) = \alpha_\infty(P) \qquad \text{(Gibbs variational principle),}$$

where the supremum must be taken with respect to all $f \in \mathcal{F}$.

d) Prove the existence of a state f_P (or a family of states) such that

$$s(f_P) - \varrho(f_P, P) = \alpha_\infty(P) .$$

The states f_P are, by definition the equilibrium states associated to the given interaction P .

e) Prove that this notion of equilibrium states agrees with the definition given by means of the equilibrium equations of the Dobrushin, Lanford, Ruelle type discussed in the next section.

We take $\mathcal{F}$ to be the class of translation invariant, weakly tempered states. In this section we prove b) and c) and in the next section we will give the available partial results about problem e) . Unfortunately, at the present stage of development of the theory, it is not possible to answer d) , since very little is known about the infinite volume states as measures in general. Therefore our results are far from definitive, but they strongly support the idea that a variational principle can be used to characterize the infinite volume equilibrium states in Euclidean field theory.

Definition 18. Given a p-smooth ($p>1$) translation invariant state f and an interaction polynomial P , we define the mean interaction associated to the compact region Λ by

$$\varrho_\Lambda(f,P) = |\Lambda|^{-1}\int_Q U_\Lambda f_\Lambda \, d\mu .$$

Proposition 19. $\varrho_\Lambda(f,P)$ has a value, $\varrho(f,P)$, independent of Λ .

Proof. If Λ_1 and Λ_2 are two regions with intersection of zero Lebesgue measure, then, for $\Lambda = \Lambda_1 \cup \Lambda_2$, we have

$$|\Lambda|\varrho_\Lambda = \int_Q U_{\Lambda_1} f_\Lambda \, d\mu + \int_Q U_{\Lambda_2} f_\Lambda \, d\mu = |\Lambda_1|\varrho_{\Lambda_1} + |\Lambda_2|\varrho_{\Lambda_2} ,$$

where we have used the locality of U_Λ and the compatibility condition for f . Since $\Lambda \to \varrho_\Lambda$ is continuous in Λ , using translation invariance, it easily follows that ϱ_Λ is indipendent of Λ .

The step b) of the previous discussion is now straightforward.

Theorem 20. (Gibbs variational inequality). For any semibounded interaction polynomial P and any weakly tempered translation invariant state f , we have

$$s(f) - \varrho(f,P) \leq \alpha_\infty(P) .$$

Proof. By Jensen's inequality we have

$$S_\Lambda(f) - |\Lambda|\,\rho(f,P) = \int_Q f_\Lambda \log(e^{-U_\Lambda} f_\Lambda^{-1})\,d\mu \le \log \int_Q e^{-U_\Lambda}\,d\mu = \log Z_\Lambda ,$$

dividing by $|\Lambda|$ and taking $\Lambda \to \infty$ we conclude the proof.

Let us now consider a class of non trivial translation invariant smooth states, very near to equilibrium states for a given interaction, defined as follows. Consider two orthogonal families of parallel lines at distance ℓ, partitioning the plane R^2 in squares of side ℓ, $\Lambda_1, \Lambda_2, \ldots$. Call σ the region of R^2 made by these lines. Then the free Markov field can be written as a countable sum of independent fields

$$\varphi = \varphi_\sigma + \varphi_1 + \varphi_2 + \ldots$$

Here φ_σ describes the degrees of freedom on σ and is given by $\varphi_\sigma(f) = \varphi(e_\sigma f)$, where e_σ is the projection in N on the subspace N_σ associated to σ as explained in Section 4. On the other hand each φ_i is defined as

$$\varphi_i(f) = \varphi((1 - e_{\Lambda'_i}) f) = \varphi^0_{\Lambda'_i}(f)$$

and describes the residual degrees of freedom in each square Λ_i . Recall that, as in Section 4 , Λ'_i is the closed complement of Λ_i , therefore φ_i is the Dirichlet field in Λ_i and is zero outside Λ_i . For the probability space (Q, Σ, μ) of the free Markov field we have

$$Q = Q_\sigma \times Q_1 \times Q_2 \times \ldots ,$$

$$\Sigma = \Sigma_\sigma \times \Sigma_1 \times \Sigma_2 \times \ldots ,$$

$$\mu = \mu_\sigma \times \mu_1 \times \mu_2 \times \ldots ,$$

in terms of the probability spaces $(Q_\sigma, \Sigma_\sigma, \mu_\sigma)$ and (Q_i, Σ_i, μ_i) associated to φ_σ and φ_i respectively, $i = 1, 2, \ldots$.

Consider now the state f given by the measure

$$\bar\mu = \mu_\sigma \times \bar\mu_1 \times \bar\mu_2 \times \ldots ,$$

where $\quad d\bar\mu_i = Z_1^{-1}\, e^{-U_i}\, d\mu_i$

Here U_i is the Dirichlet interaction in the region Λ_i

$$U_i = \int_{\Lambda_i} :P(\varphi_i(x)):\,dx ,$$

and $Z_1 = \int_{Q_1} e^{-U_1}\,d\mu_1$ is the Dirichlet partition function for a square of side ℓ .

Proposition 21. The state f introduced above is p-smooth for any $p < \infty$ and tempered.

Proof. In fact consider $\Lambda = \bigcup_{i=1}^{n} \Lambda_i$, so that $|\Lambda| = n\ell^2$, then

$$f_\Lambda = Z_1^{-n} \, e^{-\sum_{i=1}^{n} U_i}$$

Since the fields φ_i are independent, by translation invariance we have

$$\log \| f_\Lambda \|_p = c_p |\Lambda| \qquad \text{with}$$

$$c_p = \ell^{-2} Z_1^{-1} \| e^{-U_1} \|_p \, .$$

Proposition 22. For the state f introduced above the entropy density $s(f)$ is expressed as

$$s(f) = \ell^{-2} \log Z_1 + \ell^{-2} Z_1^{-1} \int_{Q_1} U_1 e^{-U_1} \, d\mu_1$$

and the mean interaction is given by

$$\rho(f, P) = \ell^{-2} Z_1^{-1} \int_{Q_1} U_1 \, e^{-U_1} d\mu_1$$

Proof. The expression of $s(f)$ follows from straightforward calculation. For ρ we must take into account that

$$U_\Lambda = \int_\Lambda : P(\varphi(x)) : dx = \sum_{i=1}^{n} U_{\Lambda i} \qquad \text{and}$$

$$\int_{Q_6} U_{\Lambda_i} \, d\mu_6 = U_i \qquad , \; i = 1, 2, \ldots ,$$

as explained in Section 4.

The state f defined above is invariant only for lattice translations of amplitude ℓ, but a translation invariant state $\bar{f}$ can be easily obtained by averaging over translations, according to a well known procedure, in such a way that

$$E^{(\bar{f})}(A) = \int_{\Lambda_1} E^{(f)}(T_x A) \, dx \, ,$$

where A is a Q space function and T_x, $x \in R^2$, is the operator implementing the translation $R^2 \to R^2 + x$. As in statistical mechanics [33] this averaging procedure does not change the entropy density and the mean interaction.

Theorem 23. (Gibbs variational principle). For any semibounded polynomial P

$$\alpha_\infty (P) = \sup \left(s(f) - \rho(f, P) \right) ,$$

where the supremum is taken over all smooth, translation invariant, tempered states.

Proof. For the state $\bar{f}$ constructed before we have

$$s(\bar{f}) - \rho(\bar{f}, P) = \ell^{-2} \log Z_1 \, ,$$

by Proposition 22. But we know from Theorem 10 that, for ℓ large enough, the right hand side can be taken as near to α_∞ as we like and the theorem is established.

Consider now the anharmonic oscillator with Hamiltonian $H = H_0 + P(q)$, where P is a polynomial bounded below, and let Ω be the ground state and E the ground state energy. The free Markov field in the Euclidean (imaginary time) region for the harmonic oscillator is a Gaussian stochastic process $q(t)$, $t \in R$, defined by the expectations $E(q(t))=0$ and $E(q(t)q(t'))=exp(-|t-t'|)$ Using the Feynman-Kac-Nelson formula it is very simple [18] to write the Euclidean infinite volume state associated to the giveninteraction P . For each finite interval $[a,b]$ the system of compatible densities is given by

$$f_{[a,b]} = (J_a \Omega)(J_b \Omega)\, e^{-\int_a^b P(q(t))dt}\, e^{(b-a)E}$$

where J_t are the embedding operators introduced in Section 2. It is simple to verify that $f_{[a,b]}$ is a positive, normalized, $L^\infty(Q, \Sigma_{[a,b]}, \mu)$ function on the Q space of the free process $q(t)$. Here $\Sigma_{[a,b]}$ is the σ-algebra generated by $q(t)$ for $a \leq t \leq b$. . Moreover for large $b-a$ we have

$$\log \| f_{[a,b]} \|_p \leq c_p (b-a),$$

therefore the state f is tempered.

Now by explicit simple computations we can prove the following theorem.

Theorem 24. For the one-dimensional Markov theory (the anharmonic oscillator in the Euclidean region) we have

$$\alpha_\infty(P) = -E\ ,\quad \rho(f,P) = \langle \Omega, P\Omega \rangle\ ,\quad s(f) = -\langle \Omega, H_0 \Omega \rangle,$$

therefore the variational principle for the entropy is verified, because $(H_0+P)\Omega = E\Omega$ and moreover it is equivalent to the usual Rayleigh-Ritz variational principle for the ground state energy.

This example shows that the entropy density is the Euclidean counterpart of the free Hamilton expectation value. In the case of non-relativistic quantum mechanics general results of this type have also been obtained by W. Crutchfield [2]. The one dimensional example suggests the following conjecture for the two-dimensional Euclidean theory.

Conjecture 25. The variational principle for the entropy density in the two-dimensional Euclidean theory is equivalent to a Rayleigh-Ritz variational principle for the energy density in the Hamiltonian theory.

From this point of view the infinite volume states of the Hamiltonian theory associated to a given interaction would be characterized as those states for which the total energy density takes the minimum value $-\alpha_\infty(P)$.

7. THE EQUILIBRIUM EQUATIONS

In classical statistical mechanics Dobrushin [3,4,5] and Lanford and Ruelle [21] introduced a set of relations for the infinite volume states, which express in a compact way the fact that the state is an equilibrium state for a system with some given interaction at some fixed temperature. Following our interpretation of Euclidean field theory as classical statistical mechanics we will introduce equations of the same type, according to the basic intuitive idea, following from Markov property, that a change of interaction outside a bounded region Λ can affect the state restricted to Λ only through a multiplicative factor concentrated at the boundary.

Definition 26. A measure $\hat{\mu}$ on the measurable space (Q, Σ) of the free Markov field is called Gibbsian for the interaction $\mathcal{P}$ ($\mathcal{P}$-Gibbsian) if and only if, for every compact region Λ in R^2, we have

i) the restriction of $\hat{\mu}$ to Σ_Λ is absolutely continuous with respect to the free measure μ, i.e. $d\hat{\mu} = f_\Lambda d\mu$ on Σ_Λ, with f_Λ in $L^1(Q, \Sigma_\Lambda, \mu)$,

ii) for any bounded, Σ_Λ measurable Q space function A we have

$$\hat{E}_{\partial\Lambda}(A) = E_{\partial\Lambda}(A\, e^{-U_\Lambda}) \,/\, E_{\partial\Lambda}(e^{-U_\Lambda}) \, ,$$

where $\hat{E}_{\partial\Lambda}$ and $E_{\partial\Lambda}$ denote conditional expectations with respect to $\Sigma_{\partial\Lambda}$ for the measures $\hat{\mu}$ and μ respectively.

As a motivation for the assumption that $\mathcal{P}$-Gibbsian states are the relevant physical states associated to a given interaction, we consider the case in which the measure $\hat{\mu}$ is obtained through a limiting procedure from the approximate measures

$$d\mu_{\Lambda'} = Z_{\Lambda'}^{-1} \; e^{-U_{\Lambda'}} \, d\mu \, ,$$

corresponding to volume cut off interacting theories, taking $\Lambda' \to R^2$. Using the Markov property then it is easy to verify that $\mu_{\Lambda'}$ is $\mathcal{P}$-Gibbsian in every region Λ contained in Λ', therefore $\hat{\mu}$ will be $\mathcal{P}$-Gibbsian in general.

In classical statistical mechanics of lattice systems the equilibrium equations are very powerful, since they allow to find all properties of the equilibrium states [3,4,5]. But in Euclidean field theory the situation is similar to the case of classical continuous systems studied by Ruelle [34]. In fact it is very easy to see that the equilibrium equations of Definition 26 are not sufficient to characterize completely the "physical" states, but must be supplemented by some restriction on the allowed behaviour of states at infinity, otherwise unphysical spurious states can appear as solutions of the equilibrium equations. A similar situation is found in quantum statistical mechanics in the study of states satisfying the KMS conditions [22].

We give an example of the occurrence of spurious solutions in Euclidean field theory and then propose that the right boundary condition on the states is weak temperedness.

For our example we consider the case of zero interaction for the one-dimensional Markov field on the real line, characterized by the free covariance

$$E\big(q(t)\,q(t')\big) = \exp\left(-|t-t'|\right)$$

The same method works for a linear or quadratic interaction and in several dimensions.

Proposition 27. For any interval $[a,b]$ consider the system of densities

$$f_{[a,b]} = \exp\left(-\tfrac{1}{2}c^2 e^{2b}\right) \exp\left(c e^b q(b)\right)$$

for real c . This system is normalized and consistent, moreover it is Gibbsian for zero interaction.

Proof. A simple computation.

Since the "physical" state for zero interaction must be the free state (corresponding to $c=0$) we see that all cases $c \neq 0$ must be considered spurious even though they satisfy the equilibrium equations. On the other hand by explicit computation we have

$$\log \| f_{[a,b]} \|_p = \tfrac{1}{2} c^2 e^{2b} (p-1)$$

therefore only the case $c=0$ gives a weak tempered state.

The example strongly suggests the idea that weak temperedness is the right boundary condition to be imposed to $\mathbf{P}$-Gibbsian states in order to characterize completely the equilibrium states associated to a given interaction. Further support comes from the following (partial) discussion about the possible equivalence of the two notions of equilibrium state given by the variational principle and the $\mathbf{P}$-Gibbsian condition.

First of all let us remark that for a $\mathbf{P}$-Gibbsian state we have the following form of the densities

$$f_\Lambda = \tilde{Z}_{\partial\Lambda}\, e^{-U_\Lambda} f_{\partial\Lambda} ,$$

where $\tilde{Z}_{\partial\Lambda} = E_{\partial\Lambda}(e^{-U_\Lambda})$ is the conditional partition function (see [18]).

Proposition 28. For any Gibbsian state

$$|\Lambda|^{-1} S_\Lambda(f) = \rho_\Lambda(f,P) + |\Lambda|^{-1} S_{\partial\Lambda}(f) + E(f_{\partial\Lambda} \tilde{p}_{\partial\Lambda}) ,$$

where $\tilde{p}_{\partial\Lambda} = |\Lambda|^{-1} \log \tilde{Z}_{\partial\Lambda}$ is the conditional pressure.

Proof. We have

$$\log f_\Lambda = -U_\Lambda + \log f_{\partial\Lambda} + |\Lambda|\, \tilde{p}_{\partial\Lambda} .$$

Multiplying by $-|\Lambda|^{-1} f_\Lambda$ and integrating we get the proposition.

Now it is easily proved, using the properties of the entropy and the basic hypercontractive estimate, that for any weakly tempered, translation invariant state we have $|\Lambda|^{-1} S_{\partial\Lambda}(f) \to 0$ as $\Lambda \to \infty$ (see [18]). The conclusion is that for any weakly tempered, translation invariant, Gibbsian state we have

$$\lim_{\Lambda \to \infty} E(f_{\partial\Lambda} \tilde{p}_{\partial\Lambda}) = S(f) - \rho(f,P) \leq \alpha_\infty(P) .$$

Therefore if it is possible to prove that the conditional pressure $\tilde{p}_{\partial\Lambda}$ converges to α_∞ in a suitable sense as $\Lambda \to \infty$, then the following conjecture would be verified.

Conjecture 29. Every weakly tempered, translation invariant, $\mathcal{P}$-Gibbsian state satisfies the variational equality.

We would like to conclude this section by referring to recent announced results of Dobrushin and Minlos [6], which show that a complete control of the equilibrium equations can be extremely important for the actual construction of the theory and the verification of the conventional wisdom [43,20,1] about dynamical instability and broken symmetry.

Theorem 30. (Dobrushin and Minlos, announced in [6]). For all polynomials $\mathcal{P}$ bounded below, the set of all Euclidean invariant, $\mathcal{P}$-Gibbsian states $\hat{\mu}$, for which the integrals $\int_Q \varphi(f)^n d\hat{\mu}$ exist for any integer n and depend continuously on $f \in \mathcal{S}(R^2)$, is a non empty convex set $W(P)$. The extremal elements of $W(P)$ are ergodic with respect to the Euclidean translations and $W(P)$ is a Choquet simplex, i.e. any element of $W(P)$ can be represented as the baricenter of some probability measure on the extremal elements of $W(P)$.

Let $P(X) = \lambda \bar{P}(X)$, $\lambda \geq 0$, then there is a $\lambda(\bar{P}) > 0$ such that for all $\lambda \leq \lambda(\bar{P})$ the set $W(P)$ consists of exactly one element. If $\bar{P}$ is even then there is a $\lambda'(\bar{P}) < \infty$ such that $W(P)$ for $\lambda \geq \lambda'(\bar{P})$ contains at least two different extremal states.

REFERENCES

[1] R. BAUMEL, Princeton University Thesis, 1973.

[2] W. CRUTCHFIELD, Princeton University Senior Thesis, 1973.

[3] R.L.DOBRUSHIN, Gibbsian Random Fields for Lattice Systems with Pairwise Interactions, Funct. Anal. Applic. 2 (1968) 292.

[4] R.L. DOBRUSHIN, The Problem of Uniqueness of a Gibbsian Random Field and the Problem of Phase Transitions, Funct. Anal. Applic. 2 (1968) 302.

[5] R.L. DOBRUSHIN, Gibbsian Random Fields, The General Case, Funct. Anal. Applic. 3 (1969) 22.

[6] R.L. DOBRUSHIN and R.A. MINLOS, Construction of a One Dimensional Quantum Field Via a Continuous Markov Field, Moscow Preprint, 1973.

[7] I.I. GIKHMAN and A.V. SKOROKHOD, Introduction to the Theory of Random Processes, W.B. Saunders Co., Philadelphia, 1969.

[8] J. GLIMM, These Proceedings.

[9] J. GLIMM and A. JAFFE, The $\lambda(\varphi^4)_2$ Quantum Field Theory Without Cutoffs.III. The Physical Vacuum, Acta Math. 125 (1970) 203.

[10] J. GLIMM and A. JAFFE, Quantum Field Theory Models, in Statistical Mechanics and Quantum Field Theory, Les Houches 1970, C. DE WITT, R. STORA, Editors, Gordon and Breach, New York, 1971.

[11] J. GLIMM and A. JAFFE, Boson Quantum Field Models, in Mathematics of Contemporary Physics, R. STREATER, Editor, Academic Press, New York, 1972.

[12] J. GLIMM and A. JAFFE, Positivity of the φ^4_3 Hamiltonian, Fort. der Physik, 21 (1973) 327.

[13] J. GLIMM, A. JAFFE and T. SPENCER, The Wightman Axioms and Particle Structure in the $P(\varphi)_2$ Quantum Field Model, New York University Preprint, 1973.

[14] R. GRIFFITHS, Rigorous Results and Theorems, in Phase Transitions and Critical Phenomena, vol. I, ed. C. Comb and M.S. Green, Academic Press, London and

New York, 1972.

[15] F. GUERRA, Uniqueness of the Vacuum Energy Density and Van Hove Phenomenon in the Infinite Volume Limit for Two-Dimensional Self-Coupled Bose Fields, Phys. Rev. Lett. 28 (1972) 1213.

[16] F. GUERRA, L. ROSEN and B. SIMON, Nelson's Symmetry and the Infinite Volume Behavior of the Vacuum in $P(\varphi)_2$, Commun.math.Phys. 27 (1972) 10.

[17] F. GUERRA, L. ROSEN and B. SIMON, Statistical Mechanics Results in the $P(\varphi)_2$ Quantum Field Theory, Phys.Lett. 44B (1973) 102.

[18] F. GUERRA, L. ROSEN and B. SIMON, The $P(\varphi)_2$ Euclidean Quantum Field Theory as Classical Statistical Mechanics, Ann. Math., to appear.

[19] F. GUERRA, L. ROSEN and B. SIMON, in preparation.

[20] A. JAFFE, These Proceedings.

[21] O. LANFORD and D. RUELLE, Observables at Infinity and States with Short Range Correlations in Statistical Mechanics, Commun.math.Phys. 13 (1969) 194.

[22] R.P. MOYA, Equilibrium States for the Infinite Free Bose Gas, University of London, Preprint, 1973.

[23] T. NAKANO, Quantum Field Theory in Terms of Euclidean Parameters, Prog. Theor. Phys. 21 (1959) 241.

[24] E. NELSON, Quantum Fields and Markoff Fields, in Proceedings of Summer Institute of Partial Differential Equations, Berkeley 1971, Amer. Math. Soc., Providence, 1973.

[25] E. NELSON, Construction of Quantum Fields from Markoff Fields, J.Funct.Anal. 12 (1973) 97.

[26] E. NELSON, The Free Markoff Field, J.Funct.Anal. 12 (1973) 211.

[27] E. NELSON, These Proceedings.

[28] E. NELSON, in preparation.

[29] C. NEWMAN, The Construction of Stationary Two-Dimensional Markoff Fields with an Application to Quantum Field Theory, J.Funct.Anal., to appear.

[30] M. REED, These Proceedings.

[31] L. ROSEN, These Proceedings.

[32] D. RUELLE, A Variational Formulation of Equilibrium Statistical Mechanics and the Gibbs Phase Rule, Commun.math.Phys. 5 (1967) 324.

[33] D. RUELLE, Statistical Mechanics, Benjamin, New York, 1969.

[34] D. RUELLE, Superstable Interactions in Classical Statistical Mechanics, Commun.math.Phys. 18 (1970) 127.

[35] J. SCHWINGER, On the Euclidean Structure of Relativistic Field Theory, Proc.Nat.Acad.Sci. 44 (1958) 956.

[36] B. SIMON, Correlation Inequalities and the Mass Gap in $P(\varphi)_2$. I. Domination by the Two Point Function, Commun.math.Phys. 31 (1973) 127.

[37] B. SIMON, Correlation Inequalities and the Mass Gap in $P(\varphi)_2$. II. Uniqueness of the Vacuum for a Class of Strongly Coupled Theories, Toulon University Preprint, 1973.

[38] B. SIMON, These Proceedings.

[39] B. SIMON and R. Griffiths, Griffiths-Hurst-Sherman Inequalities and a Lee-Yang Theorem for the $(\varphi^4)_2$ Field Theory, Phys. Rev. Lett. 30 (1973) 931.

[40] B. SIMON and R. GRIFFITHS, The $(\varphi^4)_2$ Field Theory as a Classical Ising Model, Commun.math.Phys., to appear.

[41] B. SIMON and R. HOEGH-KROHN, Hypercontractive Semigroups and Two Dimensional Self-Coupled Bose Fields, J. Funct.Anal. 9 (1972) 121.

[42] K. SYMANZIK, Euclidean Quantum Field Theory, in Local Quantum Theory, R. JOST, Editor, Academic Press, New York, 1969.

[43] A.S. WIGHTMAN, Constructive Field Theory. Introduction to the Problems, Coral Gables Lectures, 1972.

BOSE FIELD THEORY AS CLASSICAL STATISTICAL MECHANICS. II. THE LATTICE APPROXIMATION AND CORRELATION INEQUALITIES

Lon Rosen*

Mathematics Department

University of Toronto

Toronto, Canada M5S 1A1

These lectures are devoted to the idea that $P(\phi)_2$ is nothing but a model of classical statistical mechanics. Francesco Guerra has already described to you the variational principle and equilibrium equations, and I wish now to discuss the role of correlation inequalities. My lectures consist of three parts:

1. The lattice approximation
2. Proof of correlation inequalities
3. Applications

The purpose of the lattice approximation is to exhibit the ferromagnetic nature of Euclidean Bose field theories, this being the critical ingredient in the proof of correlation inequalities. It turns out that it is the free theory which determines the ferromagnetic properties and so these results are essentially model independent. Nor do they depend on the number of space dimensions, at least on a formal level.

The correlation inequalities that have been established in Bose field theories are of types G-I, G-II, FKG, and GHS . Here G-I and G-II refer to Griffiths' first and second inequalities respectively [7,4], FKG to Fortuin, Kasteleyn and Ginibre [2], and GHS to Griffiths, Hurst and Sherman [9]. In the statistical mechanics

* Research partially supported by USNSF under grant GP39048.

context, the first three types (G-I, G-II, FKG) hold for quite general Ising models (in particular, continuous spins and fairly arbitrary single spin distributions) while the GHS inequalities seem to apply only to "classical Ising models" (i.e. spin $\frac{1}{2}$ ferromagnetic pair interactions). A further approximation (the "classical Ising approximation") is needed in order to bring GHS to Bose field theories and this will be the subject of Barry Simon's lectures.

The main reference for these lectures will be the paper of Guerra-Rosen-Simon [10]. Since it will not be possible for me to supply all of the technical details in the time available, I hope that by referring freely to [10] I can concentrate on the main ideas. Much of the relevant material and some of the results that I quote can be found in Ed Nelson's lectures. Since these lectures are devoted exclusively to correlation inequalities, the reader should refer to the lectures of Guerra and Jaffe for the general statistical mechanics setting and to Jaffe's lectures for a comprehensive survey of recent results.

1. The Lattice Approximation

$\phi(x)$ denotes the Euclidean Bose field over $\mathcal{H}_{-1}(\mathbb{R}^d)$, as described in Nelson's lectures, and $d\mu$ is the corresponding free measure on Q-space (see also [13,10]). Briefly, the lattice approximation is a type of ultraviolet cutoff in which Euclidean space $\mathbb{R}^d$ is replaced by the lattice $\delta Z^d = \{n\delta \mid n \in Z^d\}$ with spacing $\delta > 0$ and the Laplacian Δ in the Euclidean propagator is approximated by a finite difference operator. By restricting to a finite dimensional subspace of Q-space, we obtain, instead of the full measure $d\mu$, a Gaussian

$$d\mu \to (2\pi)^{-N/2} |B|^{1/2} e^{-\frac{1}{2}\vec{q}\cdot B\vec{q}} \, d^N q \tag{1.1}$$

in terms of a finite number of variables $q_1, \ldots, q_N$ each associated with a lattice point. Here B is a positive definite $N \times N$ matrix which is <u>ferromagnetic</u> in the sense that its off-diagonal elements are nonpositive, and $|B| = \det B$.

The ferromagnetic property can be understood intuitively as follows. Formally, the free field measure is just the Gaussian

$$d\mu \quad = \quad \text{const. } e^{-\frac{1}{2}(\phi,(-\Delta+m^2)\phi)}d\phi \quad . \tag{1.2}$$

Now $-\Delta$ is positive "on-diagonal" and negative "infinitesimally off-diagonal" as is evident from its finite difference approximation $(\delta>0)$:

$$(-\Delta_\delta f)(n\delta) \quad = \quad \delta^{-2}[2df(n\delta) - \sum_{|n'-n|=1} f(n'\delta)] \tag{1.3}$$

where we norm Z^d by $|(n_1,\ldots,n_d)| = |n_1| +\ldots+ |n_d|$ so that the sum over $n' \in Z^d$ takes place over the $2d$ nearest neighbours of n

To convert this remark to a rigorous theorem, we introduce the Fourier transform from $\ell^2(Z^d)$ to $L^2(T_\delta{}^d)$ where $T_\delta = [-\pi/\delta, \pi/\delta]$:

$$\hat{h}_\delta(k) \quad = \quad \left(\frac{\delta}{2\pi}\right)^{d/2} \sum_{n\in Z^d} h(n)e^{-ik\cdot n\delta} \quad . \tag{1.4}$$

If we regard $A = (-\Delta_\delta+m^2)$ as an operator on $\ell^2(Z^d)$ by means of (1.3), then we see that it is a convolution operator $(Ah)(n) = \sum_{n'} a(n-n')h(n')$, where

$$a(n) \quad = \quad \begin{cases} m^2 + 2d\delta^{-2} & n = 0 \\ -\delta^{-2} & |n| = 1 \\ 0 & \text{otherwise.} \end{cases} \tag{1.5}$$

Therefore its image $\hat{A}$ on $L^2(T_\delta^d)$ is multiplication by $(2\pi/\delta)^{d/2}\hat{a}_\delta(k)$. By a simple computation

$$(2\pi/\delta)^{d/2}\hat{a}_\delta(k) \quad = \quad 2\delta^{-2}\left(d-\sum_{i=1}^{d}\cos(\delta k_i)\right) + m^2$$

$$\equiv \quad \mu_\delta(k)^2 \quad .$$

Note that as $\delta \to 0$, $\mu_\delta(k) \to \mu(k) = (k^2+m^2)^{1/2}$.

We wish the cutoff field $\phi_\delta(n)$ to have covariance matrix $C = \delta^{-d}A^{-1}$; i.e.,

$$C_{nn'} \equiv \int \phi_\delta(n)\phi_\delta(n')d\mu = (2\pi)^{-d}\int_{T_\delta^d} e^{ik\cdot(n-n')\delta}\mu_\delta(k)^{-2}dk. \quad (1.6)$$

Accordingly, we make the following definitions:

Definitions 1.1. The lattice cutoff field $\phi_\delta(n)$ is defined in terms of the field $\phi(x)$ by $\phi_\delta(n) = \phi(f_{\delta,n})$ where

$$f_{\delta,n}(x) = (2\pi)^{-d}\int_{T_\delta^d} e^{ik\cdot(x-n\delta)} \frac{\mu(k)}{\mu_\delta(k)} dk .$$

(For $C_{nn'} = \langle f_{\delta,n} , f_{\delta,n'} \rangle_{\mathcal{H}_{-1}}$ and the $\mu(k)$'s cancel with the $\mu(k)^{-2}$ in the definition of the inner product, yielding (1.6).)

Wick powers of $\phi_\delta(n)$ are defined in the usual way, e.g.,

$$:\phi_\delta(n)^2: = \phi_\delta(n)^2 - C_{00} \quad (1.7)$$

and smeared powers by

$$:\phi_\delta^r:(g) = \sum_{n\in Z^d} \delta^d :\phi_\delta(n)^r: g(n\delta)$$

where $g \in C_0^\infty(\mathbb{R}^d)$.

For the $P(\phi)_d$ theory the (smeared) Schwinger functions with space cutoff Λ and lattice cutoff δ are defined by

$$S_{\Lambda,\delta}(h_1,\ldots,h_r) = \frac{\int \phi_\delta(h_1)\ldots\phi_\delta(h_r)e^{-U_{\Lambda,\delta}}d\mu}{\int e^{-U_{\Lambda,\delta}}d\mu} . \quad (1.8)$$

Here Λ is a bounded region in $\mathbb{R}^d$, $h_1,\ldots,h_r \in C_0^\infty(\mathbb{R}^d)$, and

$$U_{\Lambda,\delta} = \sum_{n\delta\in\Lambda_\delta} \delta^d : P(\phi_\delta(n)) :$$

where P is a semibounded polynomial, and $\Lambda_\delta = \Lambda \cap (\delta Z^d)$ denotes the lattice points enclosed by Λ .

What have these definitions accomplished? In the first place the integrands in (1.8) involve only a finite number of field variables, namely $q_n \equiv \phi_\delta(n)$ for $n\delta \in \Lambda_\delta$ (where we assume for convenience that $\text{supp}\, h_j \subset \Lambda$). Thus by the definition of $d\mu$ (see Reed's lectures), the numerator in (1.8) reduces to a sum of terms of the form

$$\text{const.}\int q_{n_1}\cdots q_{n_r}\, e^{-\sum\delta^d :P(q_n):}\, d\mu_{\Lambda,\delta}(q) \tag{1.9}$$

where the Gaussian measure

$$d\mu_{\Lambda,\delta}(q) \;=\; (2\pi)^{-N/2}|C_\Lambda|^{-\frac{1}{2}}\, e^{-\frac{1}{2}q\cdot C_\Lambda^{-1}q}\, d^N q \quad . \tag{1.10}$$

Here $N = |\Lambda_\delta|$, the number of points in Λ_δ , and C_Λ is the $N\times N$ covariance matrix of the q's with entries defined by (1.6), but restricted to the indices in Λ . Similarly for the denominator in (1.8).

Secondly, this approximation is <u>locality preserving</u> in the sense that the interaction is approximated by a sum $\sum\delta^d :P(q_n):$ in which each term involves the field q_n at only a single lattice point. This is not the case for the usual ultraviolet cutoff procedure.

But most critical for our purposes is the structure of the matrix C_Λ^{-1} which occurs in the Gaussian density. By construction, the infinite matrix C has a particularly simple inverse, $C^{-1} = \delta^d A$ with entries defined by (1.5). Now of course $(C_\Lambda)^{-1} \neq (C^{-1})_\Lambda = \delta^d A_\Lambda$, but equality almost holds:

<u>Theorem 1.2</u> [10] If $\Lambda \subset \mathbb{R}^d$ is bounded, then $C_\Lambda^{-1} = \delta^d(A_\Lambda - B_{\partial\Lambda})$ where $B_{\partial\Lambda}$ is a positive semi-definite matrix with nonnegative elements that is "concentrated on the boundary $\partial\Lambda_\delta$ ".

By this we mean the following: given a set Λ_δ of lattice points we let $\Lambda_\delta^{ext} = \delta Z^d \setminus \Lambda_\delta$, $\Lambda_\delta^{int} = \{n\delta \in \Lambda_\delta | n'\delta \in \Lambda_\delta$ if $|n-n'| = 1\}$, and $\partial\Lambda_\delta = \Lambda_\delta \setminus \Lambda_\delta^{int}$. A matrix B is said to be concentrated on a set R if $B_{nn'} = 0$ unless $n\delta, n\delta' \in R$. The proof of the theorem is based on the representation

$$e^{-\frac{1}{2}q^\Lambda \cdot C_\Lambda^{-1} q^\Lambda} = \text{const.} \lim_{R\to\infty} (2\pi)^{-|R_\delta|/2} |A_R|^{1/2} \int e^{-\frac{1}{2}q^R \cdot \delta^d A_R q^R} dq^{R\setminus\Lambda} \quad (1.11)$$

where q^R stands for the variables on R_δ . Since A_R links only nearest neighbours, it is clear that when the variables $q^{R\setminus\Lambda}$ are integrated out in (1.11) the elements of A_R on Λ^{int} remain unchanged; consequently, $C_\Lambda^{-1} - \delta^d A_\Lambda$ is concentrated on $\partial\Lambda_\delta$.

From the theorem and the definition of A we see that C_Λ^{-1} has nonpositive off-diagonal elements. We call C_Λ^{-1} ferromagnetic in analogy with the Ising ferromagnet [14]. Moreover, except for the boundary variables, C_Λ^{-1} , like A , links only nearest neighbours. This is the Markov property in the lattice setting.

What is the significance of the matrix $B_{\partial\Lambda}$? Different choices of $B_{\partial\Lambda}$ correspond to different choices of boundary condition. The covariance (1.6) of Theorem 1.2 is that of free B. C. On the other hand, if we take $B_{\partial\Lambda} = 0$, we obtain the lattice theory with Dirichlet B.C. This statement is justified by Theorem 1.3 below (see also [10, §IV.3]) and is based on the observation that the finite difference approximation to the operator $(-\Delta_\Lambda + m^2)$ with Dirichlet B.C. on $\partial\Lambda$ is just A_Λ since the effect is to ignore the variables immediately outside Λ_δ . We obtain free B.C. by "integrating out" the variables in Λ_δ^{ext} and Dirichlet B.C. by setting the variables in Λ_δ^{ext} equal to zero. The simple expression for the Dirichlet covariance on the lattice

$$(C_\Lambda^D)^{-1} = \delta^d A_\Lambda \quad (1.12)$$

shows that Dirichlet B.C. are in some sense more natural than free B.C. Note also that free B.C. on Λ_δ are obtained by restricting the covariance operator $\left(C_\Lambda = C\restriction \ell^2(\Lambda_\delta)\right)$ whereas Dirichlet B.C. are obtained by restricting the <u>inverse</u> of the covariance operator $\left((C^D_\Lambda)^{-1} = C^{-1}\restriction \ell^2(\Lambda_\delta)\right)$. The usefulness of Dirichlet B.C. lies in the fact that the variables in Λ_δ are decoupled from those in Λ_δ^{ext} (as has been emphasized already in Glimm's lectures). We shall not have occasion to employ the other types of B.C. obtained by more general choices of the boundary matrix $B_{\partial\Lambda}$.

So far I have only been talking about the non-interacting or free theory and now I wish to include the interaction, paying some care to the meaning of the Wick dots. Let us denote the free Dirichlet measure by

$$d\mu^D_{\Lambda,\delta} = (\delta^d/2\pi)^{N/2}|A_\Lambda|^{1/2}e^{-\frac{1}{2}q\cdot\delta^d A_\Lambda q}\, d^N q \quad . \tag{1.13}$$

There are two logical possibilities for the definition of Wick ordering: we can order with respect to $d\mu^D_{\Lambda,\delta}$, e.g. (assuming $n\delta \in \Lambda$),

$$:q_n^2:_{D,\Lambda} = q_n^2 - \int q_n^2\, d\mu^D_{\Lambda,\delta} = q_n^2 - (C^D_\Lambda)_{nn} , \tag{1.14}$$

or with respect to $d\mu_{\Lambda,\delta}$ as in (1.7), i.e.,

$$:q_n^2: = q_n^2 - \int q_n^2\, d\mu_{\Lambda,\delta} = q_n^2 - C_{00} , \tag{1.15}$$

and similarly for higher powers. The first choice is perhaps more natural and gives rise to what we call the <u>$P(\phi)_d$ Dirichlet lattice measure with space cutoff Λ</u> :

$$d\nu^D_{\Lambda,\delta} = \frac{\exp\left(-\sum_{n\delta\in\Lambda}\delta^d :P(q_n):_{D,\Lambda}\right)d\mu^D_{\Lambda,\delta}}{\int\exp\left(-\sum\delta^d :P(q_n):_{D,\Lambda}\right)d\mu^D_{\Lambda,\delta}} ; \tag{1.16}$$

whereas we call the second choice the <u>half-Dirichlet</u> lattice measure:

$$d\nu^{HD}_{\Lambda,\delta} = \frac{\exp\left(-\sum_{\Lambda} \delta^d \ :P(q_n):\right) d\mu^D_{\Lambda,\delta}}{\int \exp\left(-\sum \delta^d \ :P(q_n):\right) d\mu^D_{\Lambda,\delta}} \ . \tag{1.17}$$

The advantage of the half-Dirichlet state is that the definition of Wick ordering does not change with Λ (see (1.14) and (1.15)). We shall return to this point in the discussion of the infinite volume limit in §3. The lattice Schwinger functions are defined in the obvious way, e.g.

$$S^D_{\Lambda,\delta}(h_1,\ldots,h_r) = \int \phi_\delta(h_1)\ldots\phi_\delta(h_r)\, d\nu^D_{\Lambda,\delta} \tag{1.18}$$

where $h_j \in C_0^\infty(\Lambda)$.

Let me restate the essential properties of the lattice theories in the following way: whether we consider free, Dirichlet, half-Dirichlet, or more general B.C., the free lattice theory is an array of Gaussian spins with interactions of ferromagnetic type that are nearest neighbours except possibly for the boundary variables. The Gibbs factor arising from the polynomial P is local and does not couple spins but only mediates the interactions of spins by perturbing the distribution of each uncoupled spin. From this point of view, it is the free theory which determines the basic properties of the model. This realization of the $P(\phi)_d$ theory as an Ising ferromagnet leads to the correlation inequalities of the next section.

In closing this section I ought to justify the word "approximation" in the title. We need something to which the lattice theories can converge, and so we take $d \leq 2$ in order to soften the ultraviolet divergences. The convergence of the theory with free B.C. as $\delta \to 0$ is then just a variation of the standard semi-boundedness proof and removal of an ultraviolet cutoff (see Nelson's lectures). In particular, objects like $\phi_\delta(h)$ and $e^{-U_{\Lambda,\delta}}$ converge in $L^p(Q,d\mu)$ for any $p < \infty$ so that by Hölder's inequality, the smeared Schwinger functions converge. In the case of Dirichlet and half-Dirichlet B.C. some further arguments are required and we must impose some mild regularity conditions on Λ . The issue for half-Dirichlet B.C. is to show that $e^{-U_\Lambda} \in L^p(Q, d\mu_\Lambda^D)$ in spite of the Wick ordering being "wrong" In two dimensions this is so because the differences in Wick ordering involve coefficients with at most logarithmic singularities. For details see [10]. Thus:

Theorem 1.3. [10] Suppose $d = 1$ or 2 and let P be a semibounded polynomial. Then as $\delta \to 0$ the smeared Schwinger functions with free, Dirichlet or half-Dirichlet B.C. converge to the corresponding Schwinger functions of the continuum theories.

In principle, these techniques and the attendant correlation inequalities should hold when $d > 2$. Of course one must add appropriate counterterms depending on δ and then prove convergence as the ultraviolet cutoff $\delta \to 0$. We expect that when $d > 2$ half-Dirichlet states will not exist since the required counterterms will be Λ dependent.

2. Correlation Inequalities

I wish next to outline the proof of G-I and G-II for the lattice model. By virtue of the approximation theorem (Theorem 1.3) these inequalities extend immediately to the spatially cutoff continuum $P(\phi)_2$ theory and to the infinite volume $P(\phi)_2$ theories for which the convergence of the Schwinger functions is known. Our proof is adapted directly from Ginibre's [4] elegant analysis of Griffiths' inequalities. G-I was first proved for the special case $(\phi^4)_2$ by Symanzik [17] using somewhat different methods.

Motivated by the lattice measures of the previous section, we define a ferromagnetic measure on $\mathbb{R}^N$ to be of the form

$$d\mu(q) = F_1(q_1) \ldots F_N(q_N) e^{-q \cdot Aq} d^N q$$

where A is an $N \times N$ positive-definite matrix with non-positive off-diagonal elements and the F_i are positive, bounded, continuous functions on $\mathbb{R}$. If each F_i is even we say that μ is <u>even</u>. We write $\langle f \rangle_\mu \equiv \int f(q) d\mu(q) / \int d\mu(q)$. Following Ginibre we let $\mathcal{G}_i$ be the set of functions of q_i of the form $f(q_i) = \varepsilon(q_i) g(|q_i|)$ where g is a positive, increasing, polynomially bounded function on $\mathbb{R}_+ = (0,\infty)$ and $\varepsilon(q) = 1$ or $\operatorname{sgn} q$. We let $\mathcal{G}$ be the set of functions on $\mathbb{R}^N$ that are sums of products $f_1(q_1) \ldots f_N(q_n)$ where $f_i \in \mathcal{G}_i$. An example of a function in $\mathcal{G}$ is the monomial $q_1^{i_1} \ldots q_N^{i_N}$ with $i_1, \ldots, i_N$ non-negative integers. Note also that $\mathcal{G}$ is closed under addition and multiplication. We shall prove:

<u>Theorem 2.1</u>. If μ is an even ferromagnetic measure on $\mathbb{R}^N$, and if $f, g \in \mathcal{G}$, then

$$\langle f \rangle_\mu \geq 0 \tag{2.1}$$

$$\langle fg \rangle_{\mu,T} \equiv \langle fg \rangle_\mu - \langle f \rangle_\mu \langle g \rangle_\mu \geq 0 . \tag{2.2}$$

Typically, the "interaction" F_i is of the form $e^{-\beta P_i(q_i)}$, P_i a polynomial, as in the lattice field theories of the previous section. The temptation is to expand this exponential (after the standard proofs of Griffiths' inequalities). This does not work. The correct point of view for correlation inequalities is to regard F_i as part of the free measure and to think of the <u>coupling terms in $d\mu$ as the interaction</u>. Thus we shall expand the <u>off-diagonal</u> part of the Gaussian. Let A_0 be the diagonal part of A and let $A_\beta = A_0 + \beta(A - A_0)$ interpolate between A_0 and A for $\beta \in [0,1]$. Define

$$d\mu_\beta(q) = F_1 \ldots F_N e^{-q \cdot A_\beta q} d^N q . \tag{2.3}$$

An important role is played by:

Lemma 2.2. For μ an even ferromagnetic measure and $f \in \mathcal{G}$, define $E(\beta) = \int f(q)d\mu_\beta$. Then the Taylor series for $E(\beta)$ at $\beta = 0$ has non-negative coefficients and radius of convergence greater than 1.

Proof. The nth derivative $E^{(n)}(0) = \int \left[\sum_{i \neq j} (-A_{ij})q_i q_j\right]^n f(q)d\mu_0$.

Since each $A_{ij} \leq 0$ for $i \neq j$, the integrand is a function in $\mathcal{G}$. Because $d\mu_0$ factors and is even, $\int g d\mu_0 \geq 0$ for any $g \in \mathcal{G}$; hence $E^{(n)}(0) \geq 0$. Now an analytic function with nonnegative Taylor coefficients at the origin must have its nearest singularity on the positive real axis, and $E(\beta)$ is clearly analytic in a neighbourhood of $[0,1]$.

In his general analysis of Griffiths' inequalities, Ginibre [4] isolated a condition (Q3) that the measure and observables should satisfy to yield correlation inequalities. In our case, (Q3) was essentially proved by Ginibre by the following argument:

Lemma 2.3. Let $d\mu_0$ be the even measure given by (2.3) and let $f_1,\ldots,f_n \in \mathcal{G}$. Then for any choice of the plus or minus signs

$$\iint \prod_{i=1}^{n} \left(f_i(q) \pm f_i(p)\right) d\mu_0(q)d\mu_0(p) \geq 0 \quad . \tag{2.4}$$

Proof. Because $d\mu_0$ factors we first reduce (2.4) to the case $N = 1$ by repeated use of the identity

$$h_1(q_1)h_2(q_2) \pm h_1(p_1)h_2(p_2) = \tfrac{1}{2}\left(h_1(q_1)+h_1(p_1)\right)\left(h_2(q_2)\pm h_2(p_2)\right)$$
$$+ \tfrac{1}{2}\left(h_1(q_1)-h_1(p_1)\right)\left(h_2(q_2)\mp h_2(p_2)\right) \quad . \tag{2.5}$$

Secondly, we think of $\mathbb{R}$ as the product space $\{\pm 1\} \times \mathbb{R}_+$ so that a function $f_i \in \mathcal{G}$ has the form $f_i = \varepsilon_i(\sigma)g_i(q)$ for $\sigma \in \{\pm 1\}$, $q \in \mathbb{R}_+$

where $\varepsilon_i(\sigma) \equiv 1$ or $\varepsilon_i(\sigma) = \sigma$ and g_i is a positive increasing function on $\mathbb{R}_+$. Then the left side of (2.4) becomes

$$\int_{\mathbb{R}_+}\int_{\mathbb{R}_+} d\mu_0(q)d\mu_0(p) \sum_{\substack{\sigma=\pm 1\\ \tau=\pm 1}} \prod_{i=1}^{n} \left(\varepsilon_i(\sigma)g_i(q) \pm \varepsilon_i(\tau)g_i(p)\right) \quad . \tag{2.6}$$

A further application of (2.5) separates out the variables so that (2.6) becomes a sum of 2^n terms of the form

$$2^{-n}\left[\sum_{\sigma,\tau} \prod_i \left(\varepsilon_i(\sigma) \pm \varepsilon_i(\tau)\right)\right] \cdot \left[\int_{\mathbb{R}_+}\int_{\mathbb{R}_+} d\mu_0(q)d\mu_0(p) \prod_i \left(g_i(q) \pm g_i(p)\right)\right] \quad . \tag{2.7}$$

But each of the factors in (2.7) is nonnegative because ε_i and g_i are increasing functions. Consider, for example, the second factor: By the change of variables $q \lt;-\gt; p$ we need only consider the case of an even number of minus signs. But in this case the integrand is clearly nonnegative since each g_i is increasing.

<u>Proof of Theorem 2.1.</u> (2.1) follows immediately from Lemma 2.2 since $E(1) = \sum_0^\infty E^{(n)}(0)/n! \geq 0$. As for (2.2) we must show that

$$2\left(\int d\mu\right)^2 \left(\langle fg \rangle_\mu - \langle f \rangle_\mu \langle g \rangle_\mu\right)$$

$$= \iint \left(f(q)-f(p)\right)\left(g(q)-g(p)\right) d\mu(q)d\mu(p) \geq 0 \quad . \tag{2.8}$$

By Lemma 2.2, the integral (2.8) can be evaluated by expanding the off-diagonal part of the Gaussian. But each term in this expansion

$$\frac{1}{n!}\iint \left(f(q)-f(p)\right)\left(g(q)-g(p)\right)\left[\sum_{i\neq i'} (-A_{ii'})(q_i q_{i'} + p_i p_{i'})\right]^n d\mu_0(q)d\mu_0(p)$$

is nonnegative by Lemma 2.3.

It is of some interest to ask how essential is the assumption in Theorem 2.1 that μ be even. It is not hard to give counter-examples for both (2.1) and (2.2) when μ is not even; e.g. take

$F_i(x) = e^{-ax}$, $a > 0$ and f and g monomials. On the other hand, it is easy to see that (2.1) holds if F_i even is replaced by the assumption that $F_i = G_i \sum_k f_{i,k}$ where G_i is even and $f_{i,k} \in \mathcal{G}_i$; and, by Lemma 2.3, (2.2) holds if products of the F's satisfy

$$\prod_{j=1}^{n} F_j(q_j)F_j(p_j) = \prod_{j=1}^{n} G_j(q_j)G_j(p_j) \sum_k \prod_{i=1}^{n_k} \left(g_{k,i}(q) \pm g_{k,i}(p)\right)$$

where G_j is even and $g_{k,i} \in \mathcal{G}$.

In particular, we obtain this result by expanding the exponentials e^{Q_i} and appealing to the arguments of Lemmas 2.2 and 2.3:

Corollary 2.4. Let μ be a ferromagnetic measure on $\mathbb{R}^N$ with $F_i(q_i) = e^{-P_i(q_i)+Q_i(q_i)}$ where P_i is an even semibounded polynomial and Q_i is an odd polynomial with $\deg Q_i < \deg P_i$ that is positive and increasing on $\mathbb{R}_+$. Then the correlation inequalities (2.1) and (2.2) are valid.

As I mentioned before, these correlation inequalities, in the case of observables f,g that are polynomials in q , go over from the lattice to the continuous models by virtue of Theorem 1.3. For instance when $d = 1$:

Theorem 2.5. Consider the $P(\phi)_1$ Euclidean Markov theory with $P = P_e - P_0$ where P_e is even and P_0 is odd and positive and increasing on $\mathbb{R}_+$ and with Schwinger functions

$$S_\Lambda^\nu(t_1,\ldots,t_r) = \frac{\int d\nu\, \phi(t_1)\ldots\phi(t_r)\, e^{-\int_\Lambda P(\phi(s))ds}}{\int d\nu\; e^{-\int_\Lambda P(\phi(s))ds}}$$

where $d\nu$ is the free measure with free, Dirichlet, Neumann, or periodic B.C. on $\partial\Lambda$ where $\Lambda \subset \mathbb{R}$ is a finite interval. Then $S_\Lambda^\nu(t_1,\ldots,t_r) \geq 0$ and

$$S_\Lambda^\nu(t_1,\ldots,t_{r+s}) \geq S_\Lambda^\nu(t_1,\ldots,t_r)\, S_\Lambda^\nu(t_{r+1},\ldots,t_{r+s}) .$$

Note that in the case $d = 1$, Wick ordering is not necessary and that $S_\Lambda(t_1,\ldots,t_r)$ is a well-defined function of $t_1,\ldots,t_r$ without the need for smearing. When $d = 2$, we must Wick order and as a result the only odd term that we can allow in P is the linear term; for when $r > 1$, $:q_n^r:$ (essentially the rth Hermite polynomial) is not positive increasing in $q_n \geq 0$.

Theorem 2.6. Consider the $P(\phi)_2$ Euclidean Markov theory with $P(x) = P_e(x) - \lambda x$ where P_e is even and $\lambda \geq 0$. Let $S_\Lambda^\sigma(x_1,\ldots,x_r)$ be the corresponding Schwinger functions in the bounded regular (see [10]) region Λ where σ denotes free, Dirichlet or half-Dirichlet B.C. Then $S_\Lambda^\sigma(x_1,\ldots,x_r) \geq 0$ and

$$S_\Lambda^\sigma(x_1,\ldots,x_{r+s}) \geq S_\Lambda^\sigma(x_1,\ldots,x_r)\, S_\Lambda^\sigma(x_{r+1},\ldots,x_{r+s}) .$$

Actually by $\phi \to -\phi$ covariance we retain the correlation inequalities when $\lambda < 0$ in Theorem 2.6 or when $a_i > 0$ for i odd in Theorem 2.5, but there may be a sign change; e.g., $(-1)^r S_\Lambda^\sigma(x_1,\ldots,x_r) \geq 0$. It would be desirable to have correlation inequalities involving Wick powers but the only obvious one is for $:\phi^2: = \phi^2 - \infty$ since the Wick renormalization is a constant:

Corollary 2.7. Consider the $P(\phi)_2$ theory with $P = P_e - \lambda x$, $\lambda \geq 0$ and B.C. σ (= free, Dirichlet, or half-Dirichlet). Then

$$\langle \phi(x_1)\ldots\phi(x_r):\phi^2(x):\rangle_\Lambda^\sigma \geq \langle \phi(x_1)\ldots\phi(x_r)\rangle_\Lambda^\sigma \ \langle :\phi^2(x):\rangle_\Lambda^\sigma .$$

As for the FKG inequalities, the interacting measure satisfies condition (A) of Fortuin-Kasteleyn-Ginibre [2], and so we obtain Theorem 2.8 below (for details see [10]). There is _no restriction on the odd terms of P_. The FKG inequalities are expressed in terms of the cone I of _increasing functions of the fields_; I consists of elements of the form $F(\phi(h_1),\ldots,\phi(h_r))$ where $h_1,\ldots,h_r \geq 0$ are in $C_0^\infty(\mathbb{R}^2)$ and $F : \mathbb{R}^r \to \mathbb{R}$ is increasing in the sense that $F(x) \geq F(y)$ if $x_i \geq y_i$, $i = 1,\ldots,r$. For technical reasons we also assume that F is continuous and polynomially bounded on $\mathbb{R}^r$. Then:

Theorem 2.8. Let P be an arbitrary (semibounded) polynomial and let $\Lambda \subset \mathbb{R}^2$ be bounded. If $F, G \in I$ are increasing functions of the fields, then the truncated expectation $\langle FG\rangle_{\Lambda,T}^\sigma \geq 0$ where σ denotes free, Dirichlet, or half-Dirichlet B.C.

Remark. The inequalities of Theorems 2.6 and 2.8 extend to the infinite volume theories once it is known that the various expectations involved converge as $\Lambda \to \infty$ (see [6], Nelson's lectures and Corollary 3.5 in the next section for results along these lines).

It should be noted that both G-II and FKG are both statements of positive correlation but for somewhat different classes of observables. The appropriate class for the $P(\phi)_2$ Griffiths' inequalities is the cone of polynomials in the fields with non-negative test functions. But clearly a product like $\phi(h_1)\phi(h_2) \notin I$ since x_1x_2 is not an increasing function on $\mathbb{R}^2$. In order to produce some nontrivial examples of observables in I we introduce, following Simon [15], the analogues $\rho(f)$ of the occupation number variables used in the lattice gas case [2]. Define the cutoff function $Y(x) = x$ if $|x| \leq 1$, $Y(x) = \operatorname{sgn} x$ if $|x| > 1$.

Definitions 2.9. $\sigma(f) = Y(\phi(f))$, $\rho(f) = \frac{1}{2}(1 + \sigma(f))$
$\Pi_A = \prod_{i\in A} \rho(f_i)$, and $\Sigma_A = \sum_{i\in A} \rho(f_i)$, where f, f_i are nonnegative test functions in $C_0^\infty(\mathbb{R}^2)$ and A is a finite index set.

Note that the occupation number variable $\rho(f)$ is a non-linear function of the field taking values between 0 and 1 . It is easy to verify that:

Lemma 2.10. The following are all in I : $\phi(f)$, $\sigma(f)$, $\rho(f)$, $\phi(f) - \sigma(f)$, Π_A, Σ_A, $\Sigma_A - \Pi_A$.

3. Applications

Some of the traditional applications of correlation inequalities in statistical mechanics are as follows [8, 14]:

(i) monotonic convergence of correlation functions in the infinite volume limit;

(ii) bounds on correlations in terms of the two point function [11];

(iii) monotonic behaviour of correlation lengths as the interaction is made more ferromagnetic;

(iv) persistence of phase transitions if an interaction is made more ferromagnetic.

We now discuss the field theoretic translations of these statements. For $P(\phi)_2$ "more ferromagnetic" means that the co-efficient of ϕ^2 has been decreased; this amounts to a decrease in the bare mass. The analogue of correlation length is m_{phys}^{-1} . The occurrence of a phase transition or spontaneous magnetization means that $\lim_{\mu\to 0^+} \langle\phi(0)\rangle_\mu > 0$ where $\langle . \rangle_\mu$ denotes expectation in the infinite volume theory with $P(x) = P_e(x) - \mu x$.

The Griffiths' inequalities lead immediately to monotonicity statements for the Schwinger functions, some of which we collect in the following lemma:

Lemma 3.1. (i) Consider the $P(\phi)_1$ theory of Theorem 2.5. The Schwinger functions $S_\Lambda^\nu(t_1,\ldots,t_r)$ are decreasing functions of the coefficients $a_1,\ldots,a_{2n}$.

(ii) Suppose that P is even with nonnegative coefficients. Then the $P(\phi)_2$ Schwinger functions $S_\Lambda^{free}(t_1,\ldots,t_r)$ are decreasing functions of the interval Λ.

(iii) Consider the spatially cutoff $P(\phi)_2$ theory with $P(x) = P_e(x) + a_2x^2 + a_1x$, $a_1 \leq 0$, as in Theorem 2.6. Then the Schwinger functions $S_\Lambda^\sigma(x_1,\ldots,x_r)$ are decreasing functions of a_1 and a_2.

Proof. All the assertions are proved in the same way; as an example, we prove the monotonicity in a_2 of (iii). It is easy to see that $S_\Lambda^\sigma(x_1,\ldots,x_r)$ is differentiable in a_2 and that

$$\frac{\partial S_\Lambda^\sigma}{\partial a_2}(x_1,\ldots,x_r) = -\int_\Lambda dx[\langle\phi(x_1)\ldots\phi(x_r):\phi^2(x):\rangle_\Lambda^\sigma - \langle\phi(x_1)\ldots\phi(x_r)\rangle_\Lambda^\sigma \cdot \langle:\phi^2(x):\rangle_\Lambda^\sigma]$$

which is nonpositive by Corollary 1.

We turn to application (i) and describe the circumstances in which monotone convergence of the Schwinger functions is known. For the other cases (high temperature and large magnetic field) in which convergence has been proved see the lectures of Glimm and Jaffe.

1. First notice how much better the situation is in one dimension than in two, owing to our lack of control over Wick powers higher than two. We conclude immediately from Lemma 3.1(ii) that the $P(\phi)_1$ Schwinger functions S_Λ^{free} (which are nonnegative by G-I) decrease monotonically to an infinite volume limit, at least when P is even with nonnegative coefficients. Actually, we know by the transfer matrix method [10] that there is (not necessarily monotonic) convergence for general P.

2. As for $d = 2$ with free B.C. there is only this meagre result:

Corollary 3.2. For the $(\phi^2)_2$ theory, the Schwinger functions S_Λ^{free} converge monotonically downward to an infinite volume limit as $|\Lambda| \to \infty$.

3. The above two convergence results were all that GRS could deduce from the Griffiths' inequalities. The general case of $P(\phi)_2$ has, happily, been salvaged by Nelson who discovered that

Theorem 3.3 (Nelson). Let $P = P_e - \lambda x$, $\lambda \geq 0$. Then the half-Dirichlet Schwinger functions S_Λ^{HD} converge monotonically upward to an infinite volume limit.

Nelson's argument is based on G-II as he has described in his lectures.

4. It is doubtful that there are any monotonicity results for free or Dirichlet B.C. when $\deg P > 4$. The difficulty with Dirichlet B.C. is that the definition of Wick ordering changes with Λ (see (1.14)). However, for the special case of ϕ^4 the change in Wick ordering occurs as a quadratic term (and a trivial constant term) which we can control by Lemma 3.1(iii). The result is:

Theorem 3.4. Consider the $P(\phi)_2$ theory with $P(x) = ax^4 + bx^2 - \mu x$, $a > 0$, $\mu \geq 0$. Then the Schwinger functions S_Λ^D converge monotonically upward to an infinite volume limit.

Part of the argument in Theorem 3.4 and in Nelson's Theorem consists of obtaining an upper bound on the Schwinger functions. A bound for the Schwinger functions with non-coincident arguments was noted in [10] on the basis of the Glimm-Jaffe linear bound [5]. Recently Frohlich [3] (see also Simon's lectures) obtained bounds for general arguments. It is sufficient to bound the S_Λ in the case of free B.C. It is here that the sign of $B_{\partial\Lambda}$ of Theorem 1.2 enters, for using G-II we deduce via the lattice approximation that

$$S_\Lambda^{HD} \leq S_\Lambda \ .$$

5. Frohlich's basic results are best expressed in terms of the generating functional $J_\Lambda^\sigma(f) = \langle e^{\phi(f)} \rangle_\Lambda^\sigma$ for the Schwinger functions: Let σ = free or HD with $P = P_e - \lambda x$, $\lambda \geq 0$, or $\sigma = D$ with $P = ax^4 + bx^2 + cx$, $c \leq 0$, and let Λ_0 be a bounded region. Then there are constants c_1 and c_2 such that for any Λ and for any f with supp $f \subset \Lambda_0$,

$$|J_\Lambda^\sigma(f)| \leq c_1 \exp(c_2 \| f \|_2^2) . \tag{3.1}$$

Moreover, in the situations of Theorems 3.3 and 3.4

$$J^\sigma(f) = \lim_{\Lambda\to\infty} J_\Lambda^\sigma(f) \tag{3.2}$$

exists and is continuous in f in L^2 norm. (3.2) follows from Theorems 3.3 and 3.4, the bound (3.1), and Vitali's Theorem.

By approximating by exponentials (the Fourier transform theorem) we can deduce the following convergence result from (3.1) and (3.2):

Corollary 3.5. Consider the situations of Theorems 3.3 or 3.4. Let F be a continuous, exponentially bounded function on $\mathbb{R}^r$, $|F(x_1,..,x_r)| \leq \exp(c\,\Sigma|x_i|)$. Then for real-valued $h_1,..,h_r$ in $L^2(\mathbb{R}^2)$, $\langle F(\phi(h_1), \ldots, \phi(h_r) \rangle_\Lambda^\sigma$ converges as $\Lambda \to \infty$.

Thus we see that the Griffiths and FKG inequalities transfer to the infinite volume limit in these cases.

6. Before leaving application (i), I wish to mention one more result of this type due to Albeverio and Hoegh-Krohn [1]. They consider the $(e^\phi)_2$ model with interaction $U_\Lambda = \int d\nu(\alpha) \int_\Lambda dx{:}e^{\alpha\phi(x)}{:}$ where $d\nu(\alpha)$ is a positive measure of compact support on $(-4/\sqrt{\pi}, 4/\sqrt{\pi})$. If, moreover, $d\nu$ is even then U_Λ contains only even powers; indeed, because

$$:\exp(\alpha\phi_h(x)): = \exp(-\tfrac{1}{2}\alpha^2 \| h\|_{-1}^2) \exp(\alpha\phi_h(x))$$

where $\phi_h(x) = \int h(x-y)\phi(y)dy$, $h \geq 0$ an ultraviolet cutoff, it follows that $U_{\Lambda,h}$ with an ultraviolet cutoff is a power series in ordinary even powers ϕ_h^{2n} with nonnegative coefficients. We conclude, as in Corollary 3.2, that the (nonnegative) Schwinger functions are decreasing functions of Λ , and thus converge in the infinite volume limit.

Application (ii) is due to Simon [15] and is patterned after Lebowitz' proof that, for Ising spin systems satisfying the FKG inequalities, a spin-spin correlation of any order can be dominated by the two point spin correlation [11]. In our case the estimate is expressed in terms of the variables of Lemma 2.10 (see Defn. 2.9; A, B are finite index sets).

Lemma 3.6. [15] Let $\langle . \rangle$ denote the expectation in a Euclidean field theory for which the FKG inequalities hold. Then we have

$$0 \leq \langle \Pi_A \Pi_B \rangle_T \leq \langle \Sigma_A \Sigma_B \rangle_T = \sum_{i \in A} \sum_{j \in B} \langle \rho(f_i)\rho(f_j) \rangle_T =$$

$$= \frac{1}{4} \sum_{i \in A} \sum_{j \in B} \langle \sigma(f_i)\sigma(f_j) \rangle_T \leq \frac{1}{4} \sum_{i \in A} \sum_{j \in B} \langle \phi(f_i)\phi(f_j) \rangle_T . \qquad (3.3)$$

The first inequality is just the FKG inequality applied to the increasing functions Π_A and Π_B . The second inequality follows from adding the pair of FKG inequalities, $\langle \Sigma_B(\Sigma_A - \Pi_A) \rangle_T \geq 0$ and $\langle (\Sigma_B - \Pi_B)\Pi_A \rangle_T \geq 0$. Similarly for the last inequality in (3.3).

Simon has exploited Lemma 3.6 to prove that in $P(\phi)_2$ theories "the field couples the vacuum to the first excited state", or in crude terms, that the mass gap must show up in the two point function. In general, consider a Euclidean Markov field theory over $H_{-1}(\mathbb{R}^2)$ satisfying Nelson's axioms including the sharp time Axiom A [12, 10] and the FKG inequalities. The relativistic

Hilbert space $\mathcal{H}$ is obtained in Nelson's construction as the "time"-zero subspace of $L^2(Q,d\mu)$; $\mathcal{H} = E_0 L^2(Q,d\mu)$, E_0 being the orthogonal projection onto vectors supported at $t = 0$. If H is the corresponding Hamiltonian operator on $\mathcal{H}$ and Ω its unique vacuum, then the Feynman-Kac-Nelson formula asserts that for $t_1 \leq t_2 \leq \ldots \leq t_n$,

$$(\Omega, \phi_0(g_1) e^{-(t_2-t_1)H} \phi_0(g_2) \ldots e^{-(t_n-t_{n-1})H} \phi_0(g_n)\Omega)_{\mathcal{H}} = \langle \phi(g_1^{t_1}) \ldots \phi(g_n^{t_n}) \rangle, \tag{3.4}$$

where ϕ_0 is the time zero field on $\mathcal{H}$, $g_j \in S(\mathbb{R}^1)$ and $g^t(x_1,x_2) = g(x_1)\delta(x_2-t)$. Using (3.4) and the convergence of $\lambda^{-1}\sigma(\lambda g)$ to $\phi(g)$ as $\lambda \to 0$, Simon has shown:

Lemma 3.7. Consider a Euclidean Markov field theory over $\mathcal{H}_{-1}(\mathbb{R}^2)$ satisfying Nelson's axioms including Axiom A for sharp time fields. Then the set of vectors $S = \{E_0 \rho(g_1^{t_1}) \ldots \rho(g_n^{t_n}) | g_j \in S(\mathbb{R}^1)$, $g_j \geq 0$, $0 < t_1 < t_2 < \ldots < t_n\}$ is total in $\mathcal{H}$.

Now suppose for simplicity that the next point in $\sigma(H)$ above 0 is an eigenvalue $E_1 > 0$. If Ω_1 is a corresponding eigenvector, then by Lemma 3.7 there is some $\psi = E_0\rho(g_1^{t_1}) \ldots \rho(g_n^{t_n}) \in S$ such that $(\psi,\Omega_1) \neq 0$. Let $f_j = g_j^{t_j}$, $f_j^t = g_j^{t_j+t}$, $\Pi_A = \prod_{j=1}^{n} \rho(f_j)$ and $\Pi_A^t = \Pi \rho(f_j^t)$. By (3.4) and the spectral theorem, we have for $t > 0$

$$\begin{aligned} \langle \Pi_A \Pi_A^t \rangle_T &= (\psi, e^{-tH}\psi) - |(\Omega,\psi)|^2 \\ &\geq |(\psi, \Omega_1)|^2 e^{-tE_1} . \end{aligned}$$

Now Lemma 3.6 extends to these sharp time f_j and we deduce from (3.3) that

$$|(\psi,\Omega_1)|^2 e^{-tE_1} \le \frac{1}{4} \sum_i \sum_j < \phi(f_i)\phi(f_j^t) >_T$$

$$= \frac{1}{4} \sum_i \sum_j [(\phi_0(g_i)\Omega, e^{-(t+t_j-t_i)H} \phi_0(g_j)\Omega) \qquad (3.5)$$

$$- (\Omega,\phi_0(g_i)\Omega)(\Omega,\phi_0(g_j)\Omega)]$$

assuming $t \ge t_i - t_j$, for all i, j. We conclude from (3.5) that for some g_i and some eigenvector ψ_1 of H with eigenvalue E_1, $(\phi_0(g_i)\Omega, \psi_1) \neq 0$. If E_1 is in the continuous spectrum, a similar argument using spectral measures carries through, and so we have proved:

Theorem 3.8. Consider a Euclidean Markov field theory over $\mathcal{H}_{-1}(\mathbb{R}^2)$ satisfying Nelson's axioms and the FKG inequalities. Then the vectors $\{\phi_0(g)\Omega \mid g\in S(\mathbb{R}^1), g \ge 0\}$ are coupled to the first excited state of the Hamiltonian H.

Remarks 1. A related result has been obtained by Glimm, Jaffe, and Spencer (see [6] and Jaffe's lectures).

2. An almost identical argument works for the spatially cutoff $P(\phi)_2$ theory (see Simon [15]).

3. In the case where H is invariant under the symmetry $\phi \to -\phi$, (say $P(\phi)_2$ with P even) it is a corollary of Theorem 3.8 that there is an eigenvector Ω_1 corresponding to E_1 which is odd under the symmetry $\phi \to -\phi$.

In this latter case $\langle\phi(f)\rangle = 0$ so that we can replace truncated expectations by ordinary expectations:

Corollary 3.9. Consider a Euclidean Markov field theory over $\mathcal{H}_{-1}(\mathbb{R}^2)$ which satisfies Nelson's axioms and the FKG inequalities and for which $\langle\phi(f)\rangle = 0$. Then the gap ΔE between the two lowest points in the spectrum of H is given in terms of the two point Schwinger function by

$$\Delta E = - \sup_{\substack{f \geq 0 \\ f \in S(\mathbb{R}^1)}} \lim_{t \to \infty} \frac{1}{t} \log \int S(x,t;y,0)\; f(x) f(y) dx\, dy . \qquad (3.6)$$

For a spatially cutoff $P(\phi)_2(g)$ theory with P even, formula (3.6) holds where S is the Schwinger function with cutoff $h(x,s) = g(s)$. Below the critical point, formula (3.6) of course tells us nothing because $\Delta E \equiv 0$.

We next turn to application (iii) : the analogue of the correlation length for Ising ferromagnets is the inverse of the mass gap $(\Delta E)^{-1}$.

1. Combining formula (3.6) with the monotonicity of Lemma 3.1 we immediately obtain:

Theorem 3.10. Consider any of the $P(\phi)_1$ theory with P even, $P(x) = \sum_1^n a_{2i} x^{2i}$; a spatially cutoff or infinite volume $P(\phi)_2$ theory with P even, $P(x) = P_e(x) + a_0 x^2$. Then the mass gap ΔE is an increasing function of the parameters $a_2, .., a_{2n}$, or a_0 .

2. There are a number of other monotomicity results for ΔE based on the correlation inequalities. Theorem 3.10 suggests, for instance, that ΔE is an increasing function of the bare mass m . Such a result is complicated by the fact that a change in m changes the definition of Wick ordering. In the special case of $a\phi^4 + b\phi^2$ this change can be controlled (as in Theorem 3.4) and it turns out that ΔE is an increasing function of m [10] above the critical point.

3. When there is a linear term in P the GHS inequality and Theorem 3.8 still permit a monotonicity result:

Theorem 3.11. [16] Consider the infinite volume theory corresponding to $P = ax^4 + bx^2 - \mu x$ defined by Theorem 3.4. Then the mass gap

ΔE is an increasing function of $|\mu|$

This result expresses the idea that we get closer to a phase transition as $|\mu| \to 0$. There is a preliminary discussion of phase transitions (application (iv)) in GRS [10] and this topic will be elaborated on in Simon's lectures.

References

1. S. Albeverio and R. Hoegh-Krohn, The Wightman Axioms and the Mass Gap for Strong Interactions of Exponential Type in Two-Dimensional Space-Time, University of Oslo preprint.

2. C. Fortuin, P. Kasteleyn and J. Ginibre, Correlation Inequalities on Some Partially Ordered Sets, Comm. Math. Phys. 22 (1971) 89-103.

3. J. Frohlich, Schwinger Functions and their Generating Functionals, University of Geneva preprint.

4. J. Ginibre, General Formulation of Griffiths' Inequalities, Comm. Math. Phys. 16 (1970) 310-328.

5. J. Glimm and A. Jaffe, The $\lambda(\phi^4)_2$ Quantum Field Theory Without Cutoffs. IV. Perturbations of the Hamiltonian, J. Math. Phys. 13 (1972) 1568-1584.

6. J. Glimm, A. Jaffe, and T. Spencer, The Wightman Axioms and the Particle Structure in the $P(\phi)_2$ Quantum Field Theory, preprint.

7. R. Griffiths, Correlation in Ising Ferromagnets, I, II, III, J. Math. Phys. 8 (1967) 478-483, 484-489, Comm. Math. Phys. 6 (1967) 121-127.

8. ———————, Phase Transitions, in Statistical Mechanics and Quantum Field Theory, Les Houches 1970, C. De Witt, R. Stora, Editors, Gordon and Breach, New York, 1971.

9. R. Griffiths, C. Hurst, and S. Sherman, Concavity of Magnetization of an Ising Ferromagnet in a Positive External Field, J. Math. Phys. 11 (1970) 790-795.

10. F. Guerra, L. Rosen, and B. Simon, The $P(\phi)_2$ Euclidean Quantum Field Theory as Classical Statistical Mechanics, Ann. Math., to appear.

11. J. Lebowitz, Bounds on the Correlations and Analyticity Properties of Ferromagnetic Ising Spin Systems, Comm. Math. Phys. 28 (1972) 313-321.

12. E. Nelson, Construction of Quantum Fields from Markoff Fields, J. Funct. Anal. 12 (1973) 97-112.

13. ———, The Free Markoff Field, J. Funct. Anal. 12 (1973) 211-227.

14. D. Ruelle, Statistical Mechanics, Benjamin, New York, 1969.

15. B. Simon, Correlation Inequalities and the Mass Gap in $P(\phi)_2$. I. Domination by the Two Point Function, Comm. Math. Phys., to appear.

16. B. Simon and R. Griffiths, The $(\phi^4)_2$ Field Theory as a Classical Ising Model, Comm. Math. Phys., to appear.

17. K. Symanzik, Euclidean Quantum Field Theory, in Local Quantum Theory, Proceedings of the International School of Physics "Enrico Fermi", Course 45, R. Jost, Editor, Academic Press, New York, 1969.

BOSE FIELD THEORY AS CLASSICAL STATISTICAL MECHANICS. III. THE CLASSICAL ISING APPROXIMATION

Barry Simon*,†
Departments of Mathematics and Physics
Princeton University

§1. Introduction and General Strategy

The techniques which have been developed initially or solely for Ising ferromagnets fall generally into two broad categories. One group, which includes correlation inequalities of GKS and FKG type, holds for general kinds of ferromagnets with more or less arbitrary (even) single spin distributions and with many body (ferromagnetic) forces allowed. The other group, which includes the correlation inequalities of GHS and the zero theorem of Lee-Yang, has been proven directly only for spin 1/2 ferromagnets (each spin takes the values ±1 with equal probability in the non-interacting systems) with pair interactions. In fact, counter examples exist with four-body interactions and spin 1/2 or with pair interactions and spins taking the values ±2,0 (but with 0,±2 having different weightings).

The lattice approximation of Guerra, Rosen, and Simon (1973) discussed already in these lectures by Nelson and by Rosen, approximates $P(\phi)_2$ by general Ising models and thus obtains GKS and FKG inequalities. Here, we wish to discuss a further approximation of Simon and Griffiths (1973) [henceforth SG] which approximates $(\phi^4)_2$ theory by "classical Ising models", i.e. systems with spin-1/2 spins and pair interactions. SG thereby obtain GHS and Lee-Yang theorems for certain $P(\phi)_2$ theories with deg P = 4. In the interests of emphasizing the main ideas we propose only to discuss the Lee-Yang theorem and one of its main applications. We will also not give certain technical details. The reader interested in further details and applications and in the GHS inequalities should consult the original papers of SG and of Simon (1973b) or the lectures of Simon (1974).

In the remainder of this introduction we want to state the Lee-Yang (1952) circle theorem and explain the general strategy of Griffiths (1970) for extending this theorem from spin-1/2 spins to more complicated situations.

<u>Theorem 1</u> Let $a_{ij} \geq 0$ for $1 \leq i < j \leq n$. Let P be the polynomial in $z_1,\dots,z_n$ of degree 1 in each z_i with 2^n terms given by:

$$P(z_1,\dots,z_n) = \sum_{\sigma_1=\pm 1,\dots,\sigma_n=\pm 1} \exp\Big(\sum_{i<j} a_{ij}\sigma_i\sigma_j\Big) z_1^{\frac{1}{2}(\sigma_1+1)} \cdots z_n^{\frac{1}{2}(\sigma_n+1)}$$

Then if each $z_i \in D \equiv \{z \mid |z| < 1\} \cup \{1\}$, then $P \neq 0$.

---

* A. Sloan Foundation Fellow

† Research partially supported by USAFOSR under Contract F44620-71-C-0108 and USNSF under Grant GP39048

Remarks

1. The connection with Ising ferromagnets is the following

$$e^{-\beta(h_1+\ldots+h_n)} P\left(e^{2\beta h_1},\ldots,e^{2\beta h_n}\right)$$

and represents the partition function of an Ising ferromagnet if spin σ_i is in a magnetic field h_i. Of course, it is not a priori clear why zeros of the partition function are important. This idea of Yang-Lee (1952) is further discussed in §3 below.

2. Since $P(z_1,\ldots,z_n) = (z_1\ldots z_n)P(z_1^{-1},\ldots,z_n^{-1})$ by spin-flip symmetry ($\sigma_i \to -\sigma_i$), P is also non-zero if each $z_i \in D^{-1}$ and in particular $P(z,z,\ldots,z) = 0$ can only happen if $|z| = 1$, i.e. $P(z,\ldots z)$ has its roots on the unit circle, hence the name "circle theorem".

3. There are various proofs of this theorem: the original Lee-Yang (1952) proof found also in Ruelle (1969); an unpublished proof of S. Sherman found in Simon (1974); a proof of Asano (1970) described also in Simon (1974); and a proof of Newman (1973).

By combining Remarks 1 and 2, Theorem 1 is easily seen to be equivalent to:

Theorem 1' Let $a_{ij} \geq 0$ be fixed for $1 \leq i < j \leq n$ and let

$$Z(h_1,\ldots,h_n) = \sum_{\sigma_1=\pm1,\ldots,\sigma_n=\pm1} \exp\left(\sum a_{ij}\sigma_i\sigma_j + \sum h_i\sigma_i\right) \tag{1}$$

Then $Z \neq 0$ if each $h_i \in \tilde{D} = \{h \mid \mathrm{Re}\, h > 0\} \cup \{h = 0\}$.

* * *

Griffiths (1970) proposed a very simple and beautiful way of extending Theorem 1' to more complex situations. As a typical case, consider a spin 1 ferromagnet, i.e. each spin s can take the values $0,\pm2$ with equal probability. We thus seek a zero theorem for the function

$$\tilde{Z}(h_1,\ldots,h_n) = \sum_{s_i=\pm2,0} \exp\left(\sum a_{ij}s_is_j + \sum h_is_i\right) \tag{2}$$

Griffiths suggests first looking at a two spin, spin 1/2 ferromagnet with $a_{12} = 1/2 \ \ln z$. Thus:

$$\text{prob}\ (s \equiv \sigma_1 + \sigma_2 = +2) = \sqrt{2}/\text{Normalization} = \text{prob}\ (s = -2)$$

$$\text{prob}\ (s = 0) = \binom{2}{0}\left(1/\sqrt{2}\right)/\text{Normalization} = \sqrt{2}/\text{Normalization}$$

That is, s looks like a spin-1 spin. In particular

$$2^{n/2}\tilde{Z}(h_1,\ldots,h_n) = \sum_{\substack{\sigma_{i1}=\pm1\\ \sigma_{i2}=\pm1}} \exp\left(\sum_{i,j} a_{ij}(\sigma_{i1}+\sigma_{i2})(\sigma_{j1}+\sigma_{j2}) + \sum_i 1/2 \ln z\ \sigma_{i1}\sigma_{j1}\right) \exp\left(\sum_i h_i(\sigma_{i1}+\sigma_{i2})\right) \tag{3}$$

by replacing the sum over $\sigma_{i1} = \pm 1$, $\sigma_{i2} = \pm 1$ by a sum over $\sigma_{i1} + \sigma_{i2} = \pm 2, 0$ doing the sum over the other degrees explicitly. On account of (3), $Z \neq 0$ if $h_i \in \widetilde{D}$.

§2. The Improved DeMoivre-Laplace Limit Theorem

It is now clear how to go about trying to prove a Lee-Yang theorem for $P(\phi)_2$. First approximate $P(\phi)_2$ by the lattice approximation, i.e. by Ising ferromagnets with pair interactions and single spin distributions $e^{-Q(q)}dq$ where $\deg Q = \deg P$ and Q is even if P is even (which we will suppose). Thus we need only obtain $e^{-Q(q)}dq$ as the output probability distribution for the total spin of an Ising ferromagnet with spin-1/2 spins and pair distributions. More accurately, we need only obtain it as the limit of suitably rescaled output distributions. This is because the Lee-Yang theorem in the form of Theorem 1' is preserved under limits on account of the following consequence of the argument principle: If $f_n(z)$ is a sequence of functions analytic and non-zero in a connected region $D \subset \mathbb{C}$ and if $f_n \to f$ uniformly on compacts of D, then f is either identically zero or non-vanishing in D. To apply this limit theorem, one needs uniform bounds on the approximating distributions as well as pointwise convergence. Below we will only prove pointwise convergence; the extra bounds (which require higher order terms in Stirling's formula) can be found in SG.

Consider first the case $\deg Q = 2$. It is easy to handle this case, for take N uncoupled spin-1/2 spins. Then the probability that $s = \Sigma s_i$ is μ is just $2^{-N}\binom{N}{(N+\mu)/2}$ where $\binom{a}{b}$ is a binomial coefficient. The DeMoivre-Laplace limit theorem asserts that the binomial distribution for large N looks like a Gaussian, explicitly:

$$\binom{N}{\frac{N+\mu}{2}} \sim D_N \exp(-\mu^2/2N) \tag{4}$$

for a suitable constant D_N. What (4) means is that if s is fixed then

$$D_N^{-1} \binom{N}{\frac{N+\mu_N(s)}{2}} \to \exp(-s^2/2)$$

as $N \to \infty$ where $\mu_N(s)$ is defined by

$$\frac{N+\mu_N(s)}{2} = \left[\frac{N+s\sqrt{N}}{2}\right] .$$

and $[x]$ = greatest integer less than x. To prove (4), one needs Stirling's formula:

$$\log n! \sim n \log n$$

from which

$$\log \binom{N}{\frac{N+\mu}{2}} \sim C_N - Nh(\mu/N)$$

with C_N a suitable constant and

$$h(x) = 1/2[(1+x)\log(1+x) + (1-x)\log(1-x)] \quad .$$

For x small, $h(x) = 1/2\, x^2 + 1/12\, x^4 + O(x^6)$. Thus for $s = \mu/\sqrt{N}$ fixed,

$$\log \binom{N}{\frac{N+\mu}{2}} \sim C_N - N(1/2(\mu/N)^2 + O(\mu/N)^4) = C_N - 1/2\, s^2 + O(1/N)$$

from which (4) follows.

Next consider $\deg Q = 4$; in fact suppose $Q(q) = q^4$. It is clear how to modify the Gaussian behavior above to get out q^4 ; just cancel the Gaussian and re-scale; i.e.

$$\binom{N}{\frac{N+\mu}{2}} e^{\mu^2/2N} \sim D_N\, e^{-\mu^4/12N^3} \tag{5}$$

For if we fix $s = \mu/N^{3/4}$:

$$\log \binom{N}{\frac{N+\mu}{2}} e^{\mu^2/2N} \sim C_N - N\,[1/12(\mu/N)^4 + O(\mu/N)^6]$$
$$= C_N - 1/12\, s^4 + O(1/N^{\frac{1}{2}})$$

But $\binom{N}{\frac{N+\mu}{2}} e^{\mu^2/2N}$ is the unnormalized probability distribution for an N spin-1/2 Ising magnet with energy $H = -\, 1/2N(\Sigma s_i)^2$ which is ferromagnetic. A similar argument works for any $Q(q) = aq^4 + bq^2$ with $a > 0$.

We are thus able to conclude the following basic theorem from SG:

Theorem 2 (Lee-Yang theorem for $(\phi^4)_2$) Let $\langle \cdot \rangle$ denote a spatially cutoff expectation value with free, Dirichlet or half-Dirichlet boundary conditions (see Guerra etal. (1973)) and $P(x) = ax^4+bx^2$; $a > 0$. Let $h \geq 0$ be in $L^\infty \cap L^1(\mathbb{R}^2)$. Then

$$F(z) = \langle \exp(z\phi(h)) \rangle$$

is an entire analytic function whose zeros lie on the axis $\mathrm{Re}\, z = 0$.

For $\deg Q \geq 6$, we have the following negative situation (SG): There are definitely sixth degree Q's which are not the limit of spin-1/2 pair-interacting(*) ferromagnets and for which the Lee-Yang theorem fails. Thus, in the lattice approximation, the Lee-Yang theorem fails for certain Q's . This suggests, but certainly does not prove, that the Lee-Yang theorem is false for some $P(\phi)_2$ theories with $\deg P = 6$.

(*) Of course if four-body interacting is allowed, there is no problem in approximating sixth degree Q's as a $1/12(\Sigma s_i)^4/N^3$ term in the energy plus re-scaling leads to $\exp(-1/30\, s^6)$.

§3. Clustering of the Schwinger Functions of $P = ax^4 + bx^2 - \mu x$ $(\mu \neq 0)$

Theorem 2 is a striking looking but, at first sight, apparently not very powerful theorem. That it is intimately connected with analyticity of the pressure is a discovery of Yang-Lee (1952) translated to $(\phi^4)_2$ by SG. That it implies strong bounds and falloff is a discovery of Lebowitz-Penrose (1968), developed in $(\phi^4)_2$ by Simon (1973b). We wish to indicate these ideas in this section. We discuss the case of Dirichlet boundary conditions although similar results hold for half-Dirichlet B.C.

Fix a,b and for μ real and $\Lambda \subset \mathbb{R}^2$, bounded, let

$$\alpha_\Lambda(\mu) = \frac{1}{|\Lambda|} \ln \int [\exp(- \int_\Lambda :a\phi^4 + b\phi^2 - \mu\phi :_D)] d\mu^D_{0,\Lambda}$$

Then, by a result of Guerra et al. (1973), for any real μ $\alpha_\Lambda(\mu) \to \alpha_\infty(\mu)$ as $\Lambda \to \infty$ (Fisher). The main point of the Lee-Yang theorem (Theorem 2) is that $\alpha_\Lambda(\mu)$ has an analytic continuation to the right half-plane, $\mathrm{Re}\,\mu > 0$. Moreover, it is clear that

$$\mathrm{Re}\,\alpha_\Lambda(\mu) \leq \alpha_\Lambda(\mathrm{Re}\,\mu) \ .$$

Let $f_\Lambda(\mu) = \exp(\alpha_\Lambda(\mu))$. We thus see that $|f_\Lambda(\mu)|$ is uniformly bounded on compacts of $\{\mu | \mathrm{Re}\,\mu > 0\}$ and converging for $\mu \in \mathbb{R}$ and thus, by the Vitali convergence theorem, convergent on $\{\mu | \mathrm{Re}\,\mu > 0\}$. Since the $f_\Lambda(\mu)$ are non-vanishing there and f_∞ is not identically zero, it is never zero, so $\alpha_\Lambda(\mu) \to \alpha_\infty(\mu)$ for all μ with $\mathrm{Re}\,\mu > 0$. We summarize by:

Theorem 3 $\alpha_\infty(\mu)$ has an analytic continuation to the entire right half-plane and $\alpha_\Lambda(\mu) \to \alpha_\infty(\mu)$ uniformly on compacts of the right half-plane.

In particular, $\sup_\Lambda d^2\alpha_\Lambda/d\mu^2 < \infty$, for any fixed $\mu > 0$. In terms of the Dirichlet state expectation value $< \ >_\Lambda$ for the $ax^4 + bx^2 - \mu x$ theory:

$$\frac{d^2\alpha_\Lambda}{d\mu^2} = \frac{1}{|\Lambda|} [<\phi(\chi_\Lambda)\phi(\chi_\Lambda)>_\Lambda - <\phi(\chi_\Lambda)>_\Lambda <\phi(\chi_\Lambda)>_\Lambda]$$

so by the Lee-Yang result $\sup_\Lambda d^2\alpha_\Lambda/d\mu^2 < \infty$, there is a D with

$$[<\phi(\chi_\Lambda)\phi(\chi_\Lambda)>_\Lambda - <\phi(\chi_\Lambda)>_\Lambda <\phi(\chi_\Lambda)>_\Lambda] \leq D|\Lambda| \tag{6}$$

Now the two-point truncated Schwinger function

$$S_2^T(x-y) = <\phi(x)\phi(y)>_\infty - <\phi(x)>_\infty <\phi(y)>_\infty$$

is positive and monotone decreasing as $|x-y| \to \infty$. If the limit is some $C > 0$, then

$$[<\phi(\chi_\Lambda)\phi(\chi_\Lambda)>_\infty - <\phi(\chi_\Lambda)_\infty <\phi(\chi_\Lambda)>_\infty] \geq C|\Lambda|^2 \ . \tag{7}$$

(6) and (7) are not directly contradictory although they clearly almost are and a further argument [Simon (1973b)] show they are: we conclude that $S_2^T \to 0$ as

$|x-y| \to \infty$. But then, by a result of Simon (1973a) (described in Rosen's lectures), all the truncated Schwinger functions go to 0 . We summarize:

Theorem 4 In the infinite volume $P(\phi)_2$ Dirichlet state with $P(x) = ax^4 + bx^2 - \mu x$ ($\mu \neq 0$) , the Schwinger functions obey clustering, i.e.

$$\int \phi(f_1 \otimes \delta_t)\ldots\phi(f_k \otimes \delta_t)\phi(f_{k+1} \otimes \delta_0)\ldots\phi(f_n \otimes \delta_0)d\nu \to [\int d\nu\, \phi(f_1 \otimes \delta_0)\ldots\phi(f_k \otimes \delta_0)] [\int d\nu\, \phi(f_{k+1} \otimes \delta_0)\ldots\phi(f_n \otimes \delta_0)] \quad \text{as} \quad t \to \infty .$$

§4 The Wightman Axioms

The status of the basic Euclidean objects for Dirichlet and half-Dirichlet states when $P(x) = ax^4 + bx^2 - \mu x$ ($\mu \neq 0$) are given by the following table:

TABLE

	Dirichlet	Half-Dirichlet
($\Lambda \to \infty$) Schwinger Functions	Yes (2)	Yes (1)
($\Lambda \to \infty$) Pressure	Yes (2)	Yes (3)
($\ell < \infty$) Transfer Matrix	?	Yes (3)
($\Lambda = \infty$) $S_2^T \to 0$	Yes (4)	Yes (4)
($\Lambda = \infty$) OS Axioms	Yes	Yes

where (1) = Nelson (these lectures), (2) = Guerra et al. (1973) (3) = Guerra et al. (1974), (4) = Simon (1973a). If the Osterwalder-Schrader (1973) reconstruction theorem is valid (there is presently a gap in the proof), all the Wightman axioms hold for the infinite volume D and HD theories. And in any event, on account of the existence of a transfer matrix, the Wightman axioms do hold for the HD theory [see Simon (1974)].

§5. α_∞ and Dymanical Instability

The field theoretic analog of a phase transition is the notion of dynmaical instability (Wightman (1969)), i.e. the existence of more than one infinite volume theory associated to a fixed interaction by some mechanism for associating infinite volume theories to interactions, e.g. the DLR equations described in Guerra's lectures. The expected picture for $\phi^4 + b\phi^2 - \mu\phi$ theories has been described in Jaffe's lecture. In this section we want to supplement the picture given by Jaffe explaining the connection between dynamical instability and the Fock space energy per unit volume, α_∞ , of Guerra (1972). As Guerra explained in his lectures, α_∞ is just the pressure. There is an old idea in field theory associated with the name of Bogoliubov that dynamical instability is present precisely when $<\phi(0)>_{\phi^4+b\phi^2-\mu\phi}$ is discontinuous in μ. In statistical mechanical language, Bogoliubov is saying that the phase transition is first order and has the field as long range order parameter. This picture is supported by the following which combines results from SG and Simon (1973b):

Theorem 5 Fix b and let $\alpha_\infty(\mu)$ denote the pressure for the $\phi^4 + b\phi^2 - \mu\phi$ and let $< >_\mu$ denote the infinite volume (Dirichlet) state for this theory. Consider the statements:

(A) There is a mass gap in the $< >_{\mu=0}$ theory.

(B) $\alpha_\infty(\mu)$ is differentiable at $\mu = 0$.

(C) The "magnetization" $<\phi(0)>_\mu$ is continuous at $\mu = 0$.

(D) There is a unique vacuum in the $< >_{\mu=0}$ theory.

Then (A) => (B) <=> (C) => (D).

Remarks

1. We emphasize that (A) is a statement about the $< >_{\mu=0}$ theory and not its decomposition into unique vacuums.

2. Suppose the picture described in Jaffe's lectures holds. Then there is a critical value b_c. When $b > b_c$ we expect (A) to hold so (B),(C) hold. When $b < b_c$, we expect (D) to fail for the following reason: The Wightman theories for $\mu = 0$ with unique vacuum (there should be two such theories!) have $<\phi(0)> \neq 0$. But by $\phi \to -\phi$ symmetry in the Dirichlet B.C. theories the value of $<\phi(0)>_\mu = 0$. Thus the $<\cdot>_{\mu=0}$ theory should not have a unique vacuum. Since (D) fails so do (A),(B). Thus away from the critical point: differentiality of the pressure should be a sensitive test of dynamical instability. At the critical point one expects (B)-(D) to hold on the basis of most stat. mech. models although we emphasize that there are stat. mech. models where (B),(C) fail at the critical point.

3. By general arguments $\alpha_\infty(\mu)$ is convex in μ and so continuous. By Lee-Yang it is analytic away from $\mu = 0$.

Sketch of proof (A)=>(B). We need only a bound on

$$\frac{d^2\alpha_\Lambda}{d\mu^2} = \frac{1}{|\Lambda|} <\phi(\chi_\Lambda)\phi(\chi_\Lambda)>_{T,\mu,\Lambda}$$

uniform in Λ and $|\mu| \leq 1$. By using the GHS and GI,II inequalities one obtains a uniform bound on the falloff of $<\phi(x)\phi(y)>_{T,\mu,\Lambda}$ as $|x-y| \to \infty$ and this yields a bound on $d^2\alpha_\Lambda/d\mu^2$. See SG.

(B) <=> (C). $$\alpha_\infty(\mu) - \alpha_\infty(0) = \lim_{\Lambda\to\infty} \int_0^\mu \frac{1}{|\Lambda|} <\phi(\chi_\Lambda)>_\mu \, d\mu$$

By using the GI,II inequalities one shows that

$$\alpha_\infty(\mu) - \alpha(0) = \int_0^\mu m(\mu)\, d\mu$$

with $m(\mu) = \langle\phi(0)\rangle_\mu$ and that $m(0+) = \lim_{\mu\downarrow 0} m(\mu)$ exists. One then proves that $m(0+)$ is the right derivative of $\alpha_\infty(\mu)$ and $m(0-)$, so differentiability of $\alpha_\infty(\mu)$ at $\mu = 0$ is equivalent to continuity of m . See SG for details.

(C)=>(D). Suppose (D) fails. Then $S^T_{2,\mu=0}(x-y) \to c^2 > 0$ as $x-y \to \infty$. But by symmetry $\phi \iff -\phi$, $S^T_{2,\mu=0} = S_{2,\mu=0}$. By GII, $S_{2,\mu>0} \geq S_{2,\mu=0}$ so $S_{2,\mu>0} \to d^2 \geq c^2$ as $x-y \to 0$. But by Theorem 4, $S^T_{2,\mu>0} \to 0$ so $\langle\phi(0)\rangle_{\mu>0} \geq c$. By symmetry $\langle\phi(0)\rangle_{\mu>0} \leq c$ so $\langle\phi(0)\rangle_\mu$ is not continuous at $\mu = 0$. For details see Simon (1973b).

References

ASANO, T. (1970):J. Phys. Soc. Jap. 29, 350.

GRIFFITHS, R. (1970): J. Math. Phys. 10, 1559.

GUERRA, F. (1972): Phys. Rev. Lett. 28, 1213.

GUERRA, F., ROSEN, L. SIMON, B. (1973): The $P(\phi)_2$ Euclidean Quantum Field Theory as Classical Statistical Mechanics, Ann. Math., to appear

GUERRA, F., ROSEN, L., SIMON, B. (1974): Boundary Conditions for the $P(\phi)_2$ Euclidean Field Theory, in preparation.

LEBOWITZ, J., PENROSE, O. (1968): Commun. Math. Phys. 11, 99.

LEE, T.D., YANG, C.N. (1952): Phys. Rev. 87, 410.

NEWMAN, C. (1973): Zeroes of the Partition Function for Generalized Ising Systems, N.Y.U. Preprint.

OSTERWALDER, K., SCHRADER, R. (1973): Commun. Math Phys. 31, 83.

RUELLE, D. (1969): Statistical Mechanics, Benjamin, New York.

SIMON, B. (1973a): Commun. Math. Phys. 31, 127.

SIMON, B. (1973b): Correlation Inequalities and the Mass Gap in $P(\phi)_2$, II. Uniqueness of the Vacuum for a Class of Strongly Coupled Theories, Ann. Math., to appear.

SIMON, B. (1974): The $P(\phi)_2$ Euclidean Quantum Field Theory, Princeton Series in Physics, Princeton University Press.

SIMON, B. GRIFFITHS, R. (1973): The $(\phi^4)_2$ Field Theory as a Classical Ising Model, Commun. Math. Phys., to appear.

WIGHTMAN, A.S. (1969): Phys. Today 22 53-58.

YANG, C.N., LEE, T.D. (1952): Phys Rev. 87 404.

CONSTRUCTIVE MACROSCOPIC QUANTUM ELECTRODYNAMICS

Klaus Hepp
Department of Physics, E.T.H.
CH-8049 Zürich, Schweiz

Elliott H. Lieb
Department of Physics, M.I.T.
Cambridge, Mass. 02139, U.S.A.

§1. Introduction

After ten days of difficult lectures the audience and the lecturers need some holidays. I have chosen the subject of this last talk half for your recreation, half for exposing you to some new and exotic aspects of the quantum world of infinitely many degrees of freedom, where there are many interesting problems in mathematical physics. My lecture will be centered around the quantum electrodynamics of the laser in the thermodynamic limit, on which E.H.Lieb and I have worked for some time. It is hard to give fair references in this field with a continuous transition to applied physics. A good starting point on laser theory is the book by Haken [H1] as well as various contributions in [A1] , [G3] , [K1]. The statistical theory of instabilities in stationary nonequilibrium systems is treated in [G1] and [G4] with many references to older contributions. A general approach to nonequilibrium phase transitions in mean field models with linear dissipation has been given in [H4] , with great emphasis on the nonlinear analysis of the Heisenberg equations of motion. In this lecture I shall presenta simplified approach to these problems, using only linear functional analysis and working in the Schrödinger picture. By this method one can easily incorporate the unbounded boson operators of the quantized radiation field, and one sees better the analogy to the usual treatment of the classical limit in quantum mechanics [H5] , [M1] .

§2. Heuristic Discussion of the Laser

The Dicke Haken Lax model of the 1-mode 2-level homogeneously broadened laser starts from the following approximation to the Hamiltonian of quantum electrodynamics

$$H = \sum_{m=1}^{\infty} \gamma_m a_m^* a_m + \varepsilon \sum_{n=1}^{N} S_n^3 + V^{-1/2} \sum_{m=1}^{\infty} \sum_{n=1}^{N} \{ a_m (\lambda_{mn} S_n^+ + \mu_{mn} S_n^-) + \text{h.c.} \} . \quad (2.1)$$

Here the $a_m^\#$ are creation and annihilation operators for the discrete set of photon modes of energy γ_m of a cavity of volume V. The λ_{mn} and μ_{mn} are coupling constants for the rotating and counter-rotating interaction of the mode m with the nth atom. We assume finitely many atoms, N, in the cavity and shall later take $N = V$ in the passage to the thermodynamic limit. The atoms have two states with fermion creation and annihilation operators $b_{+n}^\#$ and $b_{-n}^\#$ for the upper and lower level and no translational degrees of freedom. Then

$$S_n^+ = b_{+n}^* b_{-n} = (S_n^-)^* , \quad S_n^3 = (b_{+n}^* b_{+n} - b_{-n}^* b_{-n})/2 \quad (2.2)$$

satisfy SU_2 commutation relations

$$[S_n^3, S_n^\pm] = \pm S_n^\pm , \quad [S_n^+, S_n^-] = 2S_n^3 , \quad [S_n^i, S_m^j] = 0 \quad \text{for} \quad m \neq n . \quad (2.3)$$

Little is known in general about the system described by (2.1), except for thermodynamic stability with hard cores and instability without a sufficiently strong repulsion at short distances [H3]. For finitely many modes, an equilibrium phase transition from a normally radiating to a superradiant phase can be established [H2].

In these lectures we are interested in the nonequilibrium behaviour of the system, and we shall restrict ourselves to one mode and, for notational convenience, to the rotating wave approximation :

$$H_N^S = \gamma a^* a + \varepsilon S_N^3 + \lambda N^{-1/2} (S_N^+ a + a^* S_N^-) , \quad (2.4)$$

where only the total spin operators enter :

$$S_N^i = \sum_{n=1}^{N} S_n^i . \quad (2.5)$$

The total Hamiltonian H_N of the laser cavity with $N \sim V$ atoms and photons coupled to atomic pumping devices and photonic loss mechanisms is of the form $H_N = H_N^S + H_N^R$, where the reservoir part will be given later, (4.1), (4.6), once we have acquired a qualitative understanding of the laser action. Since the S_N^i give a representation of SU_2, it is natural to consider the Heisenberg equations of motion for the five operators $S_N^k(t) = \exp(iH_N t)\, S_N^k \exp(-iH_N t)$ and $a_N^{\#}(t)$:

$$\begin{aligned} \dot{a}_N(t) &= -i\gamma\, a_N(t) - i\lambda N^{-1/2} S_N^-(t) + i\left[H_N^R, a_N(t)\right], \\ \dot{S}_N^-(t) &= -i\varepsilon\, S_N^-(t) + 2i\lambda N^{-1/2} S_N^3(t) a_N(t) + i\left[H_N^R, S_N^-(t)\right], \\ \dot{S}_N^3(t) &= i\lambda N^{-1/2}\left(a_N^*(t) S_N^-(t) - a_N(t) S_N^+(t) \right) + i\left[H_N^R, S_N^3(t)\right]. \end{aligned} \tag{2.6}$$

The very successful semiclassical theory of the laser [L1] suggests that H_N^R should be chosen in such a way that

$$\begin{aligned} i\left[H_N^R, a_N(t)\right] &= -\varkappa\, a_N(t) + g_N(t), \\ i\left[H_N^R, S_N^-(t)\right] &= -\gamma\, S_N^-(t) + F_N^-(t), \\ i\left[H_N^R, S_N^3(t)\right] &= -\delta\left(S_N^3(t) - \eta N \right) + F_N^3(t). \end{aligned} \tag{2.7}$$

Here $\varkappa > 0$ and $\gamma > 0$ are damping constants for the photon amplitude and the atomic polarisation, while $\delta > 0$ and $-1/2 \leq \eta \leq 1/2$ should describe the pumping of the atoms into a mean inversion $\eta \approx S_N^3/N$. Of course, the purely dissipative terms alone on the r.h.s. of (2.7) are inconsistent with the selfadjointness of H_N. However, one hopes that in a suitable topology the additional fluctuation forces $g_N(t)$ and $F_N^k(t)$ become negligible in the limit $N \to \infty$. Assume that $g_N(t) = O(1)$ and $F_N^k(t) = O(N^{1/2})$, by some law of large numbers. Then it is plausible that the intensive observables

$$\begin{aligned} \alpha_N^{\#}(t) &= a_N^{\#}(t) N^{-1/2}, \\ \sigma_N^k(t) &= S_N^k(t) N^{-1}, \end{aligned} \tag{2.8}$$

which at $t = 0$ have $O(N^{-1})$ commutators, become c-numbers in the limit $N \to \infty$, and satisfy the ordinary differential equations

$$\dot{\alpha} = -(i\nu + \varkappa)\alpha - i\lambda\sigma^- ,$$

$$\dot{\sigma}^- = -(i\varepsilon + \gamma)\sigma^- + 2i\lambda\sigma^3\alpha , \tag{2.9}$$

$$\dot{\sigma}^3 = -\delta(\sigma^3 - \eta) + i\lambda(\sigma^-\alpha^* - \sigma^+\alpha) .$$

These equations have remarkable properties which correspond quite well to the qualitative picture of laser action. It can be rigorously shown [H4] that they have global solutions in (2.10) for all physical initial conditions

$$\alpha \in \mathbb{C} , \qquad \sigma^- \in \mathbb{C} , \tag{2.10}$$
$$\sigma^3 \in \mathbb{R} , \qquad (\sigma^3)^2 + |\sigma^-|^2 \leq 1/4 .$$

There exists a unique stationary solution

$$\sigma^3 = \eta , \qquad \alpha = \sigma^- = 0 , \tag{2.11}$$

and a 1-parameter family of periodic solutions, $\sigma^-(t) = \sigma^- \exp -i\omega t$, $\alpha(t) = \alpha \exp -i\omega t$, $\sigma^3(t) = \sigma^3$, if and only if $1/2 \geq \eta > \eta_c$, where

$$\eta_c = \frac{\varkappa\gamma}{2\lambda^2}\left(1 + \frac{(\varepsilon - \nu)^2}{(\varkappa + \gamma)^2}\right) , \qquad \sigma^3 = \eta_c ,$$

$$|\alpha|^2 = \delta(\eta - \eta_c)/2\varkappa , \qquad \omega = \frac{\varkappa\varepsilon + \gamma\nu}{\varkappa + \gamma} , \tag{2.12}$$

$$\lambda\sigma^- = (i\varkappa + \omega - \nu)\alpha .$$

At $\eta = \eta_c$, a Hopf bifurcation [H6] occurs : below threshold the laser is damping dominated, and the mean photon number per atom, $|\alpha|^2$, goes to zero from any initial state, while above threshold the stationary value $|\alpha|^2 = \delta(\eta - \eta_c)/2\varkappa$ is attained for $t \to \infty$, starting from almost any initial condition. These equations suggest that a nonequilibrium phase transition occurs at laser threshold and that the attracting self-oscillatory state is one of the dissipative structures which have been discussed by Prigogine and coworkers [G1]. In the following we shall give a Hamiltonian model H_N^R and an asymptotic expansion in $N^{1/2}$ in which this heuristic picture holds rigorously in $O(N)$.

§3. The Classical Limit for Atoms and the Radiation Field

There are several topologies in which the infinite volume limit can be studied. A very concrete and rather elementary discussion is possible in the Schrödinger picture, by passing with $N \to \infty$ along a sequence of photon and atomic coherent states [G2] using the Trotter theory of approximating Hilbert spaces [T2]. These coherent states exhaust all physical initial conditions for the intensive observables, and hence they can be considered as the true pure classical states of the infinite system. However, if one measures finer properties of the system by looking at the fluctuation observables, then one finds [H2] that the physical stationary states are not coherent, but that they can be described in the Hilbert space generated from a coherent state by applying all polynomials in fluctuation operators. In this section we shall consider the classical limit for the closed system of atoms and radiation. The interaction H_N^S has the form (2.4), and we shall first construct initial states $|\alpha, \underline{\sigma}\rangle_N = \Omega_N$ which are coherent in $a^{\#}$ and S_N^k.

Let $\alpha \in \mathbb{C}$, $U(\alpha) = \exp(\alpha a^* - \alpha^* a)$ and $|0\rangle$ be the ground state of $a^* a$. Then the sequence of photon coherent states $|N^{1/2}\alpha\rangle = U(N^{1/2}\alpha)|0\rangle$ satisfies

$$\langle N^{1/2}\alpha | \prod (a^{\#} - N^{1/2}\alpha^{\#}) | N^{1/2}\alpha\rangle = \langle 0 | \prod a^{\#} | 0\rangle , \tag{3.1}$$

$$\lim_{N\to\infty} \langle N^{1/2}\alpha | \prod (a^{\#} N^{-1/2}) | \alpha N^{1/2}\rangle = \prod \alpha^{\#} . \tag{3.2}$$

The analogue to (3.1) and (3.2) for fermions can be constructed using the representation theory of SU_2 [G2]. Any $\underline{\sigma} \in \mathbb{R}^3$ with $\|\underline{\sigma}\| = \sigma \leq 1/2$ has the form $\underline{\sigma} = R\,\underline{\sigma}_0$, where $R \in SO(3)$ and $\sigma_0^k = -\sigma\,\delta_0^k$, $1 \leq k \leq 3$. For simplicity, we restrict σ to a dense set of physical initial conditions :

$$\sigma = q/(2q + 4p) , \quad 0 \leq p,q \in \mathbb{Z} , \qquad q > 0 . \tag{3.3}$$

Take $A \in SU_2$ acting on the Pauli matrices as $A^* \tau^k A = \sum_{i=1}^{3} \tau^i R^{ik}$ and define new fermion annihilators by

$$(b_{+n}, b_{-n}) = (c_{+n}, c_{-n})\, A \tag{3.4}$$

Let $|\pm\rangle$ be the states for the fermion pair $b_{\pm}$, where

$$c^*_{-n}|-\rangle = c_{+n}|-\rangle = 0 \; , \quad T^+|-\rangle = |+\rangle , \tag{3.5}$$

where $T^+ = c^*_+ c_-$. Then in obvious notation

$$|\underline{\sigma}\rangle_N = 2^{-pM/2}\left[(|+\rangle|-\rangle - |-\rangle|+\rangle)^p \otimes (|-\rangle)^q\right]^{\otimes M} . \tag{3.6}$$

Here $N = M(2p+q)$ and $M = 1,2,\dots$ Let $T^i_N = \sum R^{ki} S^k_N$. Then

$$T^3_N |\underline{\sigma}\rangle_N = -\sigma N |\underline{\sigma}\rangle_N \; , \quad T^-_N |\underline{\sigma}\rangle_N = 0 \tag{3.7}$$

defines an irreducible representation of SU_2 . The fluctuation operators

$$t^+_N = N^{-1/2} T^+_N = (t^-_N)^* \; , \quad t^3_N = N^{-1/2} T^3_N + \sigma N^{1/2} \tag{3.8}$$

satisfy [W1] :

$$\lim_{N\to\infty} {}_N\langle\underline{\sigma}| t^k_N \dots t^m_N |\underline{\sigma}\rangle_N = \begin{cases} (\Omega_\infty, t^k_\infty \dots t^m_\infty \Omega_\infty) , & \text{if all } k,..m = \pm \\ 0 \; , & \text{otherwise .} \end{cases} \tag{3.9}$$

where $t^-_\infty \Omega_\infty = 0$ and $[t^-_\infty, t^+_\infty] = 2\sigma$. Let $t^{\pm}_\infty = t^1_\infty \pm i t^2_\infty$ and $t^3_\infty = 0$ and $\underline{s}_\infty = R\, \underline{t}_\infty$. Then

$$\lim_{N\to\infty} {}_N\langle\underline{\sigma}| \prod N^{1/2}(\sigma^{k(m)}_N - \sigma^{k(m)}) |\underline{\sigma}\rangle_N = (\Omega_\infty, \prod s^{k(m)}_\infty \Omega_\infty) , \tag{3.10}$$

$$\lim_{N\to\infty} {}_N\langle\underline{\sigma}| \prod \sigma^{k(m)}_N |\underline{\sigma}\rangle_N = \prod \sigma^{k(m)} . \tag{3.11}$$

Although all our results can be expressed in terms of correlation functions and their limits, the mathematics simplifies considerably by introducing a sequence of approximating Hilbert spaces $\mathcal{H}_N$ which converges against $\mathcal{H}_\infty$, the GNS Hilbert space for (3.1) and (3.10) . Let $\Omega_N = |N^{1/2}\alpha\rangle \otimes |\underline{\sigma}\rangle_N$ and $\mathcal{H}_N$ be the span of all $(a^* - N^{1/2}\alpha^*)^r (t^+_N)^s \Omega_N$. Define $I_N : \mathcal{H}_\infty \to \mathcal{H}_N$ by linear extension of

$$I_N (a^*_\infty)^r (t^+_\infty)^s \Omega_\infty = (a^* - N^{1/2}\alpha^*)^r (t^+_N)^s \Omega_N . \tag{3.12}$$

Then $\| I_N \| = 1$, and $\lim_{N\to\infty} \| I_N \varphi_\infty \| = \| \varphi_\infty \|$ for all $\varphi_\infty \in \mathcal{H}_\infty$, by (3.9)

On the finite particle subspace, $\mathcal{H}^{\circ}_{\infty}$, one has

$$\lim_{N\to\infty} \| x_N \ldots y_N I_N \varphi_\infty - I_N x_\infty \ldots y_\infty \varphi_\infty \| = 0 \quad , \tag{3.13}$$

if $x_N, \ldots y_N \in \{ (a^{\#} - N^{1/2}\alpha^{\#}), N^{1/2}(\underline{\sigma}_N - \underline{\sigma}) \}$. In a somehow different sense than in [H4] , the intensive observables $\alpha_N^{\#}, \underline{\sigma}_N$ have normal fluctuations around $\alpha^{\#}$, $\underline{\sigma}$ in the sequence Ω_N of coherent states [W1].

We shall say that a sequence of selfadjoint operators A_N on $\mathcal{H}_N$ converges in the Weyl-Trotter sense to a selfadjoint operator A_∞ on $\mathcal{H}_\infty$, $A_N \xrightarrow[WT]{} A$, if for all $s \in \mathbb{R}$ and all $\varphi_\infty \in \mathcal{H}_\infty$

$$\lim_{N\to\infty} \| \exp(iA_N s) I_N \varphi_\infty - I_N \exp(iA_\infty s) \varphi_\infty \| = 0 \ . \tag{3.14}$$

Let $\underline{\sigma}_N(t) = \exp(iH_N^S t)\, \underline{\sigma}_N \exp(-iH_N^S t)$ and $\alpha_N(t) = 2^{-1/2}(\xi_N(t) + i\pi_N(t))$. We shall prove in this section that

$$\xi_N(t) \xrightarrow[WT]{} \xi_t \ , \ \pi_N(t) \xrightarrow[WT]{} \pi_t \ , \ \underline{\sigma}_N(t) \xrightarrow[WT]{} \underline{\sigma}_t \ , \tag{3.15}$$

$$N^{1/2}(\xi_N(t) - \xi_t) \xrightarrow[WT]{} q_\infty(t) \ , \quad N^{1/2}(\pi_N(t) - \pi_t) \xrightarrow[WT]{} p_\infty(t),$$

$$N^{1/2}(\underline{\sigma}_N(t) - \underline{\sigma}_t) \xrightarrow[WT]{} \underline{s}_\infty(t) \ . \tag{3.16}$$

Here α_t and $\underline{\sigma}_t$ satisfy the classical equations (2.9) with $\varkappa = \gamma = \delta = 0$ and initial conditions $(\alpha, \underline{\sigma})$ at $t = 0$. Clearly the classical equations have always global solutions, since $|\alpha|^2 + \sigma^3$ and $(\sigma^3)^2 + |\sigma^+|^2$ are invariant under the flow. $a_\infty(t)$ and $\underline{s}_\infty(t)$ are solutions of the linearized equations around the classical orbit :

$$\begin{aligned} \dot{a}_\infty(t) &= - i\gamma a_\infty(t) - i\lambda s_\infty^-(t) \ , \\ \dot{s}_\infty^-(t) &= - i\varepsilon s_\infty^-(t) + 2i\lambda\alpha_t s_\infty^3(t) + 2i\lambda\sigma_t^3 a_\infty(t) \ , \\ \dot{s}_\infty^3(t) &= i\lambda(\alpha_t^* s_\infty^-(t) + \sigma_t^- a_\infty^*(t) \ - \ h.c.) \ , \end{aligned} \tag{3.17}$$

with initial conditions $a_\infty(0) = a_\infty$, $s_\infty(0) = s_\infty$.

Let $(\alpha, \underline{\sigma})$ be any initial state. Around the corresponding classical solution we develop H_N^S in powers of $N^{1/2}$ into a mean field and a fluctuation part, $H_N^S = H_N^M(t) + H_N^F(t)$:

$$H_N^M(t) = \varepsilon S_N^3 + \lambda\alpha_t^- S_N^+ + \lambda\alpha_t^* S_N^- + N^{1/2}(\gamma\alpha_t + \lambda\sigma_t^-)a^* +$$

$$+ N^{1/2}(\gamma\alpha_t^* + \lambda\sigma_t^+)a - N(\gamma\alpha_t^*\alpha_t + \lambda\alpha_t^*\sigma_t^- + \lambda\sigma_t^+\alpha_t) \tag{3.18}$$

is well known from the thermodynamics of the BCS [T1] or of the Dicke model [H2], with an explicit time dependence, if the initial condition is not stationary. The action of $H_N^M(t)$ is purely group theoretical, a translation of the boson operators by the photon part and a rotation of the atomic observables. We shall see that $H_N^M(t)$ describes correctly the time-evolution of the intensive observables when $N \to \infty$.

$$H_N^F(t) = \gamma(a^* - N^{1/2}\alpha_t^*)(a - N^{1/2}\alpha_t) + \lambda(a^* - N^{1/2}\alpha_t^*)N^{1/2}(\sigma_N^- - \sigma_t^-)$$

$$+ \lambda(a - N^{1/2}\alpha_t)N^{1/2}(\sigma_N^+ - \sigma_t^+) \tag{3.19}$$

leads to nonlinear Heisenberg equations for the fluctuation operators, when $N < \infty$, which linearize for $N \to \infty$. Consider

$$U_N^M(t) = T\exp -i\int_0^t ds\, H_N^M(s) . \tag{3.20}$$

Clearly this propagator exists for all t using a strongly convergent Dyson series on $\mathcal{H}_N^0$ and a unitary extension. $\hat{a}_N(t) = U_N^M(t)^*(a-N^{1/2}\alpha_t)\times \times U_N^M(t)$ satisfies $d\hat{a}_N(t)/dt = 0$ and $\hat{a}_N(0) = a - N^{1/2}\alpha$, hence

$$\hat{a}_N(t) = a - N^{1/2}\alpha . \tag{3.21}$$

Similarly, $\hat{\underline{s}}_N(t) = U_N^M(t)^* N^{1/2}(\underline{\sigma}_N - \underline{\sigma}_t)U_N^M(t)$ satisfies

$$\begin{aligned} d\hat{s}_N^-(t)/dt &= -i\varepsilon\,\hat{s}_N^-(t) + 2i\lambda\alpha_t\,\hat{s}_N^3(t) , \\ d\hat{s}_N^3(t)/dt &= -i\lambda\alpha_t\,\hat{s}_N^+(t) + i\lambda\alpha_t^*\,\hat{s}_N^-(t) , \end{aligned} \tag{3.22}$$

and has a global solution in terms of a 1-parameter family of rotations

$$\hat{\underline{s}}_N(t) = M(t)\, N^{1/2}(\underline{\sigma}_N - \underline{\sigma}). \tag{3.23}$$

Let us consider Weyl operators $W_N(t) = W_N(\alpha,\underline{\sigma},t,x,y,\underline{z})$ of the form

$$W_N(\alpha,\underline{\sigma},t,x,y,\underline{z}) = \exp i\left[x(q-N^{1/2}\xi_t)+y(p-N^{1/2}\pi_t)+\underline{z}(\underline{\sigma}_N-\underline{\sigma}_t)N^{1/2}\right]. \quad (3.24)$$

Then

$$\exp(iH_N^S t)\, W_N(t)\, \exp(-iH_N^S t) = \hat{V}_N(t,0)^* \hat{W}_N(t)\, \hat{V}_N(t,0) \;, \quad (3.25)$$

$$\hat{W}_N(t) = U_N^M(t)^* W_N(t)\, U_N^M(t) \quad , \quad (3.26)$$

$$\hat{V}_N(t,0) = U_N^M(t)^* \exp(-iH_N^S t) \quad . \quad (3.27)$$

$\hat{W}_N(t)$ is obtained from $W_N(t)$ by replacing in (3.24) $a-N^{1/2}\alpha_t$ by $a-N^{1/2}\alpha$ and $\underline{\sigma}_N-\underline{\sigma}_t$ by $M(t)(\underline{\sigma}_N-\underline{\sigma})$. $\hat{V}_N(t,0)$ is the solution of

$$d\hat{V}_N(t,0)/dt = -i\,\hat{H}_N^F(t)\,\hat{V}_N(t,0) \;,\quad \hat{H}_N^F(t) = U_N^M(t)^* H_N^F(t)\, U_N^M(t) \;. \quad (3.28)$$

The propagator

$$\hat{V}_N(t,s) = T\exp -i\int_s^t dr\, \hat{H}_N^F(r) \quad (3.29)$$

can again be defined by a Dyson series on $\mathcal{H}_N^0$, and $\hat{H}_N^F(r)$ is obtained from $H_N^F(r)$ by going over to fluctuation observables at time zero. Hence (3.29) can converge, when we pass to $N\to\infty$ along Ω_N.

<u>Theorem 3.1</u> : For every physical initial condition α, $\underline{\sigma}$ (with(3.3)), the intensive observables and the fluctuations along the classical orbit converge to (3.15) and (3.16) along $|\alpha,\underline{\sigma}\rangle_N$ for $N\to\infty$.

<u>Proof</u> : For $|t-s|$ sufficiently small, the Dyson series for

$$\hat{V}_\infty(t,s) = T\exp -i\int_s^t dr\, \hat{H}_\infty^F(r) \quad , \quad (3.30)$$

$$\hat{H}_\infty^F(r) = \gamma\, a_\infty^* a_\infty + \lambda(\, a_\infty^* s_\infty^-(r) + s_\infty^+(r) a_\infty) \;, \quad (3.31)$$

$$\underline{s}_\infty(r) = M(r)\, R\, \underline{t}_\infty \;, \quad (3.32)$$

converges in the strong topology on $\mathcal{H}_\infty^0$. Using (3.25) and the Schwarz inequality

$$\| I_N \hat{V}_\infty(t,0)^* \hat{W}_\infty(t)\, \hat{V}_\infty(t,0)\psi_\infty - \hat{V}_N(t,0)^* \hat{W}_N(t)\, \hat{V}_N(t,0)\, I_N \psi_\infty \| \le$$

$$\leq \| I_N \hat{V}_\infty(t,0)^* \hat{W}_\infty(t) \hat{V}_\infty(t,0)\psi_\infty - \hat{V}_N(t,0)^* I_N \hat{W}_\infty(t) \hat{V}_\infty(t,0)\psi_\infty \| +$$

$$+ \| I_N \hat{W}_\infty(t) \hat{V}_\infty(t,0)\psi_\infty - \hat{W}_N(t) I_N \hat{V}_\infty(t,0)\psi_\infty \| +$$

$$+ \| I_N \hat{V}_\infty(t,0)\psi_\infty - \hat{V}_N(t,0) I_N \psi_\infty \| . \tag{3.33}$$

For small $|t-s|$, the Duhamel formula can be used on $\mathcal{H}_\infty^0$:

$$\| I_N \hat{V}_\infty(t,s)\psi_\infty - \hat{V}_N(t,s) I_N \psi_\infty \| \tag{3.34}$$

$$= \| i \int_s^t dr\, \hat{V}_N(t,r) \left[\hat{H}_N^F(r) I_N - I_N \hat{H}_\infty^F(r) \right] \hat{V}_\infty(r,s)\psi_\infty \|$$

$$\leq \int_s^t dr\, \| (\hat{H}_N^F(r) I_N - I_N \hat{H}_\infty^F(r)) \hat{V}_\infty(r,s)\psi_\infty \| ,$$

since the Dyson series for $\hat{V}_\infty(r,s)$ solves all domain problems. Hence (3.34) converges to zero for small $|t-s|$ and $\psi_\infty \in \mathcal{H}_\infty^0$, and therefore on all $\mathcal{H}_\infty$, since $\| I_N \| = \| \hat{V}_N(t,s) \| = \| \hat{V}_\infty(t,s) \| = 1$. Large $|t-s|$ can be estimated stepwise, e.g. $\| \hat{V}_N(t,0) I_N \psi_\infty - I_N \hat{V}_\infty(t,0)\psi_\infty \| \leq \| \hat{V}_N(s,0) I_N \psi_\infty - I_N \hat{V}_\infty(s,0)\psi_\infty \| + \| \hat{V}_N(t,s) I_N V_\infty(s,0)\psi_\infty - I_N V_\infty(t,0)\psi_\infty \|$. By expanding the exponential on any $\psi_\infty \in \mathcal{H}_\infty^0$, one proves that

$$\| I_N \hat{W}_\infty(t)\psi_\infty - \hat{W}_N(t) I_N \psi_\infty \| \longrightarrow 0 \quad \text{for} \quad N \longrightarrow \infty , \tag{3.35}$$

and therefore on $\mathcal{H}_\infty$ by uniform boundedness. Hence all three terms on the r.h.s. of (3.33) go to zero for all t and all $\psi_\infty \in \mathcal{H}_\infty$. Now, $\hat{V}_\infty(t,0)^* \hat{W}_\infty(t) \hat{V}_\infty(t,0)$ has an exponential representation of the form (3.24) , where

$$\begin{aligned} a_\infty(t) &= \hat{V}_\infty(t,0)^* \hat{a}_\infty(t) \hat{V}_\infty(t,0) , \\ \underline{s}_\infty(t) &= \hat{V}_\infty(t,0)^* \hat{\underline{s}}_\infty(t) \hat{V}_\infty(t,0) . \end{aligned} \tag{3.36}$$

One checks that (3.36) satisfies (3.17) with $a_\infty(0) = a_\infty$, $\underline{s}_\infty(0) = \underline{s}_\infty$, and by uniqueness the WT-convergence (3.16) follows. (3.15) can be derived similarly.

<u>Corollary 3.2</u> : For all t_m, x_m, y_m, z_m, $1 \leq m \leq n$,

$$\lim_{N\to\infty} \left(\Omega_N, \prod_{m=1}^n \exp(iH_N^S t_m)\, W_N(\alpha, \underline{\sigma}, t_m, x_m, y_m, z_m) \exp(-iH_N^S t_m)\, \Omega_N \right) =$$

$$= \left(\Omega_\infty, \prod_{m=1}^{n} W_\infty(\alpha, \underline{\sigma}, t_m, x_m, y_m, z_m)\, \Omega_\infty\right), \tag{3.37}$$

and analogously for the intensive observables.

Proof : By repeated application of Theorem 3.1 one obtains WT-convergence for a product of time translated Weyl operators. This implies (3.37), since $\lim (I_N \varphi_\infty, I_N \psi_\infty) = (\varphi_\infty, \psi_\infty)$ by polarization.

It is typical for the classical limit of quantum mechanics in minimum uncertaincy states [H5] that the quantum mechanical correlation functions of intensive observables in coherent states (with amplitudes which are scaled so that the time zero expectation values converge) tend towards products of solutions of the classical equations, and that the fluctuations around the classical orbits follow linearized boson equations of motion. There is a close connection between the limit $N \to \infty$ and $\hbar$ fixed in mean field models and the limit $\hbar \to 0$ in ordinary quantum mechanics of a fixed number of degrees of freedom.

§4. The Lossy Laser

After having studied the classical limit for the closed system of atoms and radiation we shall now construct a model of a quantum mechanical reservoir with simple damping and pumping properties. A natural reservoir for photon losses in the laser is a quantized radiation field of infinitely many boson degrees of freedom, $a_w^{\#}$, which describes all the non-lasing photon modes of the world. The simplest interaction Hamiltonian is linear :

$$H^P = \int_{-\infty}^{+\infty} dw\, \varepsilon_w a_w^* a_w + \int_{-\infty}^{+\infty} dw\, k_w\, (a_w^* a + a^* a_w) , \tag{4.1}$$

where for notational simplicity we take $w \in \mathbb{R}$ and a continuum normalization

$$[a_w, a_{w'}^*] = \delta(w-w'), \quad [a_w, a_{w'}] = 0 . \tag{4.2}$$

Under the influence of H_N^S and H^P the Heisenberg equations become

$$\dot a_N(t) = -i\gamma\, a_N(t) - i\lambda\, S_N^-(t) N^{-1/2} - i\int_{-\infty}^{+\infty} k_w\, a_{wN}(t) , \tag{4.3}$$

$$a_{wN}(t) = - i\varepsilon_w a_{wN}(t) - ik_w\, a_N(t) . \tag{4.4}$$

In the initial state we shall assume no correlations between the cavity and the reservoir, $\Omega_N = \Omega_N^S \otimes \Omega_N^R$, and take the photon reservoir to be empty, $a_w \Omega_N^R = 0$ for all w. A dissipation of the form (4.1) will in general not lead to phenomenological equations of the type (2.7) with a fluctuation force $g_N(t)$ independent of the state of the small system, unless the interaction with the reservoir is sufficiently "white" to destroy all memory effects. We can solve (4.4) for $a_{wN}(t)$ in terms of the initial condition a_w and $a_N(s)$ for $0 \leq s \leq t$, insert the solution into (4.3) and pass to the limit $k_w \to k$, $\varepsilon_w \to w$. This leads for $\alpha_N(t)$, $t \geq 0$, to

$$\dot{\alpha}_N(t) = - (i\gamma + \varkappa)\, \alpha_N(t) - i\lambda \sigma_N^-(t) + \chi_N(t) , \tag{4.5}$$

where $\varkappa = \pi k^2$, $\chi_N(t) = -ikN^{-1/2}A(t)$ and where $A(t) = \int dw\, a_w\, e^{-iwt}$ is "white boson noise". Singular equations of motion of the type (4.5) require some care, but all the steps in the following, including some abusive notation, can be rigorously justified by treating H_N^R as a derivation and $\exp(iH_N^R t)(.)\exp(-iH_N^R t)$ as an automorphism of the Weyl and CAR algebra, which on a dense domain is defined by a convergent multicommutator series.

For the atomic degrees of freedom it is more complicated to find a phenomenologically acceptable and mathematically simple dissipation process. We shall couple the electrons of each atom to an electron sea of infinitely many degrees of freedom, again by a singular linear coupling :

$$H_N^A = \sum_{n=1}^{N} \sum_{\pm} H_{\pm n}^A , \tag{4.6}$$

$$H_{\pm n}^A = \int_{-\infty}^{+\infty} dw\, w(b_{\pm nw}^* b_{\pm nw} + c_{\pm nw}^* c_{\pm nw}) + \int_{-\infty}^{+\infty} dw \left[b_{\pm n}^* (f_\pm b_{\pm nw} + g_\pm c_{\pm nw}) + \text{h.c.} \right]$$

using Fermi fields with normal anticommutation relations in continuum normalization (4.2) . Under $H_N = H_N^S + H_N^A + H^P$, the time evolution of the atomic Fermi operators becomes

$$b_{+n}(t) = -(i\varepsilon + \gamma) b_{+n}(t)/2 - i\lambda \alpha_N(t) b_{-n}(t) - if_+ B_{+n}(t) - ig_+ C_{+n}(t) , \tag{4.7}$$

$$\dot{b}_{-n}(t) = -(\gamma - i\varepsilon)b_{-n}(t)/2 - i\lambda\alpha_N^*(t)b_{+n}(t) - if_-B_{-n}(t) - ig_-C_{-n}(t). \tag{4.8}$$

Here $B_{\pm n}(t) = \int dw\, b_{\pm nw} \exp(-iwt)$ and $C_{\pm n}(t)$ are "white fermion noise" operators and $\gamma = 2\pi(|g_+|^2 + |g_-|^2)$. We have assumed $|f_+|^2 + |g_+|^2 = |f_-|^2 + |g_-|^2$ and $|f_+|^2 = |g_-|^2$ in order to ensure macroscopic charge neutrality for the dissipation process, so that $\sum(b_{+n}^* b_{+n} + b_{-n}^* b_{-n})$ stays at N up to $N^{1/2}$ fluctuations.

The intensive observables obey the following differential equations

$$\dot{\sigma}_N^-(t) = -(i\varepsilon+\gamma)\,\sigma_N^-(t) + 2i\lambda\alpha_N(t)\sigma_N^3(t) + \varphi_N^-(t), \tag{4.9}$$

where the fluctuation force $\varphi_N^-(t) = \varphi_N^+(t)^*$ unfortunately depends on the observables of the small system :

$$\varphi_N^+(t) = \varphi_N^{++}(t) + \varphi_N^{+-}(t), \tag{4.10}$$

$$\varphi_N^{++}(t) = iN^{-1}\sum_{n=1}^{N}\left[f_+^* B_{+n}^*(t)b_{-n}(t) + g_-C_{-n}(t)b_{+n}(t)\right], \tag{4.11}$$

$$\varphi_N^{+-}(t) = -iN^{-1}\sum_{n=1}^{N}\left[f_- b_{+n}^*(t)B_{-n}(t) + g_+^* b_{-n}(t)C_{+n}^*(t)\right].$$

Similarly

$$\dot{\sigma}_N^3(t) = -\gamma(\sigma_N^3(t) - \eta) + i\lambda\alpha_N^*(t)\sigma_N^-(t) - i\lambda\alpha_N(t)\sigma_N^+(t) + \varphi_N^3(t), \tag{4.12}$$

$$\varphi_N^3(t) = \varphi_N^{3+}(t) + \varphi_N^{3-}(t), \qquad \varphi_N^{3-}(t) = \varphi_N^{3+}(t)^*, \tag{4.13}$$

$$\varphi_N^{3+}(t) = (2N)^{-1}\sum_{m=1}^{N}\left[ig_+C_{+n}(t)b_{+n}^*(t) - ig_-C_{-n}(t)b_{-n}^*(t) + \right.$$

$$\left. + if_+^*B_{+n}^*(t)b_{+n}(t) - if_-^*B_{-n}^*(t)b_{-n}(t)\right]. \tag{4.14}$$

In (4.12) the "unsaturated inversion" $\eta = \pi\gamma^{-1}(|g_+|^2 - |g_-|^2)$ was split off, in order to obtain for $i = \pm, 3$

$$\varphi_N^{i-}(t)\,\Omega_N^R = \varphi_N^{i+}(t)^*\,\Omega_N^R = 0, \tag{4.15}$$

if the B-reservoir is empty, $B_{+n}(t)\,\Omega_N^R = 0$, and the C-reservoir full, $C_{+n}^*(t)\,\Omega_N^R = 0$. By a suitable choice of $|g_\pm|^2$, we can reach any $-1/2 \le \eta \le 1/2$.

By generalizing the approach of Section 3 we shall study the WT convergence of the intensive and fluctuation observables along $\Omega_N = \Omega_N^S \otimes \Omega_N^R$, where Ω_N^R is the vacuum of the reservoir fields and Ω_N^S is coherent, $\Omega_N^S = |\alpha, \underline{\sigma}\rangle_N$. Given any physical initial condition $\alpha, \underline{\sigma}$, we split H_N along the classical orbit $\alpha_t, \underline{\sigma}_t$ into a mean field and a fluctuation part, $H_N = H_N^M(t) + H_N^F(t)$:

$$H_N^M(t) = H^P + \nu a^* a + \lambda N^{1/2}(a^* \sigma_t^- + \sigma_t^+ a) + \tag{4.16}$$

$$+ H_N^A + \varepsilon S_N^3 + \lambda(\alpha_t S_N^+ + \alpha_t^* S_N^-) - N(\alpha_t^* \sigma_t^- + \sigma_t^+ \alpha_t) ,$$

$$H_N^F(t) = \lambda(a^* - N^{1/2}\alpha_t^*)N^{1/2}(\sigma_N^- - \sigma_t^-) + \text{h.c.} \tag{4.17}$$

Under $H_N^M(t)$ the photons and the different atoms are uncoupled. Since the initial state is a product state with strong factorization properties, this will lead to "boson white noise" fluctuation forces for the fluctuation observables, when $N \to \infty$. In the corresponding interaction picture, $\hat{H}_N^F(t)$ will again become quadratic in boson operators. We will study the evolution of Weyl operators (3.24) which can be represented in the form (3.25,26,27) with H_N^S replaced by H_N and with $U_N^M(t)^*(.)U_N^M(t)$ considered as an automorphism. We compute $\hat{a}_N(t)$ and $\hat{s}_N(t)$ from their differential equations :

$$d\hat{a}_N(t)/dt = -(i\nu + \varkappa)\hat{a}_N(t) - ik\,A(t) ,$$

$$\hat{a}_N(t) = \exp-(i\nu + \varkappa)t\,(a - N^{1/2}\alpha) - ik\int_0^t ds\,A(s)\exp-(i\nu + \varkappa)(t-s) , \tag{4.18}$$

$$d\hat{s}_N^-(t)/dt = -(i\varepsilon + \gamma)\hat{s}_N^-(t) + 2i\lambda\alpha_t \hat{s}_N^3(t) + f_N^-(t) ,$$

$$d\hat{s}_N^3(t)/dt = -\gamma\hat{s}_N^3(t) + i\lambda\alpha_t^*\hat{s}_N^-(t) - i\lambda\alpha_t\hat{s}_N^+(t) + f_N^3(t) . \tag{4.19}$$

In terms of the propagator $P(t,s)$ of (4.19) one has

$$\hat{\underline{s}}_N(t) = P(t,0)N^{1/2}(\underline{\sigma}_N - \underline{\sigma}) + \int_0^t ds\,P(t,s)\,\underline{f}_N(s) . \tag{4.20}$$

In (4.19), the fluctuation forces $\underline{f}_N(t) = \underline{f}_N^+(t) + \underline{f}_N^-(t)$ satisfy

$$\underline{f}_N^{\pm}(t) = M^{-1/2}\sum_{m=1}^{M} \hat{f}_m^{\pm}(t) , \quad \hat{f}_m^{\pm}(t) = (2p+q)^{-1/2}\sum_{n\in\Delta(m)} F_n^{\pm}(t) , \tag{4.21}$$

where $\Delta(m) = ((2p+q)m, (2p+q)(m+1)]$. The $F_n^{\pm}(t)$ have the form (4.10,11,13,14) without N^{-1} and $\sum_{n=1}^{N}$ and with the $b_{\pm n}(t)$ evolved under $H_N^M(t)$. The $b_{\pm n}(t)$ therefore obey (4.7,8) with $\alpha_N(t)$ replaced by the c-number α_t and, as a consequence of the direct product structure of Ω_N , one has for all $1 \leq m \leq M$

$$(2p+q)^{-1} \sum_{m \in \Delta(m)} (\Omega_N, b_{+n}^*(t) b_{-n}(t) \Omega_N) = \sigma_t^+ ,$$

$$(2p+q)^{-1} \sum_{m \in \Delta(m)} (\Omega_N, (b_{+n}^*(t) b_{+n}(t) - b_{-n}^*(t) b_{-n}(t)) \Omega_N) = 2\sigma_t^3 , \tag{4.22}$$

where the σ_t^i satisfy the classical equations (2.9) . Hence the correlation functions in $(\Omega_N, \cdot \, \Omega_N) = \langle \cdot \rangle_N$ are δ-correlated and independent of m :

$$\langle \hat{f}_m^-(s) \hat{f}_m^-(t) \rangle_N = \langle \hat{f}_m^+(s) \hat{f}_m^+(t) \rangle_N = 0$$

$$\langle \hat{f}_m^-(s) \hat{f}_m^+(t) \rangle_N = \delta(s-t)\gamma(1/2 - \eta)(1 - 2\sigma_t^3)$$

$$\langle \hat{f}_m^3(s) \hat{f}_m^3(t) \rangle_N = \delta(s-t)\gamma(1/4 + \eta\sigma_t^3) \tag{4.23}$$

$$\langle \hat{f}_m^3(s) \hat{f}_m^+(t) \rangle_N = \langle \hat{f}_m^3(s) \hat{f}_m^-(t) \rangle_N^* = \delta(s-t)\gamma(1/2 - \eta)\sigma_t^+$$

$$\langle \hat{f}_m^+(s) \hat{f}_m^3(t) \rangle_N = \langle \hat{f}_m^-(s) \hat{f}_m^3(t) \rangle_N^* = -\delta(s-t)\gamma(1/2 + \eta)\sigma_t^+$$

Furthermore, since $\hat{f}_m^-(t)\Omega_N = \hat{f}_m^+(t)^*\Omega_N = 0$ and $[\hat{f}_m^{\pm}(s), \hat{f}_n^{\pm}(t)] = 0$ for $m \neq n$, one has independent of $N = M(2p+q)$

$$\langle f_N^i(s) f_N^j(t) \rangle_N = \langle \hat{f}_m^i(s) \hat{f}_m^j(t) \rangle_N \tag{4.24}$$

The 2-point function of $\hat{f}_m^i(s)$ with $A^{\#}(t)$ or any fluctuation operator at $t = 0$ vanishes. However, the higher correlation funtions of such quantities are far from Gaussian for finite N .

Let $\mathcal{H}_\infty$ be the Fock space of two boson modes $a_\infty^{\#}$ and $t_\infty^{\pm}$ and four boson fields $A_\infty^{\#}(t)$, $f_\infty^{i+}(t)$, i=1,2,3, with test function space $L^2([0,\infty))$ and cyclic vacuum Ω_∞, satisfying

$$f_\infty^{i-}(t)\Omega_\infty = a_\infty \Omega_\infty = A_\infty(t)\Omega_\infty = t_\infty^- \Omega_\infty = 0 \tag{4.25}$$

$$f_\infty^{i+}(t) = f_\infty^{i-}(t)^*, \quad t_\infty^+ = t_\infty^{-*}, \quad f_\infty^{\pm\pm}(t) = f_\infty^{1\pm}(t) \pm i f_\infty^{2\pm}(t) \ ,$$

and with c-number commutators, which all vanish except for

$$[a_\infty, a_\infty^*] = 1, \quad [t_\infty^-, t_\infty^+] = 2\sigma \ , \quad [f_\infty^{i-}(s), f_\infty^{j+}(t)] = \langle f_N^i(s) f_N^j(t)\rangle_N \ . \tag{4.26}$$

The identification mapping $I_N : \mathcal{H}_\infty \to \mathcal{H}_N$ is defined by linear extension of

$$I_N\, a_\infty^{*i} A_\infty^*(s_1)..A_\infty^*(s_j)(t_\infty^+)^k f_\infty^{n_1+}(t_1)..f_\infty^{n_r+}(t_r)\,\Omega_\infty = \tag{4.27}$$
$$= M^{-(k+r)/2} \sum{}' (a^* - N^{1/2}\alpha^*)^i A^*(s_1)..A^*(s_j) \hat{t}_{m_1}^+ ... \hat{t}_{m_k}^+ \hat{f}_{m_{k+1}}^{n_1+}(t_1)..\hat{f}_{m_{k+r}}^{n_r+}(t_r)\,\Omega_N .$$

where $\sum'$ extends over all $m_1,...m_{k+r} = 1,..M'$ with $m_k \neq m_l$ for $k \neq l$, and $t_N^+ = M^{-1/2}\sum_{m=1}^{M} \hat{t}_m^+$ as in (4.21). Let $\mathcal{H}_\infty^h \subset \mathcal{H}_\infty$ be the span of all vectors of the type (4.27) with $k + r = h$. Because of the product struture of Ω_N there exists an isometry I_N^h such that

$$I_N \upharpoonright \mathcal{H}_\infty^h = \prod_{g=1}^{h-1} (1 - g/M)^{1/2}\, I_N^h \tag{4.28}$$

Hence $\| I_N \| = 1$ and $\lim_{N\to\infty} \| I_N \varphi_\infty \| = \| \varphi_\infty \|$ for all $\varphi_\infty \in \mathcal{H}_\infty$. Now, we are ready to prove the main theorem on the time evolution of the laser :

<u>Theorem 4.1</u> : In the DHL model (2.4),(4.1,6) the intensive observables $\alpha_N(t)$, $\underline{\sigma}_N(t)$ converge in the WT sense along $\Omega_N = |\alpha, \underline{\sigma}\rangle_N \otimes \Omega_N^R$ (with (3.3)) towards the solutions α_t, $\underline{\sigma}_t$ of the classical equations (2.9) with initial conditions $\alpha_0 = \alpha$, $\underline{\sigma}_0 = \underline{\sigma}$. The fluctuation operators $a_N(t) = N^{1/2}(\alpha_N(t) - \alpha_t)$, $\underline{s}_N(t) = N^{1/2}(\underline{\sigma}_N(t) - \underline{\sigma}_t)$ converge in the WT sense towards solutions $a_\infty(t)$, $\underline{s}_\infty(t)$ in $\mathcal{H}_\infty$ of the linearized equations

$$\dot{a}_\infty(t) = -(i\nu + \varkappa) a_\infty(t) - i\lambda s_\infty^-(t) - ik\, A_\infty(t) \ , \tag{4.29}$$

$$\dot{s}_\infty^-(t) = -(i\varepsilon + \gamma) s_\infty^-(t) + 2i\lambda \alpha_t s_\infty^3(t) + 2i\lambda \sigma_t^3 a_\infty(t) + f_\infty^-(t) \ ,$$

$$\dot{s}_\infty^3(t) = -\gamma s_\infty^3(t) + i\lambda\,(\alpha_t^* s_\infty^-(t) + \sigma_t^- a_\infty^*(t) - \text{h.c.}) + f_\infty^3(t) \ ,$$

with $t \geq 0$ and initial condition $a_\infty(0) = a_\infty$, $\underline{s}_\infty(0) = \underline{s}_\infty$ in $\mathcal{H}_\infty$. Also the correlation functions converge in the sense of (3.37).

Proof : As in theorem 3.1, we consider the Weyl operators

$$\exp(iH_N t)\, W_N(\alpha,\underline{\sigma},t,x,y,z)\, \exp(-iH_N t) = \hat{V}_N(t,0)^* \hat{W}_N(t) \hat{V}_N(t,0) \,. \qquad (4.30)$$

Here $\hat{W}_N(t) = \exp(i\hat{K}_N(t))$ is obtained from (3.24) by replacing $a - N^{1/2}\alpha_t$ by $\hat{a}_N(t)$ and $N^{1/2}(\underline{\sigma}_N - \underline{\sigma}_t)$ by $\hat{\underline{s}}_N(t)$, which gives a rigorous meaning to the l.h.s. of (4.30). $\hat{V}_N(t,s) = T\exp -i\int_s^t dr\, \hat{H}_N^F(r)$ is defined by a Dyson series with the same substitution as in $\hat{W}_N(t)$. The estimate (3.33) proves convergence, if we can show for $0 \leq u \leq 1$ and $0 \leq s \leq r \leq t$ with $|t-s|>0$ sufficiently small and for every $\psi_\infty \in \mathcal{H}_\infty^0$ that

$$\begin{aligned} &\| (I_N \hat{H}_\infty^F(r) - \hat{H}_N^F(r) I_N) \hat{V}_\infty(r,s)\psi_\infty \| \,, \\ &\| (I_N \hat{K}_\infty(t) - \hat{K}_N(t) I_N) \exp(iu\hat{K}_\infty(t))\psi_\infty \| \end{aligned} \qquad (4.31)$$

are uniformly bounded in N and converge to zero for every r,u, if $N \to \infty$. Here $\hat{H}_\infty^F(r)$, $\hat{K}_\infty(t)$ and $\hat{V}_\infty(r,s)$ are analogously defined using $\hat{a}_\infty(t)$ and $\hat{\underline{s}}_\infty(t)$, the I_N-preimages of (4.18) and (4.20). The uniform boundedness and convergence of (4.31) can be easily proved by expanding $\hat{V}_\infty(r,s)\psi_\infty$ and $\exp(iu\hat{K}_\infty(t))\psi_\infty$ and by using (4.28). The limits $a_\infty(t) = \hat{V}_\infty(t,0)^* \hat{a}_\infty(t) \hat{V}_\infty(t,0)$ and $\underline{s}_\infty(t)$ are identified by the differential equation (4.29) which they satisfy.

§5. Conclusion

We have discussed in considerable detail a microscopic quantum mechanical model for a laser cavity coupled to different reservoirs and have given a precise sense in which the laser shows irreversible classical and stochastic behavior in the thermodynamic limit. We have, however, only barely prodded the sleeping giant of nonequilibrium statistical mechanics. There are many open problems in this field, both of a technical and conceptual nature. There is the immediate question, whether the choice of the reservoirs, in particular the use of regular Hamiltonians and nonlinear couplings, can qualitatively change the phase transition in this class of mean field models. A nontrivial generalization of our approach is necessary in order to treat the finite mode laser Our results on the intensive and fluctuation observables are the first two terms in an asymptotic expansion for large finite lasers. It is not clear how good this expansion is, if one is interested in the equilibrium properties of the finite non-linear system for $t \to \infty$. Hopefully, the two limits, $N \to \infty$ and $t \to \infty$, can be exchanged for certain observables away from threshold. Finally, it is clear that the really hard problems of irreversible statistical mechanics are not the construction of quantum mechanical cuckoo clocks but the understanding of continuous systems with short range and Coulomb forces.

References

[A1] F.T. Arecchi, Schulz-Dubois, E.O., "Laser Handbook", North-Holland Publ. Co., Amsterdam (1972)

[G1] P. Glanssdorff, I. Prigogine, "Thermodynamic Theory of Structure, Stability and Fluctuations", Wiley, London (1971)

[G2] R.J. Glauber, Phys. Rev. 131, 2766 (1963)
J.M. Radcliffe, J.Phys. A 4, 313 (1971)
F.T. Arecchi, E. Courtens, R. Gilmore, H. Thomas, Phys.Rev. A6, 2211 (1972).

[G3] R.J. Glauber "Quantum Optics", Varenna Lectures, Academic Press, N.Y. (1969)

[G4] R. Graham, in "Springer Tracts in Modern Physics", Vol. 66, Berlin (1973)

[H1] H. Haken, Handbuch der Physik, Vol. 25, 2c, Springer, Berlin (1970)

[H2] K. Hepp and E.H. Lieb, Annals of Physics 76, 360 (1973)

[H3] K. Hepp and E.H. Lieb, Phys. Rev. A, to appear

[H4] K. Hepp and E.H. Lieb, Helv.Phys.Acta, to appear

[H5] K. Hepp, "The Classical Limit for Quantum Mechanical Correlation Functions", in preparation

[H6] E. Hopf, Ber.d.math.phys.Kl.d.Sächs.Akad.d.Wiss., Leipzig 94, 3 (1942)

[K1] S.M. Kay, A. Maitland, "Quantum Optics", Academic Press, London(1970)

[L1] W.E. Lamb jr., Phys.Rev. 134, A 1429 (1964)

[M1] V.P. Maslov, YMN 15, 213 (1960)

[T1] W. Thirring and A. Wehrl, Comm.Math.Phys. 4, 303 (1967)

[T2] H.F. Trotter, Pacific J. Math. 8, 887 (1958)

[W1] A. Wehrl, Comm.Math.Phys. 23, 319 (1971)

[W2] W.F. Wreszinski, Thesis, ETH (1973)

PERTURBATION EXPANSION FOR THE $P(\phi)_2$ SCHWINGER FUNCTIONS

Jonathan Dimock
Department of Mathematics
SUNY at Buffalo

The Schwinger functions (imaginary time Green's functions) for a $P(\phi)_2$ theory are formally given by

(1) $$S(\lambda; x_1, \ldots, x_n) = \int_Q q(x_1) \ldots q(x_n)\, dq_\lambda$$

where $Q = \mathcal{S}'(R^2)$ is the space of real-valued tempered distributions and the measure dq_λ is given by

(2) $$dq_\lambda = \exp(-\lambda V(q))\, dq_0 \left(\int \exp(-\lambda V(q))\, dq_0\right)^{-1}$$

Here $\lambda \geq 0$ is the coupling constant, $V(q) = \int :P(q(x)): dx$, and dq_0 is the Gaussian measure with mean zero and covariance $(-\Delta + m_0^2)^{-1}$. Expressed in this form the Schwinger functions have a particularly accessible perturbation series. One has only to expand the exponentials $e^{-\lambda V}$. This is the imaginary time version of Feynman's original approach to field theory.

In this note we report the result that the perturbation series for a Schwinger function is asymptotic to all orders as $\lambda \to 0^+$. The details will be published elsewhere [1]. Such a result establishes the extent to which one can use perturbation theory to extract information about a field theory. By extending these results one might reasonably expect to prove that the S-matrix for the model has an asymptotic series. Since the series is non-trivial it would follow that scattering is non-trivial. Other possible applications of asymptotic series in $P(\phi)_2$ are sketched in [3].

Let Λ be a bounded region in R^2, define $V_\Lambda(q) = \int_\Lambda :P(q(x)): dx$, and define $dq_{\lambda,\Lambda}$ as in (2) with V replaced by V_Λ. Then $dq_{\lambda,\Lambda}$ is a well-defined measure on Q and we denote integrals by $\langle \ldots \rangle_{\lambda,\Lambda} = \int [\ldots]\, dq_{\lambda,\Lambda}$. The main input we require is the following result of

Glimm, Jaffe, and Spencer [2],[3]. Let A be a product of Wick monomials $\int :\varphi(x)^n: w(x)\,dx$ with each w bounded and having compact support. Then there exists an interval $\mathcal{I} = [0,\lambda_0]$ such that

I. $\sup_{\lambda\in\mathcal{I}} |\langle A\rangle_{\lambda,\Lambda}| \leq M_A$ for all Λ and

$\sup_{\lambda\in\mathcal{I}} |\langle A\rangle_{\lambda,\Lambda} - \langle A\rangle_{\lambda,\Lambda'}| \to 0$ as $dist(\Lambda\Delta\Lambda', 0) \to \infty$

II. $\sup_{\lambda\in\mathcal{I}} |\langle A', A_x\rangle^T_{\lambda,\Lambda}| \leq M_{AA'} e^{-m|x|}$

for some m (independent of A, A') and all Λ, x.

Here A_x is A translated by $x \in R^2$, and $\langle A,B\rangle^T = \langle AB\rangle - \langle A\rangle\langle B\rangle$ is the truncated two-point function. By I, $\langle A\rangle_{\lambda,\Lambda}$ converges as $dist(\sim\Lambda, 0) \to \infty$ to a Euclidean invariant limit functional $\langle A\rangle_{\lambda,\infty}$. Let $f_i \in C_0^\infty(R^2)$ and set $F(\varphi) = \langle\varphi, f_1\rangle \cdots \langle\varphi, f_n\rangle$. The Schwinger functions may be defined as distributions by $S(\lambda; f_1,\ldots,f_n) = \langle F\rangle_{\lambda,\infty}$. This provides a well-defined realization of (1),(2).

<u>Theorem</u>: The Schwinger functions $\langle F\rangle_{\lambda,\infty}$ are C^∞ on $\mathcal{I} = [0,\lambda_0]$ (including endpoints). Defining $a_n = D_\lambda^n \langle F\rangle_{\lambda,\infty}|_{\lambda=0}$ we have

(3) $$\lim_{\lambda\to 0^+} |\lambda|^{-N} \left|\langle F\rangle_{\lambda,\infty} - \sum_{n=0}^{N} a_n \lambda^n/n!\right| = 0$$

Equation (3) is the statement that the perturbation series is asymptotic and it follows immediately from the C^∞ property and Taylor's theorem. The a_n can be evaluated explicitly with the result being a sum of terms labeled by n^{th} order connected Feynman diagrams.

A more general theorem also holds which states that $\langle F\rangle_{\lambda,\infty}$ is C^∞ on any interval $\mathcal{I}$ where I, II hold. In the case at hand if we omit the crucial endpoint it is known that $\langle F\rangle_{\lambda,\infty}$ is real analytic on $(0,\lambda_0)$ [3].

The proof of the theorem is given in [1]. Here we only show that $\langle F\rangle_{\lambda,\infty}$ is C^1. It is straightforward to show that $\langle F\rangle_{\lambda,\Lambda}$ is

C^1 and that $D_\lambda \langle F \rangle_{\lambda,\Lambda} = -\langle F, V_\Lambda \rangle^T_{\lambda,\Lambda}$. Let Δ_j be a unit square centered on $j \in Z^2$, and let Λ be a square centered on the origin which is a union of the Δ_j. Then we may write

(4)
$$D_\lambda \langle F \rangle_{\lambda,\Lambda} = -\sum_{j \in \Lambda} \langle F, V_{\Delta_j} \rangle^T_{\lambda,\Lambda}$$

By I, II we have

(5)
$$\sup_{\lambda \in \mathcal{I}} |\langle F, V_{\Delta_j} \rangle^T_{\lambda,\Lambda}| \le M e^{-m|j|}$$

(6)
$$\sup_{\lambda \in \mathcal{I}} |\langle F, V_{\Delta_j} \rangle^T_{\lambda,\Lambda} - \langle F, V_{\Delta_j} \rangle^T_{\lambda,\Lambda'}| \to 0 \quad \text{as } |\Lambda|, |\Lambda'| \to \infty$$

From (5), (6) and the dominated convergence theorem we conclude that

$$\sup_{\lambda \in \mathcal{I}} |D_\lambda \langle F \rangle_{\lambda,\Lambda} - D_\lambda \langle F \rangle_{\lambda,\Lambda'}| \to 0 \quad \text{as } |\Lambda|, |\Lambda'| \to \infty$$

It follows that $\langle F \rangle_{\lambda,\infty}$ is C^1 on $\mathcal{I}$.

The higher derivatives are expressed in terms of truncated n-point functions. With a space cutoff we have

$$D^n_\lambda \langle F \rangle_{\lambda,\Lambda} = (-1)^n \langle F, V_\Lambda, \dots, V_\Lambda \rangle^T_{\lambda,\Lambda}$$

To prove differentiability in the limit the crucial estimate turns out to be

(7)
$$\sup_{\lambda \in \mathcal{I}} |\langle F, V_{\Delta_{j_1}}, \dots, V_{\Delta_{j_n}} \rangle^T_{\lambda,\infty}| \le M \exp(-m\delta(0, j_1, \dots, j_n)/n+1)$$

where $\delta(\cdot)$ is the diameter of the set. This estimate does not follow from I, II directly. Instead one uses:

II'. $\sup_{\lambda \in \mathcal{I}} |\langle A', A \rangle^T_{\lambda,\infty}| \le \langle |A'|^2 \rangle^{1/2}_{\lambda,\infty} \langle |A|^2 \rangle^{1/2}_{\lambda,\infty} e^{-md}$

whenever the localizations of A, A' can be separated by a strip of width d.

Then I, II' together with some elementary geometry give (7). The improved cluster property II' is obtained by showing that II implies that the Hamiltonian in the associated Hilbert space theory has a uniform mass gap, and then showing that this in turn implies II'. This argument is carried out by approximating the infinite volume theory first by a cutoff theory in which Λ is an infinite strip, and then by a theory in which Λ is a rectangle. See [1] for details.

References

[1]. J. Dimock, Asymptotic Perturbation Expansion in the $P(\phi)_2$ Quantum Field Theory, preprint.

[2]. J. Glimm, A. Jaffe, and T. Spencer, The Wightman Axioms and Particle Structure in the $P(\phi)_2$ Quantum Field Model, to appear in Annals of Mathematics.

[3]. J. Glimm, A. Jaffe, and T. Spencer, Erice lectures.

NONDISCRETE SPINS AND THE LEE-YANG THEOREM[1]

Charles M. Newman

Courant Inst. of Math. Sciences; New York, N.Y.,U.S.A.

Dept. of Mathematics; Indiana Univ.; Bloomington, Ind.,U.S.A.

The Lee-Yang Circle Theorem [Lee, Yang, 1952] is an important tool for the rigorous study of phase transitions in ferromagnetic spin systems. The technique of approximating Euclidean quantum fields by "general Ising models" [Guerra, Rosen, Simon, 1973] indicates that the Lee-Yang Theorem is also applicable to the study of symmetry breaking in quantum field theory providing that the theorem is extended from its original context of spin - ½ to the continuous "spin" variables arising in the field theory context. This extension has been accomplished for φ^4 models [Simon, Griffiths, 1973] by the method of "analog spin - ½ systems" [Griffiths, 1969].

In this note we report a result which essentially determines the entire class of spin variables to which the Lee-Yang Theorem is applicable. The proof of this result combines Griffiths' method of analog systems with the Hadamard factorization theorem; it appears in complete form together with an analysis of the other results presented here and a new elementary proof of the

[1]The research for this paper was supported in part by the National Science Foundation, Grant NSF-GP-24003, while the author was at the Courant Institute.

spin - ½ Lee-Yang Theorem in [Newman, 1973].

<u>Definition</u>. $\mathfrak{h}$ is the set of signed (real but not necessarily positive) measures μ on the real line which are either even or odd and which satisfy

(A) $\int \exp(b\,S^2)\, d|\mu(S)| < \infty$ for all $b \geq 0$, and

(B) $\int \exp(z\,S)\, d\,\mu(S) \neq 0$ when $\operatorname{Re} z \neq 0$.

We consider for a fixed choice of $\mu_1, \ldots, \mu_N \in \mathfrak{h}$ and $J_{kj} \geq 0$ (for $k, j = 1, \ldots, N$), the partition function,

$$Z(\vec{z}) \equiv \int \exp\Bigl(\sum_{j=1}^{N} z_j S_j + \sum_{j,k=1}^{N} J_{jk} S_j S_k\Bigr) d\,\mu_1(S_1) \ldots d\,\mu_N(S_N)$$

where $\vec{z} = (z_1, \ldots, z_N)$, and the correlations

$$\langle S_{j_1} \ldots S_{j_m} \rangle_{\vec{z}} \equiv (Z(\vec{z}))^{-1} \frac{\partial}{\partial z_{j_1}} \ldots \frac{\partial}{\partial z_{j_m}} Z(\vec{z})$$

for any choice of $j_1, \ldots, j_m \in \{1, \ldots, N\}$. We write $\operatorname{Re} \vec{z} > 0$ to mean that $\operatorname{Re} z_j > 0$ for all j. The extension of the Lee-Yang Theorem is then:

<u>Theorem 1</u>. $Z(\vec{z}) \neq 0$ when $\operatorname{Re} \vec{z} > 0$ (or when $\operatorname{Re} \vec{z} < 0$). In particular, when $\vec{z} = (z, \ldots, z)$, $Z(\vec{z})$ can only vanish when z is pure imaginary.

The measures corresponding to spin - $\frac{n}{2}$ and to uniformly distributed continuous spins are easily shown to belong to $\mathfrak{h}$, thus yielding a new proof of the Lee-Yang theorem for these cases

[Griffiths, 1969]. The measures, $d\mu/ds = \exp(-Q(S))$ with Q an even polynomial, which arise in quantum field models, can directly be shown to belong to $\mathfrak{h}$ when Q is fourth degree, thus reproducing the Lee-Yang results of [Simon, Griffiths, 1973]; it is not known (at least to the author) whether <u>any</u> higher degree polynomials satisfy condition (B). A new result of Ruelle [Ruelle, 1973] which may prove important for these higher degree cases involves the weakening of condition (B) to the requirement that all the zeros of $\int \exp(zS)\,d\mu(S)$ occur in a strip with a corresponding weakening of the conclusions concerning the vanishing of $Z(\vec{z})$. It can be shown that $Q(S) = \lambda \cosh(S)$ with $\lambda > 0$ yields a measure in $\mathfrak{h}$, and it would be relevant for exponential-interaction field theories to show that $Q(S) = \lambda \cosh(S) + bS^2$ for arbitrary real b also yields a measure in $\mathfrak{h}$.

A strong version of Theorem 1 is:

<u>Theorem 2</u>. For any fixed $\vec{y} \in R^N$ and $m_1, \ldots, m_N = 0,1,2,\ldots,$

$$\left(\frac{\partial}{\partial x_1}\right)^{m_1} \cdots \left(\frac{\partial}{\partial x_N}\right)^{m_N} \left| Z(\vec{x} + i\vec{y}) \right|^2 > 0 \quad \text{when} \quad \vec{x} > 0.$$

Theorem 2 and the fact that $\mu(S) \in \mathfrak{h} \Rightarrow S^{\ell}\mu(S) \in \mathfrak{h}$ together yield:

<u>Theorem 3</u>. For any $j, j_1, \ldots, j_m \in \{1,\ldots,N\}$,

$$\langle S_{j_1} \cdots S_{j_m} \rangle_{\vec{z}} \neq 0 \quad \text{and} \quad \mathrm{Re}\left\{ \frac{\langle S_j S_{j_1} \cdots S_{j_m} \rangle_{\vec{z}}}{\langle S_{j_1} \cdots S_{j_m} \rangle_{\vec{z}}} \right\} > 0$$

when $\mathrm{Re}\, \vec{z} > 0$. In addition,

$$\langle S_j S_k \rangle_{\vec{z}} - \langle S_j \rangle_{\vec{z}} \langle S_k \rangle_{\vec{z}} \geq 0$$

when $\vec{z} = i\vec{y}$ and $Z(i\vec{y}) \neq 0$.

The above result generalizes certain of the GKS inequalities [Griffiths, 1967; Kelly , Sherman, 1968] to complex $\vec{z}$. The final theorem we mention can be used to show that (certain of) the truncated Schwinger functions of even order in a φ^4 field theory alternate in sign as the order increases; it is a special case of a stronger conjecture concerning the truncated correlations (Ursell functions) of arbitrary order which has recently been proven for fourth order [Lebowitz, 1973].

<u>Theorem 4</u>. If $Z(\vec{0}) \neq 0$ and $Z_{\vec{\lambda}}(z) \equiv Z(z\vec{\lambda})$ with $\vec{\lambda} > 0$, then

$$(-1)^{m+1}\left(\frac{d^{2m}}{d z^{2m}} \log Z_{\vec{\lambda}}\right)(0) \geq 0$$

for $m = 1, 2, \ldots$.

REFERENCES

Griffiths, R.B.: J. Math. Phys. 8, 484-489 (1967).

Griffiths, R.B.: J. Math. Phys. 10, 1559-1565 (1969).

Guerra, F., Rosen, L., Simon, B.: $P(\varphi)_2$ Euclidean Field Theory as Classical Statistical Mechanics: Princeton preprint (1973).

Kelly, D. G., Sherman, S.: J. Math. Phys. 9, 466-484 (1968).

Lebowitz, J.: GHS and Other Inequalities: Yeshiva Univ. preprint (1973).

Lee, T.D., Yang, C.N.: Phys. Rev. 87, 410-419 (1952).

Newman, C.M.: Zeros of the Partition Function for Generalized Ising Systems: Courant Inst. preprint (1973) (to appear in Comm. Pure Appl. Math.).

Ruelle, D.: private communication (1973).

Simon, B., Griffiths, R.B.: Phys. Rev. Lett. 30, 931-934 (1973).

EUCLIDEAN FERMI FIELDS

Konrad Osterwalder*
Harvard University
Cambridge, Massachusetts

Introduction

Considering the great success of Euclidean methods in the construction of two-dimensional massive boson quantum field theory models, it is natural to ask whether these methods and ideas would also work in models involving both bosons and fermions. In particular we would like to know the answers to the following questions:

--Are there free Euclidean fermi fields such that the Schwinger functions of a free fermi theory are the vacuum expectation values of these Euclidean fields?

--Is there an integration theory for the Euclidean fermi fields? Are there L_p-space methods available?

--Is there something like a Markov property for Euclidean fermi fields?

--Can we introduce interaction through a (cut off) Euclidean action which is related to a (cut off) Hamiltonian via a Feynman-Kac formula?

--What are the properties of cut off Schwinger functions of a boson-fermion model? Can we use the Glimm-Jaffe-Spencer cluster expansion methods to obtain estimates and to pass to the no cutoff limit? Do the Schwinger functions satisfy correlation inequalities?

In the following I try to summarize what is known about these questions.

Free Euclidean Fermi fields

Free Euclidean spin $\frac{1}{2}$ fields have been constructed by Osterwalder and Schrader [OS 1], the case of arbitrary spin has been considered by Ozkaynak [Oz 1]; here we discuss the spin $\frac{1}{2}$ case only.[1)]

We want to express the Schwinger functions of a free fermion theory as vacuum expectation values of a free Euclidean fermi field. As the theory is free, it suffices to study the 2-point Schwinger function

$$\mathfrak{S}_{2,\alpha\beta}(x,y) = (2\pi)^{-4} \int e^{-ip(x-y)} \frac{(m_f - i\not{p}^E)_{\alpha\beta}}{m_f^2 + p^2} \, d^4p$$

1) Euclidean fermi fields were already studied by Schwinger [Sc 1,2].

*Supported in part by National Science Foundation Grant GP40354X.

where $\not{p}^E = \sum_{j=0}^{3} p_j \gamma_j^E$, $\gamma_0^E = \gamma_0$, $\gamma_j^E = i\gamma_j$,

and $p^2 = \sum p_j^2$; m_f is the bare fermion mass.

We can not find a Euclidean fermi field $\psi_\alpha(x)$ such that for $\bar{\psi}_\alpha = (\psi^* \gamma_0^E)_\alpha$

$$S_{2,\alpha\beta}(x,y) = (\Omega_E, \bar{\psi}_\alpha(x)\, \psi_\beta(y)\, \Omega_E)$$

because this would imply that $\| \sum_\alpha \psi_\alpha(f_\alpha)\Omega_E \|^2 = \sum_{\alpha\beta} (\Omega_E, \psi_\alpha(f_\alpha)^* \psi_\beta(f_\beta)\Omega_E)$ is not necessarily positive. The way out of this problem is to introduce two distinct fields ψ_α^1 and ψ_α^2 such that

$$S_{2,\alpha\beta}(x,y) = (\Omega_E, \psi_\alpha^1(x)\, \psi_\beta^2(y)\, \Omega_E) .$$

The fields ψ_α^i act in the Euclidean Fermi Fock space $\mathcal{E}_f$ and they satisfy the anticommutation relations

$$\{ \psi_\alpha^i(x), \psi_\beta^j(y) \} = 0$$

$$\{ \psi_\alpha^i(x), \psi_\beta^j(y)^* \} = 2\delta_{ij}\, \delta_{\alpha\beta}\, (2\pi)^{-4} \int e^{-ip(x-y)} (p^2+m_f^2)^{-1} d^4p .$$

The vacuum $\Omega_E \in \mathcal{E}_f$ is cyclic for the smeared fields $\psi_\alpha^i(f)$, $f \in \mathcal{H}^{-1/2}(\mathbb{R}^4)$. On $\mathcal{E}_f$ there is a unitary representation $U_1(a)$ of the translation group and a unitary representation $U_2(A)$ of the universal covering group SU(2) × SU(2) of the four dimensional rotation group SO_4 such that with $U(A,a) = U_2(A)U_1(a)$,

$$U(A,a)\,\Omega_E = \Omega_E$$

$$U(A,a)\, \psi_\alpha^1(x)\, U(A,a)^{-1} = \sum_\beta R_{\alpha\beta}(A^{-1})\, \psi_\beta^1(r(A)(x+a))$$

and similarly for $\psi_\alpha^2(x)$. Here $A \to R(A)$ is a four dimensional representation of SU(2) × SU(2) and $A \to r(A)$ is the homomorphism of SU(2) × SU(2) onto SO_4 .

Feynman Kac formula [OS 1]

In the boson case the relativistic Fock space $\mathcal{F}_b$ can be naturally embedded in the Euclidean boson Fock space $\mathcal{E}_b$ [Ne 1,2]. This does not hold for fermions and the relation between $\mathcal{E}_f$ and $\mathcal{F}_f$ is more complicated. Let $\mathcal{E} = \mathcal{E}_b \times \mathcal{E}_f$ be the Euclidean boson-fermion Fock space, $\Phi(x)$ the free Euclidean bose field of mass m_b, and define $\mathcal{E}_+$ to be the closed linear hull of vectors of the form

(1) $$X = :\prod_i \Psi^1_{\alpha_i}(f_i) \prod_j \Psi^2_{\beta_j}(g_j) \prod_k \Phi(h_k) : \Omega_E ,$$

where f_i, g_j and h_k are test functions which vanish at "negative times" i.e. for $x^0 \leq 0$. Then there is a map $W : \mathcal{E}_+ \to \mathcal{F}$ such that

$$WX = :\prod_i \hat{\psi}_{\alpha_i}(f_i) \prod_j \hat{\overline{\psi}}_{\beta_j}(g_j) \prod_k \hat{\varphi}(h_k) : \Omega_{\mathcal{F}}$$

where $\hat{\psi}_\alpha(x) = e^{-x^0 H_0} \psi_\alpha(0,\vec{x}) e^{x^0 H_0}$, $\hat{\overline{\psi}}_\alpha(x) = e^{-x^0 H_0} \overline{\psi}_\alpha(0,\vec{x}) e^{x^0 H_0}$

and $\hat{\varphi}(x) = e^{-x^0 H_0} \varphi(0,\vec{x}) e^{x^0 H_0}$, H_0 being the free Hamiltonian, $\psi_\alpha, \overline{\psi}_\alpha$ and φ the free relativistic fields. It turns out that there is a unitary involution Θ on $\mathcal{E}$ such that for all vectors X, Y of the form (1),

(2) $$(WX, WY)_{\mathcal{F}} = (\Theta X, Y)_{\mathcal{E}} .$$

Furthermore $\Theta \Omega_E = \Omega_E$,

$\Theta \Psi^1(x) \Theta^{-1} = \gamma_0 \Psi^2(-x^0,\vec{x})^*$, $\Theta \Phi(x) \Theta^{-1} = \Phi(-x^0,\vec{x})$, and for all $X \in \mathcal{E}_+$, $t \geq 0$,

(3) $$W U_1(t,\vec{0}) X = e^{-tH_0} WX ,$$

where $U_1(t,\vec{0})$ is the unitary time translation group in $\mathcal{E}$. Eq. (3) is the Feynman-Kac formula for a free theory.

Now let us consider fermions interacting with a boson field. The cutoff Euclidean action is given by

$$V_\varkappa(t,\Lambda) = \lambda \int_0^t dx^0 \int_\Lambda d^3x :\Big[\sum_\alpha \Psi^2_{\alpha\varkappa}(x) \Psi^1_{\alpha\varkappa}(x) \Phi_\varkappa(x) + P(\Phi_\varkappa(x))\Big]:$$

where $\varkappa$ denotes a properly chosen ultraviolet cutoff, Λ a finite volume in $\mathbb{R}^3$. P is a real polynomial, which is bounded below. $V_\varkappa(t,\Lambda)$ is not a symmetric operator on $\mathcal{E}$, but

(4) $$V_\varkappa(t,V)^* = -\Theta V_\varkappa(-t,\Lambda) \Theta^{-1}$$

holds, with the Θ introduced above.

For $X \in \mathcal{E}_+$ and $t \geq 0$ we now have

(5) $$W e^{-V_\varkappa(t,\Lambda)} U_1(t,\vec{0}) X = e^{-tH_{\varkappa,\Lambda}} WX ,$$

where

$$H_{\varkappa,\Lambda} = H_0 + \lambda \int_\Lambda d^3x : [\sum_\alpha \bar{\Psi}_{\alpha\varkappa}(\vec{x}) \psi_{\alpha\varkappa}(\vec{x}) \varphi(\vec{x}) + P(\varphi(\vec{x}))]:$$

is the cutoff Hamiltonian.

Equation (5) is the Feynman Kac formula.

Remarks: 1) If we leave out the Yukawa term in V and in H, then (5) becomes the well-known Feynman-Kac formula for boson theories, see e.g. [Ne 1],[Fe 1],[GRS 1].

2) Though there is no obvious way of formulating a Markov property for the Euclidean fermi fields, it appears that in many situations where one would like to use the Markov property, the Feynman-Kac formula is a good substitute for it.

Cutoff Schwinger functions

With $V = V_\varkappa(t,\Lambda)$ as above, cutoff Schwinger functions can be defined by

$$\mathfrak{S}_{\varkappa,t,\Lambda}(\underline{x},\underline{\alpha};\underline{y},\underline{\beta};\underline{z}) = \tag{6}$$
$$= \frac{1}{(\Theta e^{-V}\Omega_E, e^{-V}\Omega_E)} \left(\Theta e^{-V}\Omega_E \ , \ \prod_i \Psi^1_{\alpha_i}(x_i) \prod_j \Psi^2_{\beta_j}(y_j) \prod_k \Phi(z_k) e^{-V}\Omega_E\right)$$

In the case of $\mathcal{P}(\varphi)_2$ models, the volume cutoff Schwinger functions satisfy all the Euclidean axioms (see [GJS 1],[OS 1],[O 1]) with the exception of Euclidean convariance. (This property holds in the no cutoff limit only and follows easily if this limit is unique). We conjecture that the same is true for the cutoff Schwinger functions defined by (6). Notice that the denominator in (6) is not zero, because by (2) and (5) it is equal to $\| e^{-tH_{\varkappa,\Lambda}} \Omega \|^2$.

Because of the (anti)commutation relations of the Euclidean fields the $\mathfrak{S}_{\varkappa,t,\Lambda}$ are (anti)symmetric in their arguments: this is axiom (E3) in [OS 2] or [O 1]. The positivity axiom (E2) again follows easily from equation (2). To prove the distribution and the cluster axioms, estimates on $\mathfrak{S}_{\varkappa,t,\Lambda}$ are needed. For a Y_2 model with small coupling constant, the GJS-cluster expansion method should lead to the necessary bounds, were it not for the non hermiticity of the action V, which might cause problems.

At present there seems to be no reason to believe that the Schwinger functions of, for example, a Yukawa model satisfy correlation inequalities.

Functional integration

The problem of functional integration for fermions has been extensively discussed in the literature. There are two main approaches:
(a) The "physical" approach, which grew out of Feynman's formulation of quantum mechanics in terms of "sums over histories" [Fe 1,2], interprets fermi fields as anticommuting c-number functions. The mathematical objects resulting from a "functional integration" over fermions are Fredholm determinants depending on the bose field $\Phi(x)$! References for this approach are Matthews-Salam [MS 1,2,3], Edwards [E 1], Novožilov-Tulub [NT 1], Berezin [Be 1] and further references given in these publications.
(B) The "mathematical" approach of non-commutative integration, introduced by Segal [Se 1,2]; see also Gross [Gr 1] and Nelson [Ne 3].

The Euclidean fermi fields introduced above naturally lead to the "functional integrals" of (A). As a matter of fact, just in terms of these Euclidean fields one can understand why the c-number functions η and $\bar{\eta}$ —see [NT 1]—corresponding to the fields $\psi(x)$ and $\bar{\psi}(x)$ have to be independent functions; not each other's complex conjugates: they correspond to the uncorrelated Euclidean fields Ψ^1 and Ψ^2. It should also be remarked that the way renormalization can be dealt with in this formalism seems quite attractive (in a Y_2 model, for example).

It is an interesting but still open question whether there is a natural way to formulate a Euclidean fermi theory within Segal's framework of non commutative integration.

References

[Be 1] F. A. BEREZIN, The method of second quantization, Academic Press, New York, 1966.

[Bo 1] N. N. BOGOLUIBOV, Dokl. Akad. Nauk SSSR 99, 225 (1954).

[Ed 1] S. F. EDWARDS, Phil. Mag. 47, 758 (1954).

[F 1] J. FELDMAN, Nuclear Physics B52, 608 (1973).

[Fe 1] R. P. FEYNMAN, Rev. Mod. Phys. 20, 367 (1948).

[Fe 2] R. P. FEYNMAN, Phys. Rev. 76, 749 (1949).

[Fr 1] J. S. FRADKIN, Dokl. Akad. Nauk SSSR 98, 47 (1954).

[GJS 1] J. GLIMM, A. JAFFE, T. SPENCER, The Wightman axioms and particle structure in the $P(\varphi)_2$ quantum field model, preprint.

[Gr 1] L. GROSS, J. Functional Anal. 10, 52, (1972).

[GRS 1] F. GUERRA, L. ROSEN, B. SIMON, The $P(\varphi)_2$ Euclidean quantum field theory as classical statistical mechanics, preprint.

[MS 1] P. T. MATTHEWS, A. SALAM, Proc. Roy. Soc. A221, 128 (1953).

[MS 2] P. T. MATTHEWS, A. SALAM, Nuovo Cimento 12, 563 (1954).

[MS 3] P. T. MATTHEWS, A. SALAM, Nuovo Cimento, 2, 120 (1955).

[Ne 1] E. NELSON, Quantum fields and Markoff fields, Amer. Math. Soc. Summer Institute on PDE, held at Berkeley, 1971.

[Ne 2] E. NELSON, J. Functional Anal. 12, 97 (1973).

[Ne 3] E. NELSON, J. Functional Anal. 12, 211 (1973).

[Ne 4] E. NELSON, Notes on non commutative integration, preprint.

[NT 1] J. V. NOVOŽILOV, A. V. TULUB, Uspechi fiz. Nauk 61, 53 (1957) German translation in Fortschr. d. Physik 6, 50 (1958).

[OS 1] K. OSTERWALDER, R. SCHRADER, Euclidean Fermi Fields and a Feynman-Kac Formula for Boson-Fermion Models, to appear in Helv. Phys. Acta.

[OS 2] K. OSTERWALDER, R. SCHRADER, Commun. Math. Phys. 31, 83 (1973) and Axioms for Euclidean Green's functions II, to appear.

[O 1] K. OSTERWALDER, Euclidean Green's functions and Wightman distributions, Erice lectures 1973.

[Sc 1] J. SCHWINGER, Proc. Natl. Acad. Sci. U.S. 44, 956 (1958).

[Sc 2] J. SCHWINGER, Phys. Rev. 115, 721 (1959).

[Se 1] I. E. SEGAL, Ann. of Math. 57, 401 (1953).

[Se 2] I. E. SEGAL, Ann. of Math. 58, 595 (1953).

Lecture Notes in Physics

Bisher erschienen / Already published

Vol. 1: J. C. Erdmann, Wärmeleitung in Kristallen, theoretische Grundlagen und fortgeschrittene experimentelle Methoden. 1969. DM 20,–

Vol. 2: K. Hepp, Théorie de la renormalisation. 1969. DM 18,–

Vol. 3: A. Martin, Scattering Theory: Unitarity, Analyticity and Crossing. 1969. DM 16,–

Vol. 4: G. Ludwig, Deutung des Begriffs physikalische Theorie und axiomatische Grundlegung der Hilbertraumstruktur der Quantenmechanik durch Hauptsätze des Messens. 1970. DM 28,–

Vol. 5: M. Schaaf, The Reduction of the Product of Two Irreducible Unitary Representations of the Proper Orthochronous Quantummechanical Poincaré Group. 1970. DM 16,–

Vol. 6: Group Representations in Mathematics and Physics. Edited by V. Bargmann. 1970. DM 24,–

Vol. 7: R. Balescu, J. L. Lebowitz, I. Prigogine, P. Résibois, Z. W. Salsburg, Lectures in Statistical Physics. 1971. DM 18,–

Vol. 8: Proceedings of the Second International Conference on Numerical Methods in Fluid Dynamics. Edited by M. Holt. 1971. DM 28,–

Vol. 9: D. W. Robinson, The Thermodynamic Pressure in Quantum Statistical Mechanics. 1971. DM 16,–

Vol. 10: J. M. Stewart, Non-Equilibrium Relativistic Kinetic Theory. 1971. DM 16,–

Vol. 11: O. Steinmann, Perturbation Expansions in Axiomatic Field Theory. 1971. DM 16,–

Vol. 12: Statistical Models and Turbulence. Edited by M. Rosenblatt and C. Van Atta. 1972. DM 28,–

Vol. 13: M. Ryan, Hamiltonian Cosmology. 1972. DM 18,–

Vol. 14: Methods of Local and Global Differential Geometry in General Relativity. Edited by D. Farnsworth, J. Fink, J. Porter and A. Thompson. 1972. DM 18,–

Vol. 15: M. Fierz, Vorlesungen zur Entwicklungsgeschichte der Mechanik. 1972. DM 16,–

Vol. 16: H.-O. Georgii, Phasenübergang 1. Art bei Gittergasmodellen. 1972. DM 18,–

Vol. 17: Strong Interaction Physics. Edited by W. Rühl and A. Vancura. 1973. DM 28,–

Vol. 18: Proceedings of the Third International Conference on Numerical Methods in Fluid Mechanics, Vol. I. Edited by H. Cabannes and R. Temam. 1973. DM 18,–

Vol. 19: Proceedings of the Third International Conference on Numerical Methods in Fluid Mechanics, Vol. II. Edited by H. Cabannes and R. Temam. 1973. DM 26,–